PATTERN THEORY

Pattern Theory is a groundbreaking exploration of the concept of pattern across a range of disciplines, including science, neuroscience, psychology, and social sciences.

This book examines the meaning and implications of pattern, presenting a comprehensive body of theory that unifies concepts of form, order, and regularity and connects them to memory and perception. By challenging existing orthodoxies and linking evidence from brain and mind function, it outlines a robust theoretical framework around pattern searching and matching, pattern activation, and the continuity of pattern nexuses. This in-depth study of pattern theory and pattern thinking delves into the cognitive basis of patterns, their impact on reasoning and learning, and the social and collaborative nature of pattern recognition, expression, and representation. It also addresses philosophical issues and implications surrounding shared pattern thinking and introduces a broad conceptual basis for "pattern inquiry", providing a range of questions and methodologies for applying pattern theory. The book culminates in a manifesto for pattern theory and its application in pattern inquiry, offering 50 key principles that can be applied across various settings. Researchers, scholars, and practitioners are encouraged to explore and critique this unified theory as a lens for examining social and cognitive phenomena.

Ideal for academics and professionals seeking to challenge their understanding of the connections between mind and society, as well as for those looking to deepen their understanding of pattern as a cognitive phenomenon, as a theoretical lens, and as a meta-methodology for inquiry, this text provides a substantive foundation for ongoing development and application of pattern science across multiple fields.

Rachel H. Ellaway is a professor in Community Health Sciences, and Director of the Office of Health and Medical Education Scholarship, at the Cumming School of Medicine, University of Calgary, Alberta, Canada.

PATTERN THEORY

Memory, Interpretation, Understanding, Meaning

Rachel H. Ellaway

Routledge
Taylor & Francis Group

LONDON AND NEW YORK

Designed cover image: Rachel H. Ellaway

First published 2025
by Routledge
4 Park Square, Milton Park, Abingdon, Oxon OX14 4RN

and by Routledge
605 Third Avenue, New York, NY 10158

Routledge is an imprint of the Taylor & Francis Group, an informa business

British Library Cataloguing-in-Publication Data
A catalogue record for this book is available from the British Library

Library of Congress Cataloging-in-Publication Data
Names: Ellaway, Rachel, author.
Title: Pattern theory : memory, interpretation, understanding, meaning /
Rachel H. Ellaway.
Description: Abingdon, Oxon ; New York, NY : Routledge, 2025. | Includes
bibliographical references and index.
Identifiers: LCCN 2024035886 | ISBN 9781032895925 (hardback) | ISBN
9781032877006 (paperback) | ISBN 9781003543565 (ebook)
Subjects: LCSH: Pattern perception.
Classification: LCC BF294 .E45 2025 | DDC 152.14/23--dc23/eng/20240919
LC record available at https://lccn.loc.gov/2024035886

ISBN: 978-1-032-89592-5 (hbk)
ISBN: 978-1-032-87700-6 (pbk)
ISBN: 978-1-003-54356-5 (ebk)

DOI: 10.4324/9781003543565

Typeset in Times New Roman
by SPi Technologies India Pvt Ltd (Straive)

CONTENTS

ACKNOWLEDGEMENTS

I would like to thank the late Joanna Bates for exploring the concept of pattern with me in the context of medical education. I would also like to thank Catherine Patocka for sharing some of this journey into pattern. I would also like to thank Kent Hecker, Fil Cortese, Meredith Young, Don Boudreau, and Aliya Kassam for their feedback.

1

AN INTRODUCTION

In which I consider what pattern is.

We often find the world bewilderingly complex and diverse, but our will to make sense of things is strong. This is why we so readily devise explanations of what things are and why they are the way they are. Our understanding is augmented and directed by our experiences, our socialization, and our schooling such that we (hopefully) develop ever more sophisticated and nuanced understanding of the world and of ourselves. All of this, I argue in this book, depends on our use of patterns and pattern thinking. To be human, for good or ill, is to be inescapably entangled with patterns.

But what is *pattern*? Can we agree on what this one deceptively simple word means? Perhaps not. For example, pattern is often used synonymously with *regularity*, such as observing patterns of behaviour in a workplace. Sometimes pattern is taken to refer to *order*, such as appreciating a floral pattern on a piece of fabric. I will go on to argue that there are critical differences between different meanings ascribed to pattern, not least of which is that patterns allow us to understand orders and regularities by ascribing causes, origins, logics, and implications to them. I might argue that patterns act as our theories of regularities, and as such they are cognitive phenomena. This is reflected in the subjectivities of pattern. For instance, a group of observers may witness the same regularities but perceive them differently based on the different patterns that are triggered in their minds. Indeed, the cognitive and subjective basis of pattern is perhaps more apparent when pattern recognition goes awry.

Perceiving patterns that are not actually there is called pareidolia:

Illusory sensory perception, or *pareidolia*, is common. It occurs when external stimuli trigger perceptions of non-existent entities, reflecting erroneous matches between internal representations and the sensory inputs … Among all forms of pareidolia, face pareidolia is the best recognized … [which] suggests that our visual system is highly tuned to perceive faces, likely due to the social importance of faces and our exquisite ability to process them.

(Liu et al. 2014)

DOI: 10.4324/9781003543565-1

Consider the image in Figure 1.1. Do you see a little girl with her parents or the head of a bearded man who looks like Western depictions of Jesus, or do you see both? This is an example of a bistable image that has two or more legitimate readings (Rodríguez-Martínez & Castillo-Parra 2018). Our minds fix on one reading but can (with some effort) switch between them. Interestingly, it is nearly impossible for us to perceive more than one reading at a time. This suggests that not only do we seek pattern understanding of regularity, but when we find it, we readily anchor our perceptions in that particular reading. A related condition is *apophenia* in which patterns are not only perceived erroneously, they are seen as particularly meaningful to the point of delusion (Conrad 1958).

FIGURE 1.1 The "Jesus" photograph. Public domain image from Wikimedia Commons.

That we can see things that are not there is reflected in the many missteps and illusions in the history of science. As an example, consider the astronomer Percival Lowell who observed canals on the surface of Mars and developed an elaborate theory based on his observations. Lowell's pareidolia may be forgiven as a matter of imprecision given that he was using a late 19th-century telescope to observe an object many millions of kilometres away. It has also been suggested that Lowell was influenced by Italian astronomer Schiaparelli's observations of what he called "canali" on Mars (*canali* is Italian for channels, not canals) (Simon 2014). Whatever the basis of Lowell's perceived canals, this is just one of many plausible and yet erroneous explanations and interpretations of unfamiliar and potentially ambiguous phenomena in the history of science. Other examples include geocentrism, vitalism, alchemy, astrology, phrenology, and phlogiston. On the one hand, we value repeated observations by multiple investigators, critical thinking, and peer review to help to address the risk of pareidolia of any one scientist. On the other hand, pareidolia can pervade scientific communities and even whole cultures. This shifts the focus of pareidolia from a visual or perceptual sense to a conceptual sense, from preconscious momentary individual reactions to systemic and shared long-term beliefs and behaviours. Kuhn argued that science advances through successive paradigm shifts; as old paradigms fail or are found insufficient, they are replaced by new paradigms (Kuhn 1962). Arguably geocentrism, vitalism, and alchemy as paradigms were reasonable ideas or interpretations at their time, but they were superseded as scientific understanding changed and their pareidolic and apophenic characteristics acknowledged and (to some extent) corrected.

Not only can we perceive patterns that are not there, we may also fail to perceive patterns that are there. This is reflected in the differentiation of Type I (false positive) and Type II (false negative) errors. Shermer made this distinction in defining his concept of "patternicity":

> … the tendency to find meaningful patterns in both meaningful and meaningless noise … patternicity will occur when the cost of making a Type I error is less than the cost of making a Type II error.
>
> *(Shermer 2011)*

Following a similar line of argument, it has been argued that pareidolia is a positive evolutionary trait, as perceiving danger that is not there is less likely to prove fatal than failing to perceive danger that is there (Kumar & Wroten, 2023). This suggests that we may as a species be predisposed to Type I over Type II errors as a result of the evolution of our pattern perception capabilities.

Just because someone fails to perceive something that is there does not mean that they will necessarily do so in subsequent encounters. It could have been a momentary lapse resulting from tiredness, inattention, or distraction. We should, therefore, distinguish between inattention or mistakes in pattern recognition and more fundamental incapacities. *Agnosia*, the inability to recognise things, is a cognitive disorder, which can take different forms such as *prosopagnosia* (an inability to recognise faces). As another general principle, it would seem that our pattern perception capabilities can vary from person to person, sometimes to the point of genius, sometimes to the point of disability, and sometimes both.

I will argue that our patterns as cognitive phenomena do not just connect current experience to past memory, they suggest meanings, functions, and even support making predictions. We do not simply perceive an array of discrete things; we perceive them in relation to

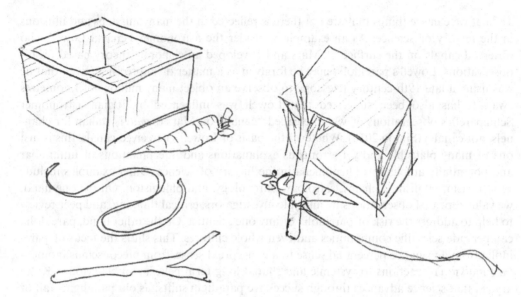

FIGURE 1.2   A box, a carrot, a twig, and a piece of string. We might view them as discrete elements (left) but when assembled in a particular configuration, a pattern of an animal trap becomes apparent (right). Other configurations reflect other patterns, such
as carrying the carrot, rope, and stick in the box.

Author's own artwork.

each other, and we can to some degree perceive how those relationships might change. Take
Figure 1.2 as an example. Not only would most of us (I hope) recognise a box, a carrot, a
stick, and a length of rope, seeing them together might lead us to also consider how they
might interact with each other or with other things. For instance, if you are a hunter (or if you
have watched a lot of Warner Brothers cartoons) you may see that these four things in a particular configuration can be used to build an animal trap. While a human might recognise the
pattern of this trap, a rabbit or a roadrunner (or whatever other creature it is intended for)
likely will not, not because they do not see it in a physical sense, but because they are unfamiliar with and, therefore, cannot recognise it for what it is or what it might do. There may be
many reasons why humans, rabbits, or roadrunners might not see the trap as a trap, these may
include a lack of experience, distractions, or sensorial or cognitive bias or impairment. That
we attempt to understand something unfamiliar (sometimes successfully and sometimes
unsuccessfully) by comparing our experiences with our memories is central to pattern theory.

Word Games

By advancing definitions of pattern and critique others' definitions of pattern, I acknowledge that I am inescapably playing Wittgensteinian "word games" (Wittgenstein 2009). I
will say that pattern means this or that, while others may advance their own definitions or
allow meaning to be inferred based on convention or context. Words enable exposition and
debate, but their meanings can, like patterns, be partial and fluid. For instance, by articulating my arguments in English, I must by necessity work within its linguistic rules. I claim no
intrinsic authority or superiority in conducting this work in my native language, simply
that it is the only language in which I am competent to do so.

How might the arguments I make translate to other languages? How well does the term *pattern* map to different languages? Are the meanings of pattern I explore in this book the same as the meanings of *mönster* in Swedish, or *pauta* in Spanish, or *muster* in German? In Japanese there are at least 16 possible translations of pattern, but I would have to defer to native Japanese speakers to suggest which of these, if any, are appropriate translations of pattern in the sense(s) in which I use it. This problem has been noted by others:

> English speakers fluently and effortlessly use the word pattern to describe regularities, exact or approximate, that they perceive in the world. However, if they wish to talk about such phenomena in French, they will soon learn, to their frustration, but there is no French word that exactly covers it ... depending on ... what they mean, they will have to choose among French words such as motif, regularité, système, style, tendance, habitude, configuration, disposition, périodicité, design, modèl, schéma, and perhaps others.
>
> *(Hofstadter & Sander 2013)*

This is not just a matter of mapping meanings as fixed symbolic entities; words can have implications and underlying meanings that we do not always intend or notice. For example, the term *pattern* is somewhat gendered given its roots in the Middle English *patron*, the Medieval Latin *patronus*, and the Latin *pater*, which all refer to father figures. Why not challenge the gendered nature of this language and talk about *matterns* rather than *patterns*?

There are also translational challenges between different scientific paradigms and discourses. What *pattern* as a term means to a biochemist and what it means to a linguist may have little in common other than that regularity is involved in some way. As Grenander observed:

> What, then, is a pattern? ... In everyday language we use the word a great deal, allowing it to mean many different things. The most usual meaning of the word is perhaps "design" or "style of marking", but it can also mean "sample" or "copy". The French "dessin", German "Muster" and Swedish "mönster" can mean design but have, in addition, a flavor of something desirable, something one ought to imitate.
>
> *(Grenander 1969)*

Perhaps I should, like any good scholar, define terms and be done. I do not do this (yet), in part because I want to consider the polysemy as part of my overall thesis, and in part because conceptual precision is perilous without control, and pattern has long since fled Pandora's box. Is pattern just an arbitrary word or does it have currency and coherence in the arguments I make? Could I just as well claim that *splonges* are the *pablics* of *mucklebies*? Resolving this challenge will take some time, it will involve exploring many meanings and concepts associated with the term *pattern*, and it will call into question whether definitions are as necessary as many scientists believe.

Structure of the Book

Chapter 2 explores the ways in which the concept of pattern has been articulated in different academic and non-academic discourses. Focusing on works that have explicitly claimed to deal with pattern or patterns, I consider five discursive spaces. The first considers philosophical perspectives that reflect tensions whether or not pattern is a phenomenon of mind.

The second focuses on *design patterns* and *pattern languages* as ways of codifying the shared thinking of a discipline or network of practice. The third considers texts that have argued that mathematics is the science of pattern. The fourth considers how scientists from many disciplines have either treated pattern as an undefined concept or as a synonym for regularity. The fifth explores the ways in which artists, designers, and other creatives have used the concept of pattern. Collectively the problems identified in these different discourses set the stage for the development of pattern theory.

My starting point for Chapter 3 is the conflation of pattern and regularity I described in Chapter 2. In considering what regularity is and how it relates to pattern, I tackle three interlinked but distinct concepts: *form, order,* and *regularity.* I explore what each of them are, what they depend on, and how they are related. Form as a noun is constituted in our perception of the appearances, structures, expressions, and performances of specific things or ideas. Form can also be a verb in the sense of the forces applied to something that result in form (as a noun). Either way, form depends on perception. Order can also refer both to the state of something and to the manipulation of states, and again it depends on perception; one person may perceive order where another does not. Order is primarily based on the perception of ordering principles that can have limits and transformations in the thing that appears to be ordered. Regularity can refer to repetitions in which case frequencies and variations matter. Regularity can also refer to governance in which case causes and controls are what matter. We tend to distinguish between form, order, and regularity because our minds try to make sense of our world by categorising and grouping similar things, all of which reflect pattern memories.

Chapter 4 develops a cognitive and neurological basis for *pattern theory* and explores some of the implications of doing so. I argue that patterns are nexuses of coherently connected and structured memories, and that perception involves searching and matching pattern memories to current experiences. I distinguish between *pattern recognition* and *pattern development,* and I describe the heuristic, transient, and adaptive nature of *pattern perception.* Building on this, I describe *pattern thinking* more broadly as being abductive, integrative, and highly efficient. I expand on this argument to note how memories of emotions that are woven into patterns form the basis of selective attention, cueing, anticipation, and our sense of significance. Patterns aggregate such that fleeting impressions (*micropatterns*) can be aggregated to form concrete perceptions (*mesopatterns*) and even abstract theories and understandings (*macropatterns*). Throughout the chapter I weave neurological and psychological evidence together to support my theories and observations. I close the chapter by considering some of the weaknesses of patterns and pattern thinking.

Pattern thinking is an umbrella term for any mental process that engages pattern in some way. In Chapter 5, I explore how pattern thinking is the basis for a range of otherwise distinct cognitive behaviours and functions. The human capacity for language is fundamentally based on pattern thinking capacities and capabilities and may be both the driver of the development of pattern thinking as well as its most notable product. I consider the concept of affordances as related to pattern thinking, arguing that Gibson's theory of affordances did not account for the anticipatory nature of pattern thinking. I also consider the interpretive nature of pattern thought and the use of hermeneutic theory to explain how this works. Pattern thinking involves reasoning to pattern and from pattern and I explore these sequential relationships. I expand on the role of affective memory in pattern thinking and I review the relationships between pattern thinking and learning, both in terms of human learning across an individual's lifespan and in terms of learning theories. I then turn to the paradoxical topic of shared pattern thinking; clearly patterns are unique to individual minds and yet humans are adept at sharing their pattern thinking. As much as pattern thinking has many strengths, I note that there are also many limitations to pattern thinking including the ways in which it can falter and fail and the consequences thereof. I close the chapter by considering artificial pattern thinking and its implications for the development of pattern theory.

Chapter 6 picks up the issue of *shared pattern thinking* introduced in the previous chapter to consider the different ways in which we externalise our patterns and pattern thinking. Given that patterns are nexuses of coherent memories, patterns cannot be shared in any absolute sense, rather we share mediated approximations of our perceptions of our patterns with each other. Although perceiving pattern instances does not externalise pattern *per se*, it can help to explain what does count as externalisation. I then consider four kinds of externalisation. *Pattern expressions* are intentional instances of a pattern while pattern representations are intended to capture the structure and meaning of certain parts or characteristics of a pattern. *Pattern systems* are collections of interlocking pattern representations that cover a wider domain than any one pattern can usefully encompass. Although *pattern languages* have been articulated in terms of idealised problem solving in a single domain, I redefine a pattern language as a pattern system that has additional grammar (rules) and syntax (idioms and styles of expression). I distinguish between five kinds of pattern systems and pattern languages: speculative (reflecting imaginary shared pattern thinking), normative (reflecting idealised shared pattern thinking), naturalistic (reflecting idiosyncratic shared pattern thinking), historical (reflecting the development of shared pattern thinking), and completist (reflecting tacit shared pattern thinking). I close the chapter by describing pattern creativity as another aspect of the repertoire of human pattern thinking.

In Chapter 7, I step back from the exposition of pattern theory over the previous chapters to consider the ontological, epistemological, and axiological characteristics of patterns and pattern thinking. From an ontological perspective, I ask whether patterns and pattern instances exist, drawing on the mind-body problem as a lens. I also consider the existential nature of different kinds of externalised patterns, noting their intrinsic disconnection from mind. I then take a functionalist perspective to ask how it is that patterns work, and I explore the kinds of ontologies that patterns can generate for us. I then switch to considering general pattern epistemologies with a particular focus on pattern knowledge. I describe four dimensions of pattern knowledge: the knowledge of what something is; the knowledge of what something means; the working pattern knowledge built from interactions between pattern memories and current experience; and the knowledge created when our pattern memories are elaborated. I then consider the use of category as a pattern index, and the *fuzzy* nature of pattern knowledge. Macropatterns have particular epistemological characteristics: aggregation, abstraction, theorisation, signification, formalisation, and systematisation. I close by considering pattern axiologies and the way values both emerge from pattern and are encoded into pattern.

In Chapter 8, I translate pattern theory into the basis for scientific inquiry. I describe how patterns and pattern thinking can contribute to the development and articulation of theory, while pattern expressions and pattern representations are ways of externalising and sharing theory. A pattern methodology is a pattern theoretical position with respect to phenomena of interest that might involve examining, generating, or elaborating patterns, or using pattern as a lens. I describe five modalities of pattern inquiry: purpose, focus, disposition, process, and product, and I describe each in turn in terms of the following pattern characteristics: cognition, constitution, logics, uniqueness, awareness, dynamism, continuum, categorisation, formation, dependency, modalities, triggering, elaboration, weaving, micro-meso-macro, layering, similarity and difference, generation, limitations, sharing, latency, expressing and representing, and medium of expression. I then consider some of the methodological precedents for *pattern inquiry* with a particular focus on typological pattern analysis, topological pattern analysis, and pattern language analysis.

Having outlined the principles of pattern inquiry, in Chapter 9 I provide practical examples of different approaches to pattern inquiry. This includes a substantive worked example of developing a pattern language of medical school admissions along with other examples from my work and that of my colleagues in this area. I use these examples to explore some strengths and limitations of different approaches to pattern inquiry, and to consider how issues of scope might be managed. I then explore different ways in which grammar and syntax might be analysed so as to transform a pattern system into a pattern language. I describe pattern analysis in terms of mapping landscapes of shared pattern thinking that can be the foci of inquiry or the basis for subsequent inquiry. Rather than engaging in pattern inquiry as an exclusive methodological approach, I describe approaches to pattern-informed inquiry that borrows from pattern theory as part of a broader framing of a research agenda. I then consider the implications of exploring and analysing real, implied, and interpreted patterns, and how this relates to integrating pattern inquiry into the social sciences and into other paradigms of inquiry. I close the chapter by considering issues of rigour and integrity in pattern inquiry.

Finally, in Chapter 10 I draw the various threads and themes developed throughout the book to outline a pattern manifesto for the social sciences. I start by outlining 50 core tenets of pattern theory, grouped by the nature of patterns, the nature of pattern thinking, the nature of shared pattern thinking, and the nature of pattern inquiry. I then consider the place of a pattern theoretical approach in the context of contemporary trends in the social sciences, I reprise the matter of word games and the issue of what pattern as a word can be taken to mean. I close the book with my thoughts on the strengths and limitations of the ideas I have presented, the idiosyncratic ways in which I have approached the topic, and the broad implications for what we may call *pattern science*.

References

Conrad K. *Die Beginnende Schizophrenie. Versuch einer Gestaltanalyse des Wahns*. Stuttgart: Thieme: 1958.

Grenander U. Foundations of Pattern Analysis. *Quarterly of Applied Mathematics*. 1969; 27(1): 2–55.

Hofstadter D, Sander E. *Surfaces and Essences: Analogy as the Fire and Fuel of Thinking*. New York Basic Books: 2013.

Kuhn TS. *The Structure of Scientific Revolutions*. Chicago, IL: University of Chicago Press: 1962.

Kumar A, Wroten M. Agnosia. In: *StatPearls* [Internet]. Treasure Island (FL): StatPearls Publishing; 2023 Jan-. Available from: https://www.ncbi.nlm.nih.gov/books/NBK493156/

Liu J, Li J, Feng L, Li L, Tian J, Lee K. Seeing Jesus in Toast: Neural and Behavioral Correlates of Face Pareidolia. *Cortex*. 2014: 53: 60–77.

Rodríguez-Martínez GA, Castillo-Parra H. Bistable Perception: Neural Bases and Usefulness in Psychological Research. *International Journal of Psychological Research*. 2018; 11(2): 63–76. DOI: 10.21500/20112084.3375

Shermer M. *The Believing Brain*. New York, NY: St Martin's Press: 2011.

Simon M. Fantastically Wrong: One Astronomer's Quest to Expose the Alien-Built Canals of Mars. Wired Magazine. May 21st 2014 - https://www.wired.com/2014/05/fantastically-wrong-martian-canals/

Wittgenstein L. *Philosophical Investigations* (4th Ed.). Hoboken, NJ: Wiley-Blackwell: 2009.

2

WE NEED TO TALK ABOUT PATTERN

In which I consider uses and misuses of the term "pattern"

Why write a book about pattern in scientific inquiry? After all, if pattern is so central to scientific thought and practice, then surely pattern as a concept is well grounded and established in scientific practice. In this chapter I explore whether that is the case in different discursive spaces by considering texts that have laid titular claim to the concept of pattern.

Philosophical Perspectives on Pattern

I start with the concept of pattern in discourses in the philosophy of science. Despite the presence of pattern in its title, Hanson's treatise on the philosophy and nature of scientific inquiry first mentioned pattern more than a third of the way in (in the context of discussing causality) when he claimed that "perceiving the pattern in phenomena is central to their being explicable" (Hanson 1958). While I agree with this in principle, quite what Hanson meant by "the pattern in phenomena" was left undefined. The closest that Hanson came to a definition was in differentiating pattern statements from detail statements:

> Pattern statements are different from detail statements. They are not inductive summaries of detail statements ... to deny a detail statement is to do something within the pattern. To deny a pattern statement is to attack the conceptual framework itself.
>
> *(Hanson 1958, pp. 87–88)*

Hanson's reading of pattern was arguably more to do with knowledge than phenomena and it was more epistemological than ontological. Indeed, Hanson's treatment of pattern seems to have been more about how we understand phenomena that consist of details rather than being about a characteristic *of* the phenomena or the details they contain.

Bateson's influential treatise on pattern in the philosophy of science (1979) argued that pattern was all about epistemic connections; patterns connect elements within a

DOI: 10.4324/9781003543565-2

phenomenon and they connect one phenomenon to other phenomena. The concepts of recognition and regularity were central to Bateson's argument: "To be meaningful – even to be recognized as pattern – every regularity must meet with complementary regularities" (Bateson 1979, p. 46). In other words, it is the regularities within a phenomenon that allow us to recognise the pattern and the ways in which it connects to other phenomena. Although we may perceive regularity in a phenomenon, it is our connecting (remembering, abstracting) the regularities of past phenomena with those of the current phenomenon that allows us to see the pattern that connects them.

The idea that pattern is in the mind was explored in Hayek's work on pattern in the context of scientific discovery (Hayek 1967). Hayek challenged the observation that pattern was often considered to be synonymous with perceptions of regularity, form, or order, by arguing that, although humans are willing and able pattern recognisers, not all patterns are apparent. Indeed, we require some level of understanding or experience of a pattern before we are able to perceive it in the world around us. In other words, we need to develop a pattern theory before we can perceive that pattern in the regularities we encounter. Hayek also noted that, because the ability to see pattern in the world around us is something is so familiar and ingrained, we tend not to notice the importance of pattern in affording these perceptions:

> It is probably the capacity of our senses spontaneously to recognize certain kinds of patterns that has led to the erroneous belief that if we look only long enough, or at a sufficient number of instances of natural events, a pattern will always reveal itself. That this often is so means merely that in those cases the theorizing has been done already by our senses.
>
> *(Hayek 1967 p. 25)*

Hayek also stressed that there is an intrinsic imprecision in pattern theory, in that it can predict: "general attributes of the structures that will form themselves, but not … the individual elements of which structures will be made up" (Hayek 1975). Acknowledging that this suggested that pattern was somehow inferior to other kinds of knowledge, Hayek stressed that pattern theory and knowledge work at different levels of complexity and at different scales:

> Compared with the precise predictions we have learnt to expect in the physical sciences, this sort of mere pattern predictions is a second best with which one does not like to have to be content. Yet … to act on the belief that we possess the knowledge and the power which enable us to shape the processes of society entirely to our liking, knowledge which in fact we do not possess, is likely to make us do much harm.
>
> *(Hayek 1975)*

Interestingly, Dennett also connected pattern, interpretation, and prediction, although he conflated pattern as a phenomenon of mind with that which a mind perceives:

> I claim that our power to interpret the actions of others depends on our power – seldom explicitly exercised – to predict them. Where utter patternlessness or randomness prevails, nothing is predictable. The success of folk-psychological prediction, like the success of any prediction, depends on there being some order or pattern in the world to exploit.
>
> *(Dennett 1991)*

Christen (2009) built on Hayek in outlining patterns as raising "scientific and theoretical complexes of questions". One complex focused on the fundamental difference between perceptions of the presence of patterns and perceptions of the causes of patterns, another complex was that pattern can change according to context. For instance, we might perceive a pattern of behaviour involving individuals thumping each other in the face but not know why they were doing it without attending to the context, such as a boxing ring, a political demonstration, or a movie studio backlot.

Interestingly, all of these writers gravitated toward pattern as a function of mind, although none of them fully committed to this conceptualisation. Hanson stressed that our understanding something is tied to our ability to *see* its pattern. Bateson took a similar position, emphasising that patterns connect perception and knowledge. Hayek also argued that past pattern experiences allow us to see patterns in current regularities, and in doing so he differentiated between different logics of pattern discovery and pattern recognition, noting that it is often assumed that pattern recognition is all that is needed. Hayek also noted the limitations of pattern knowledge and its connections with the complexity of the phenomenon under consideration, and that patterns reflect the dispositions of wholes rather than details. Christen stressed the role of context in pattern thinking, focusing on the frame of reference and the level of complexity involved. Bateson's emphasis on connection as an essential characteristic of pattern was another important observation, as was his grounding of pattern (more or less) as a cognitive phenomenon. As much as these writers wrestled with what pattern is or means, these texts serve as starting points for a pattern theoretical stance rather than explicitly expressing one.

Design Patterns and Pattern Languages

A second broad discourse around pattern has focused on the concept of *design patterns* (the recognition, codification, and deliberate use of patterns in creative activities) and *pattern languages* (collections of design patterns). Architectural theorist Christopher Alexander has been credited with introducing these concepts as being intimately related to problem-solving:

> Each pattern describes a problem which occurs over and over again in our environment, and then describes the core of the solution to that problem, in such a way that you can use the solution a million times over, without ever doing it the same way twice.
>
> *(Alexander et al. 1977)*

Alexander described a design pattern as a generalised way of solving a particular class of problem in different contexts. For instance, the design pattern of a window solves the broad problem of *how to let natural light through a wall while retaining the integrity of the wall*. There are many different kinds of walls with many different kinds of windows, but they all conform to the same broad design pattern. To that end, Alexander argued that pattern is inescapably tied to context: "a pattern of events cannot be separated from the space where it occurs" (Alexander 1979, p. 73).

Mouzon & Henderson (2004) described four realms of architectural patterns: universal patterns, national patterns, regional patterns, and location-specific patterns. Universal patterns included doors, windows, and roofs, national patterns compared variations in building styles between countries, regional patterns reflected certain geographies or climates, and location-specific patterns reflected the design of specialist structures such as lighthouses or harbours. Mouzon & Henderson described these patterns as having many uses: helping architects to solve design problems, helping builders to solve construction problems, helping inhabitants to use these buildings, and helping scholars to critique them. Alexander (1979) argued that, given that a pattern may be instantiated in many ways and in many contexts, only a few patterns are needed to generate a multitude of applications. He also argued that a pattern may have sub-patterns, it may be a part of other patterns, and it can come together with other patterns in solving problems:

> All these rules of thumb – or patterns – are part of larger systems which are languages ... Every person has a language in his mind ... It is as true of any great creative artist, as of the humblest builder.
>
> *(Alexander 1979, pp. 202–203)*

Concepts of design patterns and pattern languages have had a significant impact on thinking beyond architecture, including software development, regional development, education, and permaculture (Leitner 2015). In each translation, the concept of a design pattern has been redefined to align it with a particular community's dynamics, needs, and interests, which has only added to the polysemy of pattern:

> As a result of the diversity of this patterns community, we have had difficulty in defining the term pattern. We all think we can recognize a pattern when we see it, we think most of us would agree in most cases, but we cannot come up with a single definition.
>
> *(Fowler 1997)*

Scheurer, in contrasting Alexander's architectural design patterns with those from software design (drawing on the seminal work of Gamma et al. [1994]), noted three key variances between them. First, they varied by scale. In architecture, scale matters because "everything is based on the fundamental human dimension", whereas "software can be scaled at will, at least while it is being programmed, without any need to change the basic patterns". Second, they varied by degree of abstraction: "Alexander's patterns largely refer to concrete application problems from the perspective of the end user", while "Gamma's software design patterns are completely free from the actual application". And third, they varied in terms of their logics; Alexander's design patterns were bottom-up (taking a user's point of view), while software engineers necessarily worked top down (Scheurer 2009, pp. 41–56). Differences of scale and logic can also be seen in Schlechte's work (2019) in outlining a pattern language for composing music. Interestingly, he emphasised the practical and adaptive side of pattern thinking in music in that it can equally apply to single phrases and to whole compositions.

In summary, discourses on design patterns focused on patterns as ways of solving practical problems, which can be collected to form pattern languages where many patterns may contribute to solutions to particular problems. Notably, the focus in this discourse has been on the convergent shared patterns of communities of practice rather than on the divergent interpretations of those patterns by individuals within those communities.

Mathematics and Patterns

There is a different discourse around pattern to be found in mathematics, much of which is grounded in claims of this kind:

> Mathematics is the science of patterns. The mathematician seeks patterns in number, in space, in science, in computers, and in imagination. Mathematical theories explain the relations among patterns; functions and maps, operators and morphisms bind one type of pattern to another to yield lasting mathematical structures. Applications of mathematics use these patterns to "explain" and predict natural phenomena that fit the patterns. Patterns suggest other patterns, often yielding patterns of patterns. In this way mathematics follows its own logic, beginning with patterns from science and completing the portrait by adding all patterns that derive from the initial ones.
>
> *(Steen 1988)*

While this may be so (certainly other mathematicians have made similar claims), is this not also true of other sciences? Physicists, biologists, meteorologists, and sociologists (to name but a few) all pursue patterns (albeit in their own paradigmatic ways) in their attempts to describe, explain, and predict the phenomena they are interested in. While mathematics can play a central role in other fields, it is perhaps an ambitious assertion to claim the science of pattern exclusively for mathematics. Notably, Steen was one of the few mathematicians who both evoked pattern as a central element in their work and who also reflected on the nature of pattern, arguing that patterns can be found both in data (i.e., in the world) and in the mind (i.e., in our perceptions of the world). Mulligan & Mitchelmore (2009) also reflected on the nature of patterns, observing that "a mathematical pattern may be described as any predictable regularity, usually involving numerical, spatial or logical relationships". These examples notwithstanding, most mathematical texts on patterns I reviewed did not define what pattern was, seemingly because their authors assumed that the term needed no definition.

That is not to say that attempts have been made to generate a more inclusive mathematical approach to modelling patterns in general. Of particular note was Grenander's work in algebraic pattern analysis that led to the development of a "general pattern theory" (GPT) as a way of formally representing pattern structures. Interestingly, Grenander shared my concern that, despite occurrence of pattern concepts wherever you look, what pattern meant as a concept was all too often left undefined:

> The concept of patterns pervades the history of intellectual endeavor; it is one of the eternal followers in human thought. It appears again and again in science, taking on different forms in the various disciplines, and made rigorous through mathematical formalization. But the concept also lives in a less stringent form in the humanities, in novels and plays, even in everyday language. We use it all the time without attributing a formal meaning to it and yet with little risk of misunderstanding. So, what do we really mean by a pattern? Can we define it in strictly logical terms? And if we can, what use can we make of such a definition?
>
> *(Grenander & Miller 2007, p. 1)*

The goal of GPT was to develop formal techniques for deconstructing and analysing patterns, for representing and modelling them, and to do so in ways that could be applied in many different disciplines. To that end, GPT focused on creating "realistic representations of knowledge" of the world and Grenander and colleagues focused on discovering and modelling individual pattern rules as well as rules for combining patterns:

> Pattern theory attempts to provide an algebraic framework for describing patterns as structures regulated by rules, essentially a finite number of both local and global combinatory operations. Pattern theory takes a compositional view of the world, building more and more complex structures starting from simple ones. The basic rules for combining and building complex patterns from simpler ones are encoded via graphs and rules on transformation of these graphs.
>
> *(Brown University Pattern Theory Group 2002)*

A GPT analysis of a pattern was based on developing an appropriate "pattern theoretic model", an abstraction that reflected essential differences in a pattern's topology (Grenander 1989). Selecting, matching, and fitting a pattern theoretic model to a particular pattern instance was a somewhat abductive process but eventually this generated a "pattern template". Once the pattern's principles had been outlined, their specifics were added to the pattern template (locations, sizes, colours, etc.). The next step was to explore and define how the pattern template could be transformed while still reflecting the underlying pattern. These pattern transformations could be objective (such as rotations, reflections, deformations) or subjective (such as the angle at which something is viewed and how it is lit). This was a central concern in GPT; whether two or more apparently distinct phenomena were instances of the same underlying pattern but transformed in some way or whether they were instances of two quite different patterns. This in great part reflects the contributions of GPT to getting computers to recognise the same object from different positions and in different conditions.

Interestingly, while Grenander and colleagues made extensive use of graph theory and set theory in their work, they found no one branch of mathematics that was sufficient to encompass all aspects of GPT, rather they developed a hybrid combination of different mathematical modelling techniques:

> Pattern theory is algebraic in the sense that it manipulates symbols according to given rules. It is compositional in that some of these manipulations consist of combinations of simple parts into more complicated ones. It is metric in that it measures the plausibility (or legality) of such combinations in terms of probabilities. It is therefore a hybrid of algebra, topology, and probability theory.
>
> *(Grenander & Miller 2007, p. 202)*

Despite Grenander's wish to render patterns and pattern science more concrete and tractable, beyond the models of patterns he and his colleagues developed, quite what a pattern is or was does not seem to have been substantially addressed.

Mumford & Desolneux (2010) built on GPT in their exploration of how shapes, frequencies, distributions, and rhythms (regularities) can be detected in various signals. However, the difference between the phenomenon and the model thereof in this work was also ambiguous:

Pattern theory proposes that the types of patterns (and the hidden variables needed to describe those patterns) that are found in one class of signals will often be found in the others and that their characteristic variability will be similar … The underlying idea is to find classes of stochastic models that can capture all the patterns we see in nature, so that random samples from these models have the same look and feel as the samples from the world itself.

(Mumford & Desolneux 2010, p. 1)

Interestingly, Mumford & Desolneux at one point described patterns as being "caused by objects, processes, and laws present in the world but at least partially hidden from direct observation" such that "patterns can be used to infer information about these unobserved factors", which leans towards a cognitive basis of pattern.

Linked to general pattern theory and yet distinct from it are theories of pattern in computational science and computational logic that focus on understanding pattern recognition and pattern discovery. For example, Shawe-Taylor & Cristianini (2004, p. 3) defined patterns as "… any relations, regularities or structure inherent in some source of data". Like many others, Niemann (1990) focused on patterns as images or sounds that are perceived by machine processing, although he did acknowledge that there were other meanings that his work did not encompass. I note Niemann's work in particular because he extended the meaning of pattern recognition to include "understanding" such that patterns are representations of recurring regularities in data that are based on predictable and indicative characteristics that can be described mathematically. Regularities are mapped to abstractions of regularities (patterns) which are then mapped to symbols representing broader characteristics and associations:

If the machine is able to map a pattern … to an internal symbolic knowledge structure, representing the task domain, then it is said that the machine understands the pattern.

(Niemann 1990, p. 8)

As I will describe in subsequent chapters, although the terms that Niemann and I use are not well-aligned, this cascade of comparison, abstraction, and symbolisation is not dissimilar to cognitive processes of pattern thinking.

Interestingly, Shalizi & Crutchfield (2001) argued that using dense and elaborate mathematics to "crack" the problem of modelling pattern phenomena was doomed to failure because of an intrinsic imprecision in pattern. The strength of any given pattern model should, they argued, be understood as a balance between representing the complexity of the pattern and representing its predictive abilities; the more of one, the less of the other. Other mathematical texts used pattern to connect ideas both from within mathematics and from other epistemologies such as philosophy and metaphysics. For instance, Devlin's otherwise impressive and impassioned 1994 treatise on mathematics employed "pattern" in its title and in its central arguments but did not define what patterns were. The closest Devlin came to a definition was in describing a pattern typology:

What a mathematician does is examine abstract patterns – numerical patterns, patterns of shape, patterns of motion, patterns of behavior, and so on. Those patterns can be either real or imagined, physical or mental, static or dynamic, qualitative or quantitative, purely utilitarian or of little more than recreational interest. They can arise from the world around us, from the depths of space and time, or from the inner workings of the human mind.

(Devlin 1994, p. 3)

In summary, while much effort has gone into formal mathematical modelling and use of patterns, much (but not all) of this work has been limited by a lack of attention to what pattern actually is. If we are not sure quite what pattern is in these texts, then their claims that mathematics somehow is the *de facto* science of pattern are rather undermined and we are obliged to look elsewhere.

Dynamic Pattern Theories

Another scientific pattern discourse has used the concept of pattern as a way of describing dynamic and emergent phenomena. For instance, Izard et al. (2000) argued that, while individual emotions were hard to identify, pattern theory could be used to model the regularities of common aggregated emotional states, even when they were precipitated by different triggers and involved different interacting individual emotions. Pattern in this discourse is all about aggregation and the constitutive logic of pattern elements (features), with a particular focus on what tips an instance into or out of conformance with a pattern:

A feature F is constitutive for a pattern X if it is part of at least one set of features which is minimally sufficient for a token to belong to a type X. "Minimally sufficient" means that these features are jointly sufficient for the episode to be of type X, but if one of them would be taken away the episode would not count as an instance of type X anymore.

(Newen et al. 2015)

The suggestion of an episodic nature of pattern conformance is of particular interest, both ontologically and epistemologically. Not only can pattern recognition be momentary (now you see it, now you don't), the phenomenon that is recognised may also be momentary. This is particular apparent in considering emotion, which involves both the complex product of dynamic cognitive and physiological processes and perceptions thereof which are dependent on those processes:

[As] a pattern theory, "emotion" is a cluster concept that includes a sufficient number of characteristic features. Taken together, a certain pattern of characteristic features constitutes an emotion, although no individual feature by itself may be necessary to constitute an emotion. This means ... [that] there are borderline cases where it is not clear whether some complex cluster of aspects counts as an emotion.

(Gallagher 2013)

This suggests that patterns (whatever they are) need to be considered in terms of their persistence (how long they last), sustainability (how susceptible they are to decay or collapse), and adaptability (how they can change with or in response to external contextual changes). In this way, Gallagher (2013) accounted for differences in perceptions of emotion and feelings of emotion over time, in different contexts, and by different individuals. The argument here is not that patterns cannot be rigidly constitutive, only that they need not be. Gallagher extended this line of argument to consider how pattern theories of an individual's sense of self might work at different levels of abstraction:

> First, one can think of the pattern theory of self as operating like a meta-theory that defines a schema of possible theories of self, each of which would itself be a pattern theory … Second, however, any particular theory of self can be a pattern theory, and one pattern theory can differ from another pattern theory by specifying different aspects (from among a through g) to be included as aspects of self.
>
> *(Gallagher 2013)*

Suggesting that different pattern theories might be applied to a particular problem begs the question whether a pattern is transient because it is perceived in different ways, because the mind perceiving the pattern is changing, or because the phenomenon the pattern refers to is also in a state of flux.

Scott Kelso, in focusing on how brains work, outlined pattern dynamics in terms of circular causalities that had competing amplifying and dampening effects that shape the resulting brain patterns. Sometimes these forces are random, at other times they may be skewed or directed such that certain pattern forms predominate. Collectively, "patterns arrive spontaneously as the result of large numbers of interacting components … loss of stability gives rise to new or different patterns and/or switching between patterns" (Scott Kelso 1995, pp. 16–17). These observations notwithstanding, Scott Kelso avoided committing to any particular definitions of pattern, saying instead that:

> Everybody knows the world is made up of processes from which patterns emerge, but we seldom give pause to what this means. Maybe it's enough to know a pattern when we see it.
>
> *(Scott Kelso 1995, p. 3)*

In summary, while other discourses have proposed models of pattern elements and pattern structures, dynamic pattern theorists have focused on how patterns can emerge, change, fracture, and dissolve over time.

Pattern and Mind

There are some who have considered the tension between the idea that pattern is in the world around us and the idea that pattern is a cognitive phenomenon. For instance, when Elkins (2002) discussed patterns, he referred to perceived regularities in terms of personal patterns of understanding rather than to regularities independent of mind. Bell (2012) argued, as others have done, that patterns are everywhere in the world around us and that being able to perceive them is an essential human capability. Although he drew upon the work of pattern theorists such as Peter Stevens (who I will shortly come to), Bell's definition of pattern was couched in the form of a list of expressions:

> The dictionary offers various definitions of 'pattern', each of which is relevant to the subject of this book: an arrangement of repeated or corresponding parts, decorative

motifs, etc.; a decorative design; a style; a plan or diagram used as a guide in making something; a standard way of moving or acting; a model worthy of imitation, a representative sample.

(Bell 2012, p. 13)

Stevens' 1974 work "Patterns in Nature" has been widely cited in other pattern-related works. His core thesis was that there are a limited number of basic patterns that nature follows, which reflect "universal spatial constraints", the efficiencies of morphogenesis (the origin of form), and the way matter responds to forces such as stress, turbulence, and flow. These abstracted patterns (that I would call regularities) included spirals, meanders, explosions, trees and branches, and bubbles and cells. Although Stevens used the term "pattern" to refer to his examples from the natural world, he was careful to distinguish between the appearance of things, their underlying form, and the causes of the underlying form, but not the thinking involved in perceiving them. M'Closkey & VanDerSys also discussed pattern in the context of natural landscapes and ecologies, but focused less on the intrinsic forms and regularities of those phenomena than on the ways in which we understand them and share that understanding (M'Closkey & VanDerSys 2017).

Ball's "Patterns in Nature" (2016) was another impassioned work on general principles of pattern, but again the concepts of pattern, design, regularity, order, and symmetry were used in a fluid and rather interchangeable way without defining them or differentiating between them. As others have done, Ball hinted at pattern as our understanding of the world rather than the world itself, but did not expand on this idea:

The world is a confusing and turbulent place, but we make sense of it by finding order. We notice the regular cycles of day and night, the waxing and waning of the moon and tides, and the recurrence of the seasons. We look for similarity, predictability, regularity: those have always been the guiding principles behind the emergence of science. We try to break down the complex profusion of nature into simple rules, to find order among what might at first look like chaos. This makes us all pattern seekers.

(Ball, 2016, p. 6)

Margolis (1987) proposed that "thinking is based on recognising patterns" such that "an individual's cognitive repertoire" is based on their "available cognitive patterns and their relation to cues", which aligns with some of my arguments later in this volume. That said, Margolis did not define pattern *per se*, and he used the term variously to denote discrete things in the world, the qualities of those things, perceptions of those things, and cognitive responses to those perceptions.

Hofstadter considered pattern in several of his works, often positioning it as being inseparable from mind and yet something that mind seeks beyond itself:

The elusive sense for patterns which humans inherit from our genes involves all the mechanisms of representation of knowledge, including nested contexts, conceptual skeletons and perceptual mapping, slippability, descriptions and meta descriptions and their interactions, fission and fusion of symbols, multiple representations (along different dimensions and different levels of abstraction), default expectations, and more.

(Hofstadter 1979, p. 674)

In his later work *Metamagical Themas: Questing for the essence of mind and pattern*, Hofstadter spent a little more time reflecting on pattern phenomena (not unreasonably given it was mentioned in the title) but still without actually defining what pattern was, leaving us to infer what he meant from statements such as:

> There are inner mental patterns underlying our ability to conceive of mathematical ideas, universal patterns in human minds that make them receptive not only to the patterns of mathematics but also to abstract regularities of all sorts in the world ... mind [is] the principal apprehend of pattern, as well as the principal producer of certain kinds of pattern.
>
> *(Hofstadter 1985, p. xxv)*

Ericsson (2006) took a cognitivist perspective in discussing the ways in which our minds "chunk" knowledge in forming and accessing memories, and how experts develop and use ever more complex and elaborate pattern chunks in their thinking. However, while these chunks might or might not be equivalent to patterns, Ericsson and Harwell (2019) more typically used the term pattern to refer to regularities. Ericsson's colleagues used the concept of pattern even more liberally to refer to connections between external regularities, our perceptions of these regularities and connections, and hypothetical connections between our perceptions and our understanding. For instance, Ross et al. (2006) observed that pattern could serve as a bridge between new information and prior experience, while Chi (2006) used pattern to refer to regularities in the world, our perceptions of patterns within these regularities, and how we reason based on those pattern perceptions.

Many of the texts that used pattern as a central concept that I read in developing this book did not define what pattern meant, and many of them used pattern interchangeably with regularity. These include Dawkins (in his work on patterns of life [Dawkins 1986]), Parsons (in his work on "pattern variables" [Parsons & Shils 1951, p. 77]), King (in her work on patterns of human connection [King 2020]), and Diesing (in his work on patterns of discovery in the social sciences [Diesing 1971]). Again and again, pattern has been presented as a central construct but articulated in circuitous, ambiguous, and/or polysemic ways, usually implying much more than was said. At times it even seemed that concept of pattern was used by some as a talisman to add metaphysical authority to their knowledge claims and theses. As much as I appreciated the frisson of ontological mystery and epistemic ambiguity that this use of pattern seemed to add, this kind of rhetorical sleight of hand hardly contributes to robust scientific reasoning.

Pattern Aesthetics

One last pattern discourse to consider here has focused on the aesthetics of pattern. For instance, Jones (1868) mentioned pattern again and again in the context of things designed and intended to please through the deliberate use of regularities (repetition, symmetry, transformation), natural form (such as leaves), abstract forms (lines, squares, circles), and the use of colour. Manufactured patterns are the basis of much of the expressive arts:

> We find pattern in music whether it is a symphony or a blues song. Poetry has strong pattern qualities which may be in the repetition of words, or rhyme, or the scansion of metrical verse. Dances have their patterns, not only the tribal, magical and social ones,

but also those which are less formal, often individual. Man continually structures his ideas and in so doing organise his thoughts and actions into patterns which enable him to communicate.

(Palmer 1972, p. 43)

Pattern has been discussed in terms of single performances or artefacts, in terms of multiple performances or artefacts, and in unifying an idiom (the organisation of and connection between many instances). Moreover, there has been rather less ambiguity in this discourse as to what pattern refers to. For instance, Jirousek (1995) defined pattern as:

An underlying structure that organizes surfaces or structures in a consistent, regular manner. Pattern can be described as a repeating unit of shape or form, but it can also be thought of as the "skeleton" that organizes the parts of a composition.

The relationship between the externalities of patterns and our affective reactions to them has been a recurring concern in the psychology of art, reflected in the eternal question of whether it has some intrinsic qualities or is solely "in the eye of the beholder". Although much of the lay literature on the aesthetics of pattern does not distinguish between form and experience, when this tension is considered, it can afford additional insights. For instance, in discussing music, Hofstadter observed that:

One must distinguish ... between syntactic pattern and semantic pattern. The notion of a syntactic pattern in music corresponds to the form of structural devices used in poetry: alliteration, rhyme, meter, repetition of sounds, and so on. The notion of a semantic pattern is analogous to the pattern or logic that underlies a poem and gives it reason to exist: the inspiration, in short.

(Hofstadter 1985, p. 181)

In this case, syntactic aspects of pattern refer to its regularities, symmetries, form, etc., while its semantic aspects refer to our understanding of and responses to these regularities, implying the recurring tension between phenomena we encounter and our perceptions of those phenomena. Pattern, from an aesthetic perspective, is again about relationships. These can be relationships in form, substance, or execution, but they also include relationships between that which observes (spectators, audiences) and that which is observed (artefacts, performances etc.). Not only do the aesthetics of pattern depend on affective responses, they are often designed to stimulate them.

Some aesthetic pattern sources have considered pattern as being related to perception and mind, such as in considering the universality or cultural specificity of patterns and the transformative nature of aesthetic experiences (Winner 2018). However, most writers have taken a pragmatic position, often based on context-specific interpretations of what patterns were for. For instance, pattern is a central concern in dressmaking, but with a particular focus on physical templates (plans, blueprints, or sets of instructions) necessary to producing a particular kind of thing:

Patterns are so useful and important because they allow a designer's idea to become repeatable. Patterns allow for a correct result to be reached time and time again, and it is important to think of them in this way.

(NA 2013)

While some aesthetic pattern texts have focused on practical expressions or uses of pattern or on the ways in which pattern can be perceived or valued, others have focused on working with pattern in creative ways. For example, Alison (2019) proposed a series of narrative patterns as tools for writers (drawing on Stevens' typology of "natural patterns") to provide underlying form and structure, to make connections and associations within and beyond texts, and to suggest new and unusual forms of crafting narratives.

Clearly, pattern is an important concept in aesthetic discourses and is often grounded in applied examples and practices. Issues of the relationships between pattern and mind and between pattern and regularity are also to be found here, such that pattern as relationship is perhaps the dominant underlying metanarrative here. However, the (perfectly reasonable) focus on application and performance rather than on philosophical or scientific debate suggests that there is little here that can help to advance a pattern theoretical perspective.

Chapter Summary

Although what the concept of pattern means changes to suit the domain or paradigm at hand, I found a number of regularities across different discourses in texts that made use of the concept of pattern. For one, I found inflexibility in the tendency to seek or express absolutes or fixed representations of pattern phenomena. Hayek noted the imprecision of patterns, Alexander noted that patterns can be interpreted and combined in many ways and that no pattern language can ever be definitive, and Grenander focused on pattern transformations and probabilistic pattern conformance. And yet, despite this, they all sought to formalise and codify representations of patterns in ways that were more concrete and definitive than a philosophy of pattern (that embraces variance) might reasonably afford. This may reflect an underlying assumption that patterns equate to templates, or it may simply reflect these scientists' pursuit of clarity, definition, and solidity in understanding the universe. I also found what seemed to be a systemic and widespread inattention to pattern as a concept. I will not say that all these individual works are fatally flawed because of this inattention, as they all pursued rigour and significance within their own discourses and frames of reference. And yet, if pattern is central to their ideas, then why is it treated so ambiguously?

I was somewhat concerned about the recurring rhetorical use of pattern to confer authority to arguments and knowledge claims. When pattern "just is" such that it cannot or need not be defined, then this has more to do with rhetorical and sometimes even poetic form than it does with scientific inquiry. A more appropriate scholarly response to encountering or working with a term like pattern and the concepts it refers to is to examine the term closely and deconstruct the assumptions made and beliefs held about it. Often pattern was used in the same text and sometimes even in the same sentence to refer to regularities in the world, to perceptions of these regularities, and to the understanding and meaning that might be ascribed to those perceptions. Often there was a circularity to these arguments; regularities were explained in terms of their pattern causes even as patterns were articulated as a way of creating or directing those same regularities.

So, what can we take from this jumble of observations, theories, and descriptions? It would seem that, whatever they are, patterns are parsimonious; a few patterns can generate or reflect a multitude of instances. It would also seem that a pattern is not the same as an instance of a pattern; one pattern may have many instances, and those pattern instances may resemble each other but they are not necessarily identical. Pattern is not just "out there" in the world, there is an intrinsic cognitive character to patterns, particularly in regard of pattern perception, pattern recognition, and pattern discovery. Patterns may be static, changing, and emergent. Patterns may change quickly, slowly, or not at all. Patterns may be single occurrences, episodic, or show other spatial or temporal dynamism. Certain patterns can be linked with values such that the legitimacy of different patterns may be contested, and systems of pattern values can define paradigms, genres, and cultures. Patterns can be combined such that one pattern can depend on another, and one pattern can contain another. Patterns work at different levels and scales; a pattern may be realised in specific instances (performances, works, episodes) and across whole pattern systems (idioms, traditions, genres).

Despite these insights, we are left with many questions. For instance, are patterns phenomena in the world, are they what we perceive, recognise, or understand about a phenomenon, or are they both? Many of the things we call *patterns* in the world are what I would call *regularities*, while I (and others) would argue that patterns are better understood as the ways in which we perceive and understand these regularities. Indeed, do patterns depend on our perceptions of them, or do patterns exist independent of our perceptions, or both? The answer to this question is, I think, provisionally "both". Although patterns are grounded in perceptions and understandings of regularities then they must constitute a cognitive phenomenon. On the other hand, patterns can be recorded, codified, and shared. The links between cognitive patterns and any expressions thereof clearly needs more thought and attention. Clearly, although there are many questions and many unresolved issues regarding pattern, they provide useful starting points for developing pattern theory.

References

Alexander C, Ishikawa S, Silverstein M. *A Pattern Language: Towns, Buildings, Construction*. New York, NY: Oxford University Press: 1977.

Alexander C. *The Timeless Way of Building*. New York, NY: Oxford University Press: 1979.

Alison J. *Meander, Spiral, Explode: Design and Pattern in Narrative*. New York, NY: Catapult: 2019.

Ball P. *Patterns in Nature: Why the Natural World Looks the Way it Does*. Chicago, IL: The University of Chicago Press: 2016.

Bateson G. *Mind and Nature: A Necessary Unity*. EP Dutton. New York, NY: 1979.

Bell S. *Landscape: Pattern, Perception and Process*. London, UK: Routledge: 2012.

Brown University Pattern Theory Group overview, circa 2002. *Archived Copy on the Internet Archive Wayback Machine*: https://web.archive.org/web/20090207181519/http://www.dam.brown.edu/ptg/

Chi MTH. Laboratory Methods for Assessing Experts and Novices Knowledge. In: Ericsson KA, Charness N, Feltovich PJ, Hoffman RR (eds). *The Cambridge Handbook of Expertise and Expert Performance*. Cambridge, UK: Cambridge University Press: 2006.

Christen M. Patterns in the Brain: Neuroscientific Notes on the Pattern Concept. In: Gleiniger A, Vrachliotis G. Pattern: *Ornament, Structure, and Behaviour*. Context Architecture – Birkhauser: 2009.

Dawkins R. *The Blind Watchmaker*. New York, NY: W.W. Norton: 1986.

Dennett DC. Real Patterns. *The Journal of Philosophy*. 1991; 88(1): 27–51.

Devlin K. *Mathematics: The Science of Patterns*. New York, NY: Scientific American Library: 1994.

Diesing P. *Patterns of Discovery in the Social Sciences*. Transaction: 1971.

Elkins J. *Stories of Art*. New York, NY: Routledge: 2002.

Ericsson KA, Harwell KW. Deliberate Practice and Proposed Limits on the Effects of Practice on the Acquisition of Expert Performance: Why the Original Definition Matters and Recommendations for Future Research. *Front. Psychol.* 2019; 10: 2396.

Ericsson KA. An Introduction to Cambridge Handbook of Expertise and Expert Performance: its Development, Organisation, and Content. In: Ericsson KA, Charness N, Feltovich PJ, Hoffman RR (eds). *The Cambridge Handbook of Expertise and Expert Performance*. Cambridge, UK: Cambridge University Press: 2006.

Fowler M. *Analysis Patterns*. Indianapolis, IN: Addison Wesley: 1997.

Gallagher S. A Pattern Theory of Self. *Front. Hum. Neurosci.* 2013; 7: 43.

Gamma E, Helm R, Johnson R, Felicides J. *Design Patterns: Elements of a Usable Object-oriented Software*. Amsterdam: 1994.

Grenander U, Miller M. *Pattern Theory: From Representation to Inference*. London, UK: Oxford University Press: 2007.

Grenander U. Advances in Pattern Theory. *The Annals of Statistics*. 1989; 17(1): 1–30.

Hanson NR. Patterns of Discovery. Cambridge UK: Cambridge University Press: 1958.

Hayek FA. *Epistemology of Complexity. Studies in Philosophy, Politics and Economics*, London, UK: Routledge & Kegan Paul: 1967.

Hayek FA. The Pretence of Knowledge. *The Swedish Journal of Economics*. 1975, 77(4): 433–442.

Hofstadter DR. *Gödel, Escher, Bach: An Eternal Golden Braid*. New York, NY: Basic Books 1979.

Hofstadter DR. *Metamagical Themas: Questing for the essence of mind and pattern*. New York, NY: Basic Books: 1985.

Izard CE, Ackerman BP, Schoff KM, Fine SE. Self-organization of Discrete Emotions, Emotion Patterns, and Emotion-Cognition Relations. In S. E. Lewis and I. Granic (eds), *Emotion, Development, and Self-Organization: Dynamic Systems Approaches to Emotional Development*. New York, NY: Cambridge University Press: 2000.

Jirousek C. Pattern. In: *Art, Design, and Visual Thinking*. Cornell University: 1995. Online at http:// char.txa.cornell.edu/language/ELEMENT/PATTERN/pattern.htm

Jones O. *The Grammar of Ornament*. London, UK: Bernard Quaritch: 1868.

King M. *Social Chemistry: Decoding the Patterns of Human Connection*. New York, NY: Dutton: 2020.

Leitner H. *Pattern Theory: Introduction and perspectives on the Tracks of Christopher Alexander*. Spartanburg, SC: CreateSpace: 2015.

M'Closkey K, VanDerSys K. *Dynamic Patterns: Visualizing Landscapes in a Digital Age*. London, UK: Routledge: 2017.

Margolis H. *Patterns, Thinking and Cognition*. Chicago IL: University of Chicago Press: 1987.

Mouzon SA, Henderson S. *Traditional Construction Patterns*. New York, NY: McGraw Hill. 2004.

Mulligan J, Mitchelmore M. Awareness of Pattern and Structure in Early Mathematical Development. *Mathematics Education Research Journal*. 2009, 21(2): 33–49.

Mumford D, Desolneux A. *Pattern Theory: The Stochastic Analysis of Real-World Signals*. Natick, MA: AK Peters Ltd: 2010.

N/A. *How Patterns Work: The Fundamental Principles of Pattern Making and Sewing in Fashion Design*. London, UK: Assembil Books: 2013.

Newen, A., Welpinghus, A., and Juckel, G. Emotion Recognition as Pattern Recognition: The Relevance of Perception. *Mind Lang*. 2015; 30, 187–208. DOI: 10.1111/mila.12077

Niemann H. *Pattern Analysis and Understanding* (2nd ed.) Berlin, Germany: Springer-Verlag: 1990.

Palmer F. *Visual Awareness*. London, UK: Batsford: 1972.

Parsons T, Shils EA. *Toward a General Theory of Action*. Cambridge, MA: Harvard University Press: 1951.

Ross KG, Shafer JL, Klein G. Professional Judgements and "Naturalistic Decision-making". In: Ericsson KA, Charness N, Feltovich PJ, Hoffman RR (eds). *The Cambridge Handbook of Expertise and Expert Performance*. Cambridge: Cambridge University Press: 2006.

Scheurer F. Architectural Algorithms and the Renaissance of the Design Pattern. In: Gleiniger A, Vrachliotis G. *Pattern: Ornament, Structure, and Behaviour*. Context Architecture – Birkhauser: 2009.

Schlechte T. *A Pattern Language for Composing Music*. Dresden, Germany: PfCM Press: 2019.

Scott Kelso KA: *Dynamic Patterns: The self-organization of brain and behavior*. Cambridge MA: MIT Press: 1995.

Shalizi CR, Crutchfield JP. 2001. Computational Mechanics: Pattern and Prediction, *Structure and Simplicity*. *Journal of Statistical Physics*. 104: 816–879.

Shawe-Taylor J, Cristianini N. *Kernel Methods for Pattern Analysis*. Cambridge, UK: Cambridge University Press: 2004.

Steen LA. The Science of Patterns. *Science*. 1988; 240(4852): 611–616.

Stevens PS. *Patterns in Nature*. London, UK: Penguin: 1974.

Winner E. *How Art Works: A Psychological Exploration*. London, UK: Oxford University Press: 2018.

3

FORM, ORDER, AND REGULARITY

In which I consider foundational concepts related to pattern.

As I outlined in the previous chapter, many authors have used the terms *pattern* and *regularity* interchangeably. When Portmann (1967) discussed patterns in the morphology of animals, he referred to the regularities of colouration and camouflage and in the ordering of species, and when Capra (2021) discussed patterns, he referred to the regularities of propagating wave forms and the (not quite) repeating regularities in the form of leaves and branches. Before describing the basis for pattern theory in Chapter 4, in this chapter I will explore the concepts of form, order, and regularity, and consider their relationships with pattern and pattern thinking.

Form

Form is both a noun and a verb. As a noun, form refers to the apparent shape, structure, and/or configuration of something (its perceived morphology). As a verb, form refers to giving shape, structure, and/or configuration to something. These provisional definitions notwithstanding, what form is and what it means have long been debated by philosophers. Plato argued that there was a perfect world of which the one we perceive was only a shadow, and that *Forms* were the ideals from which the things we perceive were derived. Although agreeing that form was an important concept, Aristotle rejected much of Plato's concepts of Form, arguing instead that form reflected a thing's *essence*. Bacon rejected these kinds of metaphysical perspectives on form, emphasising that the evidence of our senses was all we could rely on. Kant further differentiated between *matter*, which was what our experiences of the world tell us, and *form* as our ability to make sense of these experiences. Sartre argued that existence preceded essence, again rejecting the idea of form as being any more than the reality of a thing (or perceptions thereof). Quine went further to argue that essences are only to be found in our understanding of phenomena.

DOI: 10.4324/9781003543565-3

I would argue that form is constituted in our perceptions of the appearances, structures, expressions, and performances of specific things. Indeed, I would suggest that everything that can be perceived, concrete and abstract, physical and mental, must have some kind of form. Even being relatively formless is a kind of form, while being entirely formless might equate to nonexistence. We can draw on Kantian ideas of *noumena* and *phenomena* in this regard. Noumena have no apparent form, but this does not mean they do not exist, only that their existence is not apparent to us. Form is not required for existence, but it is required for the perception of existence. Put another way, for something to have form it needs to be perceivable in some way or other, even if this is only obliquely through sensors, mathematical analyses, logical reasoning, or by exercising our imaginations. In perceiving form, we need to be mindful that we can both establish and dissolve perceptions of form by changing our focus of attention. Form is not continuous therefore; it can appear and disappear as our attention shifts, which in turn means that form requires an observing mind to make those distinctions. Form is not, therefore, separable from mind, even though it can clearly refer to things outside the mind (Baecker 2021).

Let me illustrate these ideas with an example. What forms do you perceive in Figure 3.1 below? How you answer this question will in part depend on the frame of reference you use. I think it likely that you will observe and maybe contrast the form of the five objects in the image, but you could instead focus on the form of the image as a whole, the form of the page the image is on, or the form of this book. Alternatively, you may focus on different forms within an object, or drill down further to pixels on a screen or the ink and fibres of a printed page. If you have focused on the five apparently discrete objects, what material forms do you perceive? You may perceive the object on the left as a rounded but uneven speckled blob. The object in the middle is another blob but with jagged edges and more unevenness than the first object. The three objects on the right likely appear to be parallelograms of different tints although each parallelogram is of a uniform tint, some of which have clusters of partial circles on them. All five objects have smudges on their right-hand sides.

Because it is likely you have previously encountered two-dimensional images of three-dimensional things, you may interpret these smudges as shadows and from this you may perceive these objects as three dimensional. If so then you likely *see* a pebble, a piece of rock, and three LEGO® bricks based on memories of other pebbles, rocks, and bricks that you have seen. From this, you might also infer their sizes, weights, temperatures, and

FIGURE 3.1   What might we learn about form and order from observing this image?

Author's own photograph.

textures, perhaps by imagining you are picking these things up and brushing them against your cheek or lips. These are all expressions of medial form, and they are all examples of pattern recognition.

However, in translating your sensory experience to perceptions of identities, other perceptions may come to you, although what they are will depend on your experiences and memories. If you are a geologist, you might recognise the pebble as being made of some kind of quartzite and the rock as being volcanic scoria. If you are a LEGOist you might perceive the bricks in terms of how they fit into the LEGO® system. You may have many memories of rocks and bricks that include where you were when you previously encountered these things, what you were doing, who you were with, and how you felt at the time.

In summary, form refers to the specifics of things but is rooted in perception such that anything that can be perceived has form. Form is not required for existence, but it is required for the perception of existence. However, form also depends on scales, levels, and frames of reference of observers such that it can appear and disappear according to context, observer, and frame of reference.

Order

Like form, *order* can be both a noun and a verb. As a noun, order refers to particular arrangements, dispositions, or collections of things. As a verb, order is about establishing or imposing a particular arrangement, disposition, or collection on things. For instance, in Figure 3.1, the five objects might be grouped into rocks and bricks or into natural and manmade things (the same grouping based on different criteria) or they might be grouped according to their utility, weight, material strength, or retail value. Alternatively, we might simply say that their spatial order is (left to right) rock, rock, brick, brick, brick. Despite this flurry of possible orders, the image of five objects is unchanged, which in turn suggests that order is also a product of mind even if it depends on things outside the mind.

Lorand (1994) noted tensions between order as a principle of organisation and order as the state of things that reflect that principle. An ordering principle is "anything that can serve, under certain circumstances, as a pattern, rule, law, model, or formula by which a set can be ordered", it is "the rule according to which the elements are organized and unified". Things can be ordered at one scale or in one way while being unordered at another (Bohm 1996) as illustrated in the classic Eames' movie "Powers of Ten" (Eames & Eames, 1977). Not only can order appear or disappear according to scale, it can also appear and disappear based on dynamic changes in a system, for instance, in water moving between states of ice (ordered), water (moderately disordered) and steam (very disordered). We could therefore consider order in terms of local reductions in entropy, or perceptions thereof.

Bohm argued that there are many orders that we cannot perceive but that nevertheless exist. He called these *implicate orders* of which the *explicate orders* (those we can perceive) are a subset (Bohm 1980). This echoes Bhaskar's concept of three domains of reality: *empirical* (that which we experience), *actual* (that which happens independent of experience), and *real* (the objects and mechanisms that constitute reality) (Bhaskar 2008). At a more practical level, we might also consider order to be contextually invisible in cases where one observer perceives order where another observer does not, where an observer initially does not perceive order but through reflection, research, training, or practical experience comes to perceive it, or where attending to some orders renders others invisible. It is not that invisible order cannot be perceived, only that it has not been perceived by particular observers in particular contexts.

While there may be many possible orders that we might observe in something, we tend to focus on one or two particular orders to the exclusion of any others (for instance in bistable images). Can we objectively account for order, or is it, as Michaelides (2008) also suggested, always in the eye of the beholder? Lorand (2000) argued that an ordering principle may be objectively present but if it cannot be perceived by some means or other then order is not subjectively present. Bohm & Peat (1987) argued that it is context that changes order more than the observer, but what is context but an observer's frame of reference? Interestingly, Bohm also argued that some kinds of order were not entirely subjective:

> ... for example, distance, time, mass, or anything else of this nature ... can be defined and communicated just as well as can be done with other qualities that are commonly recognised to be capable of an objective description.
>
> *(Bohm 1996, p. 8)*

Lorand argued that "order is neither objective nor subjective but a mutual product of mind (subject) and things (object)" (Lorand 2000, p. 25). Wikman (2013) described this as a combination of *gathering together* and *gathering apart*, while Foucault in his "The Order of Things" (Foucault, 1970) (which unfortunately had little else to say about the concept of order) observed that "comparison and order are one and the same thing". Clearly, perception is bound up with order in all of these perspectives.

In considering order, we should also consider the concept of *disorder*, not least because it does not simply mean the absence of order. Rather, disorder typically refers to a deficit in or deviation from functional behavioural ideals, healthcare, social, criminal, and military contexts ("medical disorder", "drunk and disorderly", "disorderly conduct"). Often these reflect a sense of order as being good and desirable ("law and order"), while disorder is bad and as such it needs to be challenged, mitigated, or punished. Even disorder as a state of untidiness generally fits this conceptual framing. However, noting the biases that this might introduce to a more abstract consideration of order and disorder, this is not the meaning I will pursue. Rather I shall focus on degrees of order and disorder.

Bohm & Peat (1987) argued that neither absolute order nor absolute disorder are real or achievable, that disorder is a form of order, and that we need only use the concept of order to also reflect disorder. Bohm later suggested that order and disorder are a coupled dyad:

> What is commonly called disorder is merely an inappropriate name for what is actually a certain rather complex kind of order that is difficult to describe in full detail. Our real task can, therefore, never be to judge whether something is ordered or disordered, because everything is ordered, and because disorder in the sense of the absence of every conceivable kind of order is an impossibility. Rather, what one really has to do is to observe and describe the kind of order that each thing actually has.
>
> *(Bohm 1996, pp. 10–11)*

If orders are imperfect and partial then they will of necessity have boundaries, transformations, and transitions between higher, lower, or differently ordered regions. There are whole literatures and theories of transitions, such as those describing transitions between solid, liquid, and gaseous states, and life transitions from child to adult to senescence, and finally to death. Rather than disorder being a concept of absence or lack of order, we can think of disorder in terms of transitional states of order. Transitions are, after all, a necessary aspect of order given its dependence on distinctions and boundaries. Moreover, we can consider many ways in which such transitions might occur (within levels, between levels etc.), and the characteristics of those transitions (changing degrees or kinds of ordering principles, changing levels or kinds of distinctiveness etc.).

In summary, order refers both to the state of one or more things and to the manipulation of those states. Given that we may perceive form in our own thoughts and feelings as well as in the objects and environments we encounter, those things in which we perceive order can be both concrete and abstract, and both objective and subjective in nature. However, order, like form, is dependent on perception and as such it cannot be disentangled from mind. Perceptions of order, like those of form, can change according to level, scale, and frame of reference such that the perception of order involves a complex interplay of subjectivity and objectivity. After all, we do not all need to perceive or even be able to perceive a particular order for it to exist; arguably it just takes one observer. Order connects an instance to an ordering principle but, in doing so, order is always perceived in reference to an ideal that it cannot perfectly embody.

There are many possible ordering principles: alignments (spatially or conceptually oriented to other things), symmetries (the same or similar when translated in some way or other), geometries (local structure such as tiling, stacking, or tessellation), topologies (global structure such as paths, deformations, and equivalences), sequences (spatial, chronological, abstract), rhythms (repetitions, frequencies), and hierarchies (precedence, modularity). We might add Stevens' (1974) spirals, meanders, explosions, vortices, trees, branches, meanders, cracks, bubbles, and cells to this list. We might also draw on Kontopoulos' (1993) social structuration principles that interestingly both overlap and diverge from Stevens' ordering principles: hierarchies, flows and impaired flows, self-selection and segregation, constraining identities, comparative advantage and comparative exclusion, oligarchy and oligopoly, gaming, federalism, social frustration, market reversals, tipping rules, sorting rules, matching rules, travelling, interfacing, voting, fluctuation, nucleation, differentiation, allometry and allometric speciation, islands and the logic of fission, percolation, cellular automata, fractals, mixing and merging, waves, branching/uncoupling, cascades, packing rules, surface-to-volume ratios, rules of unpacking (partitioning), informational packing, parsing, swamping, rules of pruning, eco-logics (entropic and heterarchical), totalising logics, and metaptations.

In considering order we should consider those *natural orders* that we perceive in the forms and behaviours of plants and animals, in the formation and behaviours of our world (rivers, oceans, weather), and in the universe beyond our world (orbits, aurora, eclipses). We can also perceive natural order in the structure of crystals and molecules and in the behaviours of stars and galaxies. Indeed, if we perceive an ordering principle in things that is (or appears to be) independent of human intent or artifice, then we are likely to claim that it is an example of natural order. Many of us find natural order in the world around us pleasing, whether it is in the whorls of seashells, the propagation of ripples across water or across a field of wheat, or in the sequence of colours in a rainbow. However, this is not just a matter of form or perception, identifying something as natural or unnatural often involves some kind of ethical or aesthetic stance. For instance, heterosexual, cisgender, monogamous relationships have often been portrayed as the natural order of things, with any other kind of intimate relationship painted as unnatural. Similarly, social orders of class and caste, as well as those associated with gender roles, race relations, and even religions have been considered by some to be natural orders, which in turn provides the basis for oppressing or removing examples of unnatural orders (Cartwright & Ward 2016).

Establishing, extending, and maintaining ideologies that are based on concepts of natural order often reflects in-group and out-group divisions (Taijfel 1970) that focus on protecting and extending in-group privilege, safety, or control over other groups through advancing certain natural orders and stigmatising "unnatural" orders. When this happens, the concept of natural order shifts to being something that is controlled, imposed, and replicated, such that conformance or submission to the authority and significance of order is required and often enforced (Napier 2014).

Natural orders can also be illusory. For instance, an apple in a store is likely to be the product of years of crossbreeding and perhaps genetic manipulation, as well as farmers and retailers selecting what produce gets to the shelf. The apple may be waxed to improve its appearance and longevity, and it may be permeated with various fertilisers, insecticides, and other substances. The question whether something as simple as an apple reflects a natural order may, therefore, be hard to answer.

We can only conclude that our perceptions of order are just that, they are perceptions; order is observer-generated even as it reflects real things. Moreover, perception of order can also be understood as having both cognitive and sociological aspects (Bhaskar 2008). All of this means that ordering principles are not *in* the thing before us, they are perceived, interpreted, abstracted, and they need not be logical, systematic, or even rational:

> Ordering principles do not have to be given or found through scientific methods or logical procedures, they can simply be created by the observer, be capricious and meaningless and still function as ordering principles.
>
> *(Lorand 1994)*

For instance, the order of the earth orbiting the sun is naturally occurring, but when a year starts and ends and what year it is are orders defined by arbitrary cultural conventions. Even the concepts and terms we use to distinguish between natural and artificial orders are culturally constructed (Polany 1974; Lorand 2000), such that consensus (or the lack thereof) might also be considered a kind of deliberate order. As Foucault (1970) observed, order (and discontinuities thereof) may seem natural and inevitable but still arise from and serve systems of belief.

Regularity

The word *regularity* in English is derived from the Latin *regula* meaning to rule, which is also the root of *regal* (pertaining to a ruler), *regulation* (a means of ruling), and *regulated* (being ruled), all of which reflect concepts of authority rather than form. So, does regularity refer to the thing that is regulated or to the regulation of the thing? The answer is both; regularity can refer to repetitions, and regularity can also refer to governance.

In terms of regularity referring to repetition, a regular thing repeats in space or in time, or both. Things that repeat have frequencies, both spatial and temporal, such that temporal regularity is measured as the time between repeats and sequences thereof, while spatial regularity is perceived in terms of distances, displacements, and movements (Chetverikov 2000). Another concern is whether regularity is continuous in its repetition or episodic, which in turn resurrects the issue of boundaries, translations, and transitions that I noted in the context of order. There are also regularities in things that draw on a repertoire of repeating elements, such as adenine, cytosine, guanine, and thymine forming elaborate DNA strands, and the seven notes in the diatonic scales in Western music.

As much as the facts of repetition are important, regularity as repetition is also a perceptual phenomenon that can be understood as subjective interpretations of objective events. Repetitions become regularities only when they are perceived as such, but what triggers perception or recognition? How many times does something need to repeat to be perceived as a regularity? If I have a cup of coffee twice in one day is that a regularity? Perhaps not, but if I had a cup of coffee every day for a month then that might perhaps be perceived as a regularity. If I stopped drinking coffee the following month then the regularity would be broken unless the month after that I started my once-a-day coffee drinking again. If I continued with a sequence of one month on followed by one month off, then that too could be perceived as regularity. Does the cup of coffee need to be a standardised unit to count as a regularity? Is it sufficient that I consume some kind of beverage made from roasted coffee beans, whether from an espresso machine, a cafetiere, or from a jar of instant coffee? What about other caffeinated beverages such as tea? If I only take a few sips of the coffee then does that count, or must I finish the whole cup? Is it the act of making or purchasing a cup of coffee that counts independent of whether or not I or anyone else subsequently consumes it? Clearly, what is being attended to is also an important part of regularity.

Although repeats in a regularity may be identical, they do not need to be, they may just be similar. Moreover, what I may perceive as regularity may not be perceived as such by other observers. Not only may we have different interpretive thresholds for something to be perceived as regular, we may also be attending to different putative regularities. Regularity is interpretive and it depends on subjective judgements of sufficiency and similarity in repetitions.

Regularity referring to governance is the perception that something seems to follow or obey laws, rules, or commands, it refers to something that it is regulated. Perceiving this kind of regularity does not require an examination of its causes and origins, it only notes or acknowledges the presence of some apparent guiding force. For instance, we might perceive the regularity of a train or bus service if it follows a timetable, if it observes rules of ticketing, or if it stops to let passengers on and off at bus stops. Regularity as regulation can also reflect a sense of things being "as they should". In this sense, regularity as regulation may be experienced as an absence of surprises, dissonances, conflicts, or paradoxes rather than any affirmative quality. Regularity as regulation can also be experienced or perceived in a more affirmative sense as the imposition of order in an otherwise unregulated situation. For instance, the arrival of the police to deal with a street fight or enforcing a ceasefire in a conflict might be perceived as imposed regularities.

Any consideration of regularity also requires some consideration of irregularity. Like order and disorder, regularity and irregularity can be seen as two sides of the same coin

(Diesing 1971). However, irregularity is not simply the absence of regularity; it can refer to degrees of divergence from some default state or it can reflect changes in or transitions between different regularities. Irregularity may be used to refer to flaws in something such as in "financial irregularities" or when physicians interpret irregularities in heart sounds as indicators of pathology. Irregularity may also be valued, for instance, in the *wabi sabi* qualities of handmade Japanese pottery. If we think about irregularity as breaks in regularity, we might consider irregularity as defining the boundaries and extents of regularities, the places where regularity falters and fails. We might even see irregularity as a different and unexpected or incongruous regularity within a broader regularity. Wherever we look, we find regularity and irregularity tightly entwined.

In summary, something may be perceived as having many kinds of regularity: temporal, spatial, conceptual, symbolic, and affective. Although it might be argued that temporal and spatial regularities are more objective than the others, all regularity involves some degree of subjectivity. This dependence on perception does not deny the facts of repetition, only that we engage arbitrary thresholds of how much repetition is needed to count as regularity, how much precision in repeats is required, and the scale, level, or frame of reference involved.

Chapter Summary

There have been many books (and I am sure that they will be many more) that describe the objectivities and syntactics of form, order, and regularity. Some have done so using mathematical equations, representations, and algorithms, while others have taken a more prosaic interpretive approach to describing, explaining, or even predicting how the objective characteristics of form, order, and regularity are expressed and how they interact. By taking a cognitive interpretivist stance on form, order, and regularity, and particularly on pattern, I do not discount the value or application of this work. However, I would note that, given that all of these approaches (and their products) have been created by human minds, they are themselves inextricably entangled with perception and the basis of perception.

Form arises from perceptions of specific characteristics of things. Order arises from the perception of principles in the grouping or arrangement of things. Regularity arises from the perception of repetition and regulation. Not only are all three dependent on perception (and therefore on mind), form, order, and regularity are not unconnected modalities of perception; the distinction we make between them is also a product of mind. This is apparent in the ways that the perception of one modality can slip into another, that several may be perceived at the same time, and that our attention may oscillate between them. We distinguish between form, order, and regularity because our minds try to make sense of our world by categorising and grouping similar things alongside what we think or feel about them. Indeed, humans find it difficult not to link perceptions of form, order, and regularity to each other, and to purposes, origins, meanings, and significances. Although a teleological tendency to seek and find purpose in all things may lead to illusions and false beliefs, without it human minds could not have developed as they have. And this is the crux of my thesis; pattern is not produced from perceptions of form, order, and regularity, rather these distinctions are the products of pattern thinking. After all, it is easy to think in terms of pattern, but we have to work at thinking in terms of form, order, and regularity. Is pattern our foundational cognitive modality? If so, then where does pattern come from, how does it work, and how is it that we seem so oblivious to its central role in our lives?

References

Baecker D. A Sociological Reading of George Spencer-Brown's Laws of Form. In: Kauffman LH, Compton A, Conrad L, Cummins F, Dible R, Ellsbury G, Grote F (eds.). *Laws of Form – A Fiftieth Anniversary*. Singapore: World Scientific Publishing: 2021.

Bhaskar R. *A Realist Theory of Science* (2nd ed.). London, UK: Verso: 2008.

Bohm D, Peat FD. *Science, Order, and Creativity: A Dramatic New Look at the Creative Roots of Science and Life*. New York, NY: Bantam: 1987.

Bohm D. *On Creativity*. London, UK: Routledge: 1996.

Bohm D. *Wholeness and the Implicate Order*. London, UK: Routledge: 1980.

Capra F. *Patterns of Connection: Essential Essays from Five Decades*. Albuquerque, NM. University of New Mexico Press: 2021.

Cartwright N, Ward K. *Rethinking Order: After the Laws of Nature*. London, UK: Bloomsbury: 2016.

Chetverikov D. Pattern Regularity as a Visual Key. *Image and Vision Computing* 2000; 18(12): 975–985.

Diesing P. *Patterns of Discovery in the Social Sciences*. Chicago IL: Aldine: 1971.

Foucault M. *The Order of Things*. London, UK: Tavistock: 1970.

Kontopoulos KM. *The Logics of Social Structure*. Cambridge, UK: Cambridge University Press: 1993.

Lorand R. The Concept of Order. *The Jerusalem Philosophical Quarterly* 1994; 43: 305–327.

Lorand R. *Aesthetic Order: A Philosophy of Order, Beauty and Art*. Taylor & Francis: 2000.

Michaelides EE. Entropy, Order and Disorder. *The Open Thermodynamics Journal*. 2008; 2: 7–11.

Napier JL. The Natural Order of Things: The Motivated Underpinnings of Naturalistic Explanations for Inequality. In Forgas JP, Harmon-Jones E (Eds.). *Motivation and its Regulation: The Control Within*. London, UK: Psychology Press: 2014: 299–311.

Polany M. *personal Knowledge: Towards a Post Critical Philosophy*. Chicago, IL: Chicago University press: 1974.

Portmann A. *Animal Forms and Patterns: A Study of the Appearance of Animals*. New York, NY: Schocken: 1967.

Stevens PS. *Patterns in Nature*. London, UK; Penguin: 1974.

Taijfel H. Experiments in Intergroup Discrimination. *Scientific American* 1970; 223(5): 96–102.

Eames C, Eames R (1977). *Powers of Ten* [Video]. YouTube. https://www.youtube.com/watch?v=0fKBhvDjuy0

Wikman G. The Notion of Order in Mathematics and Physics. Similarity, Difference and Indistinguishability. *Foundations of Physics* 2013; 43: 568–596.

4

PERCEPTION AND PATTERN

In which I explore the cognitive and neurological bases of pattern.

Perception is relational, it requires someone to do the perceiving, something to be being perceived, and a context within which perception occurs (Johns & Saks 2011). There are many different theories of perception, but they more or less agree that as soon as photons fall on the rods and cones in our eyes, as soon as air pressure changes are detected by our ears, as soon as changes in air chemistry are detected by our olfactory nerves, as soon as the touch or heat receptors in our skin are triggered, or as soon as some other stimulus occurs (setting aside the reactions of the peripheral nervous system) then our brains are involved, and this means that our minds are involved too.

The human brain is a complex collection of structures involving billions of interconnected neurons that acts as the coordinating part of a nervous system that extends down through the spinal column and radiates out to every part of our bodies. We can lose teeth, even limbs, or have organ transplants and we are (more or less) still ourselves; this is not true of brains. However, as conscious and self-aware individuals, we are not particularly aware of the function of our own neurons and neural structures (any more than we are of any other cells in our bodies), but we are more or less aware of what we call our own mind. Our minds are what we use when we think about things, when we recognise things or imagine them, when we reason or reflect on something, when we remember things, and when we feel things both kinetically and emotionally. Mind is intimately connected to brain, as illustrated by brain injuries that also impact aspects of mind, but quite how the translation or transition between brain and mind works has been debated for millennia, and it has spawned many conflicting theories. Dualists hold that mind is not brain-dependent and that mind could exist in other media. Materialists hold that there is a direct correlation between neuronal behaviours in the brain and the working of the mind such that they are inseparable. Idealism is a rather solipsistic belief that there are only phenomena of mind and everything else is illusory. Given the huge strides made in the neurosciences in recent years, materialism is the position I take in articulating this chapter.

DOI: 10.4324/9781003543565-4

Although I argue that brain and mind are inseparable, whether we seek to understand pattern in terms of neurons or experiences would seem to take the discussion in quite different directions. However, since many of the sources I draw on in this chapter seem not to have made much of the brain-mind issue and moved between them in a somewhat fluid fashion, I will follow their lead for now, while noting that this is an issue that I will return to later in this volume.

Patterns and Memories

There are distinct schools of thought that have linked the structure and function of human brains to pattern. There are those who have researched the nature and function of organic brains (such as Bor, Mattson, and Damasio) and those who have sought to build machine analogues of brains (such as Hawkins and Kurzweil). Hawkins (2004) described the brain as a "pattern machine", while Kurzweil's (2012) core thesis was based on a "pattern recognition theory of mind" which posited our having vast numbers of "pattern recognition circuits" in our minds. This is a bold and rather compelling assertion that suggests that pattern thinking (whatever it is and however it works) is the basis for much of our perception, thinking, and cognition in general. However, we should not assume this is a done deal as Hawkins conflated concepts of pattern to include external events, sensory stimuli, cortical behaviours, memory, and connecting sensory data with memory. Kurzweil's thesis was similar (and he acknowledged the influence of Hawkins work on his own thinking) in that he argued that patterns are how the brain makes sense of the world, which in turn affords efficient, adaptable, and recursive ways of "hierarchical thinking" (Kurzweil 2012). Essentially, pattern from Kurzweil's perspective was a function or product of our brains that allowed us to understand the world around us. However, Kurzweil did not specify what patterns were, again leaving us to infer what he meant from the ways in which he used the term.

Mattson's thesis was based on what he called "superior pattern processing", which extended beyond perception to include many other cognitive abilities, including communication, imagination, and creativity:

> Superior pattern processing (SPP) [is] the fundamental basis of most, if not all, unique features of the human brain including intelligence, language, imagination, invention, and the belief in imaginary entities such as ghosts and gods.
>
> *(Mattson 2014)*

Although Mattson also did not specify exactly what he meant by pattern, he did anchor the implied meaning of the term within concepts of cognitive processing:

> … large numbers of encoded images and sound patterns can … be recalled and mentally manipulated in ways that enable comparisons of different patterns and, at least in the human brain, the generation of new patterns that convey objects and processes that could possibly exist or are impossible or implausible.
>
> *(Mattson 2014)*

Bor used the idea of pattern to describe both the nature of sensory information and its content. He also described pattern as "capturing the repetitions [of sensory data] in a rule" (Bor 2012, p. 149). Bor did, however, acknowledge that the terms "pattern", "structure", and "regularity" were used interchangeably, and given this ambiguity, he tended to use the less contentious term "chunks" to refer to discrete neurological units or episodes of neural activity. I do not think that this resolved the issue, but at least he acknowledged it.

I argue that pattern arises from the aggregation of memory. Pattern is not synonymous with memory, it is possible after all to have memory without pattern, but we weave much of our memory together to create impressions, interpretations, and understandings, and this is pattern. As I have noted, others have approximated to this understanding of pattern as connection (Bateson), pattern as knowledge (Grenander), and pattern as problem solving (Alexander). That said, it is not enough to say that pattern is simply aggregated memory. Pattern can create and store understanding, meaning, and significance in addition to identity and interpretation. To that end, I offer three bodies of evidence to support the substance and constitution of pattern as I have suggested it. The first body of evidence is subjective and experiential, based on what I perceive to be my own patterned mind. The second is inductive and reasons from available empirical evidence towards a pattern memory theoretical model. The third involves deductive reasoning to reverse engineer how memories work and what this suggests in terms of a theory of pattern memory.

From a subjective perspective, I can recall my past experiences, feelings, and thoughts, and as a native English speaker, I call these things "memories". Having lived for many years, I have acquired a great number of memories, some of which I have retained from my early childhood. I am also aware that I do not and cannot remember everything I have experienced or thought about. Indeed, my mind seems to be quite selective in terms of what I commit to memory, and this selectivity is only partially controllable by my conscious mind. I have also found that, although I may witness the same event as someone else, their memories of that event often differ from mine. In summary, I experience my memories as partial, selective, interpretive, and subjective.

As much as I have an impressionistic sense of having memories, I cannot interrogate them directly as I would a book or a catalogue; they are only apparent to my conscious mind when I engage in recall and reflection. Sometimes I can access my memories in slightly more deliberate and focused ways, such as when I try to recognise a face or a song. I can also recall memories without any conscious intent, such as in my dreams (whether waking or in my sleep), or when unexpectedly encountering something. I find that I can consciously organise, aggregate, abstract, and generalise my memories (but only to some extent) to create categories of things, and to develop hypotheses and theories about them. I can subsequently remember these abstractions and generalisations as associated memories. Some of my memories seem highly specific and embodied, while others are highly abstract and generalised. Although some of my memories seem more discrete than others, I have a general sense that my memories are interconnected, for instance I can follow a train of thought, inference, or imagination relatively effortlessly.

It seems to me that I recognise and thereby make sense of new things by drawing on my memories of having previously encountered similar things. Often, when I recognise a thing, I do not simply identify it, I can recall its relationships, behaviours, significances, and meanings. My memories also seem to reflect my embodied perceptions and understanding of my experiences such that when I recognize something I also remember how I felt about it or about things like it. I also have some sense that my memories and how I use them seems to vary according to context. For instance, memory in the context of my work as an academic seems to me to be a lot more formal and abstract than memories of things in my personal life or from my childhood.

I have friends whose memories seem better at recall or association than mine and others whose memories seem less adept than mine. I know people whose memories seem to have significantly deteriorated or have been otherwise damaged to the point that their loss of memory has become functionally disabling. I also have a sense of a fading of my own memories over time. Sometimes I misrecognise things or I see things that are not there. Sometimes these lapses seem due to tiredness or inattention, sometimes they seem to reflect past experience and conditioning, and at other times they seem to be due to insufficient or fleeting information. I would hazard that you have similar experiences. Indeed, we might learn much from exploring convergences and divergences in our experiences of our memories. However, assuming that there are more convergences than divergences, I would propose these observations as a provisional experiential basis for the concept of pattern memory. That said, clearly experiential evidence is insufficient in and of itself to either fully imply or validate a pattern theoretical perspective, so let me turn to empirical evidence.

Let me start with "Hebbian learning" which, despite having originally been proposed in 1949, is still the dominant theory of how learning and memory have a common neurological basis. Hebb's theory was based on the idea that groups of neurons become permanently altered through the process of learning and memory formation such that:

> Any frequently repeated, particular stimulation will lead to the slow development of a 'cell-assembly', a diffuse structure comprising cells in the cortex and diencephalon (and also, perhaps, in the basal ganglia of the cerebrum), capable of acting briefly as a closed system, delivering facilitation to other such systems and usually having a specific motor facilitation. A series of such events constitutes a "phase sequence" – the thought process. Each assembly action may be aroused by a preceding assembly, by a sensory event, or – normally – by both.
>
> *(Hebb 1949)*

While Hebb's model is still broadly taken to reflect the neurological basis of memory, current understanding of memory formation at a neurological level that argues that our prefrontal cortices process our experiences into memory elements that are stored across our cortices, while our hippocampi weave these memory elements together into coherent pattern memories (Voss et al. 2017).

A long-lasting reconfiguration of specific neurons has been related to the formation of specific memories, a phenomenon that has been called an *engram*. These engrams are increasingly the focus of neuroimaging scientists, for instance, in exploring the impacts of neurological dysfunctions and the ways individuals recover from them or adapt to them (Ortega-de San Luis & Ryan 2018). In terms of pattern formation and subsequent recall, Yadav et al. (2002) noted that:

> Memory formation involves binding of contextual features into a unitary representation, whereas memory recall can occur using partial combinations of these contextual features. The neural basis underlying the relationship between a contextual memory and its constituent features is not well understood; in particular, where features are represented in the brain and how they drive recall.

This suggests that memory is only coherent when connections between memory fragments (engrams) are properly established, and that memories can be iteratively built on and around other memories. Moreover, the means to do this are also to be found in the brain. For instance, Edelman (1990) noted the importance role of the hippocampus, thalamus, cerebellum, and basal ganglia in building and connecting memory.

Coherent memories are not created in an instant; they require reinforcement and iteration in order to coalesce, and as they form, they become independent of the hippocampus. This suggests again that the hippocampus is a primary weaver of pattern memory but not essential to its ongoing existence:

> Successive reactivation of this hippocampal–cortical network leads to progressive strengthening of cortico-cortical connections (for example, by strengthening existing cortico-cortical connections or establishing new ones). Incremental strengthening of cortico-cortical connections eventually allows new memories to become independent of the hippocampus and to be gradually integrated with pre-existing cortical memories.
>
> *(Frankland & Bontempi 2005)*

Frankland & Bontempi also noted that coherent memories are woven together by the hippocampus, even if the resulting memories are stored elsewhere:

> Encoding of perceptual, motor and cognitive information initially occurs in several specialized primary and associative cortical areas. The hippocampus integrates information from these distributed cortical modules that represents the various features of an experience, and rapidly fuses these features into a coherent memory trace.
>
> *(Frankland & Bontempi 2005)*

There is also neurological evidence to substantiate perceptions that not all of our experience is woven into memory, and that only selected experiences become a part of our longer-term memories:

> Most memories disappear within minutes, but those that survive the fragile period strengthen with time. Long-term memories are formed in a two-way conversation between the hippocampus and the cortex, until the hippocampus breaks the connection and the memory is fixed in the cortex – which can take years.
>
> *(Medina 2008, p. 147)*

The formation of new memories can be disrupted, but generally speaking, the longer they persist the more robust they become: "This process of strengthening and stabilization is known as memory consolidation" (Bisaz et al. 2014). Moreover, memory can become more fluid when it is reactivated but will later reconsolidate:

> A memory that has become insensitive to molecular interferences can again become labile if it is reactivated, for example by retrievals. This post-retrieval fragility, like the one that occurs after acquisition, is temporally limited and the memory returns to a stable state through a process known as reconsolidation.
>
> *(Bisaz et al. 2014)*

Another source of empirical evidence we can draw on is that pertaining to clinical and pathological dysfunctions. For instance, amnesias of various kinds (retrograde, anterograde, etc.) with various causes (stroke, head injury, pharmacological, etc.) all reflect loss of memory. There are different causes of memory loss. There may be damage to the substrate for memory such that specific memories are lost but not the capacity for memory, the connections and associations connecting specific memory elements may be damaged or lost such that memory loses its coherence, or sometimes a mind simply loses the ability to access memories that are still presumably intact (Mayes 1995). Memories that can be accessed but lack coherence are considered fragmented, which can be the result of traumatic experiences (Brewin 2016). Schacter (2022) described forgetting in terms of transience (memory decays), absent-mindedness (inattentive to memory), and blocking (lost access to memory), to which he added four "sins of commission": misattribution, suggestibility, bias, and persistence (intrusive memory). All of these "sins" can be linked to pattern memory in ways that I will describe later in this volume.

Assembling evidence for a theory of pattern memory reflects the approaches used in developing other theories of cognition, such as those that dichotomised short- and long-term memory (Baddeley & Warrington 1970) and those that argued for a dual process theory of mind (Wason & Evans 1974). This kind of approach is about establishing a plausible and grounded basis for a body of theory, but not necessarily proof. This is not to suggest that there is no contradictory evidence for pattern memory and pattern theory, only that I was unable to identify any. Perhaps others will generously help to address this gap. I will not belabour my point that research evidence aligns well with (although it neither proves nor disproves) a pattern theoretical stance, in great part because this book is not about neuroscience or psychology but rather social theory that is congruent with these fields.

A third body of evidence I offer in support of a pattern theoretical stance is based on deductive arguments that suggest the necessity of pattern memory. Given that we can store experiences, feelings, and thoughts in our minds, there must be mechanisms by which active mental events (including sensoria (sight, sound, touch, etc.), feelings and emotions, and analytical and abstract thought) are encoded and stored. Given that we assemble a great number of memories over time, these mechanisms need to have significant capacity and efficiency, and they need to be able to retain memory over many years.

Given that we make sense of novel experiences by recalling memories of similar past experiences and comparing the two, our minds must have mechanisms for searching our memories, assessing the similarity of these memories to current experiences, selecting those memories that are sufficiently similar, and activating or pulling those memories into active thought.

Given that memory can be used to establish identities we must have mechanisms that organise our memories by (variously) aggregating, abstracting, and generalising them to create categories, models, and theories, which we can also store and subsequently access as memories. This apparent aggregation and abstraction means that our memories are deeply interconnected, which in turn explains how memory recall does more than establishing identity, it also suggests relationships, behaviours, significances, and meanings associated with a particular identity.

Given that we can recall knowledge of similar things by using different associations reinforces the proposition that memory is profoundly based on connections and associations between memory elements. Moreover, given that, in reactivating a memory, we recall things as we experienced them, suggests that our memories are embodied, which again supports the principle of memories being made up of many cognitive phenomena such as perceptions, feelings, and abstract thoughts.

Given that, when our minds can find no identity matches, we again draw on our memories and appraise their similarities and differences to establish new understanding (that can then also be retained as memories) means that not only are there mechanisms involved in accessing memories, there are others that create or edit them. Moreover, given that we can progressively develop associations, categories, models, theories related to our memories over time (and that these can also be stored as related memories) means that we can continuously elaborate our memories. This is also supported by the observation that new experiences can be added to existing memories without wiping or replacing them. It also means that memories are not just records of percepts or thoughts, they can become abstract representations of how to act, think, and behave.

Given that parts of a memory can be recalled and that parts of memories can be edited and elaborated independent of others means that memories are stored as networks of parts rather than as single indissoluble units. This in turn means that there must be some way of connecting these parts to keep track of them and to recreate a sense of a coherent memory on recall. That we can forget or lose memories in other ways also means that memory is not permanent in any absolute sense even though it can be long lasting. Moreover, not everything we experience is memorised, which means that there must be mechanisms that select which memories will be created and which will not. That we can retain memories but lose the ability to access them means that mechanisms of memory creation and recall function independently of the memories they work with such that certain parts might be altered or compromised without changing others. There is an apparent efficiency to this in that our minds can make significant use of existing memory parts and associations.

That sometimes we can recall memories when we try suggests that our memories are at least partially responsive to conscious thought. However, that at other times we recall memories without any conscious intent, including when we are asleep, means that our memories and the mechanisms by which we access them are working constantly and independently of our conscious minds. That we cannot always recall a memory that we are nevertheless aware of having means we have some means of indexing or cross-referencing our memories independent of the memories themselves. That our memories are partial, selective, interpretive, and subjective reinforces the individual and cognitive nature of memory, while the fact that we are often unaware or inattentive to these characteristics of memory means that our memories and their associated mechanisms are deeply integrated with conscious thought to the point that the two are effectively inseparable.

That different individuals witnessing or participating in the same events form similar but unique memories reinforces the subjective and individual nature of memory. Moreover, that we cannot directly experience or share our memories with others means that memories are intrinsically cognitive and personal. Although we cannot access each other's memories directly, that we can express and share abstractions of our memories means that we can access our own memories in ways that allow us to create meaningful representations thereof (although not necessarily complete), that those abstractions are meaningful to others, and that we can build new memories based on abstractions of memories that have been shared with us. That we, as a social species, set so much store in expressing and sharing memories with each (for instance, in the form of stories, images, literature, performances) illustrates both the shared experience of memory and our use thereof, and the dependence we have on memory and memory exchange as the basis of cultures, societies, and other collective activities.

In summary, whether we approach the issue of memory from the point of view afforded by experiential knowledge, theoretical-empirical knowledge, or knowledge from deductive reasoning, the proposition that we have pattern memories that we access through various forms of pattern thinking stands. There are many implications of this, not least in terms of the many things that we do with our pattern memories; building, consolidating, editing, elaborating, and organising them, and using them in support of reasoning, feeling, signification, and learning, all of which I cover in Chapter 5. Although I fully accept that there may be other plausible explanations, I have not encountered integrative theory with the same scope and focus as I have attempted to outline in this book. One metatheory is more than enough, at least for now. I now outline the use of pattern memory in the context of pattern perception.

Pattern Perception

I provisionally define *pattern perception* as the way we make sense of ourselves and the world around us based on matching pattern memory to current experience. The processes that produce perception have been identified as mostly, but not exclusively, taking place in the cerebral cortex, a folded sheet of nerve fibres that overlies the rest of the human brain. The cortex has regions (lobes) that tend to govern different functions (such as motor functions and planning towards the front of the brain, and vision and bodily sensation towards the rear) and it has two hemispheres (left and right) that are connected through the corpus callosum (Ashwell 2019).

The cortex has layers with columns of nerve cells connecting them. Hawkins (2004) and Kurzweil (2012) independently described the neural "traffic" in these cortical columns in terms of sensory data arriving at one end and decisions and actions emerging at the other. Although Mattson focused less on neuroanatomy in describing pattern thinking than Hawkins or Kurzweil, he also argued that the cortex is the seat of pattern thinking (Mattson 2014). Cumulatively, it would seem to be the sheer scale and interconnectedness of these structures that provides the capacity for pattern thinking (Kurzweil 2012). This is reinforced by Shermer's (2011) observations of flows and processing of sensory inputs that work at different scales and in different regions of the brain.

Although human brains have a much larger and more elaborated cortex than other species, cortical columnar structures are found in the brains of primates and in those of many other species, which suggests that they too have some capacity for pattern thinking. That said, human pattern thinking seems to be many orders more complex and nuanced than its expression in other species, involving, as it does, abstraction, stratification, and systematisation, and the pursuit and exploration of meaning (M'Closkey & VanDerSys 2017). The function of these cortical columns seems to vary between species as well as between individuals of the same species, suggesting that these cortical structures are not pre-programmed to hold or process particular kinds of patterns, rather, they seem to provide a powerful and adaptive resource for pattern thought (Mattson 2014). This neuroplasticity also suggests that each brain develops its own approaches to pattern perception rather than being hard-wired to them from birth (acknowledging that many creatures do seem to have some "baked-in" pattern thinking dispositions). That said, this is not simply about structure; biochemistry would also seem to play an important role in pattern thinking, a point that I will return to in the context of the role of emotion in pattern thinking.

If, as it seems, minds have both the capacity and disposition for pattern perception, then how does it work and what can this tell us about pattern more broadly? Kurzweil (2012) argued that perception involves comparing sensory input to patterns stored in our memories, selecting them based on familiarity or common elements or contexts or connections, while Damasio argued that recall depends on pattern memory such that:

> The brain must have a way of storing the respective patterns, somehow, somewhere, and must retain a path to retrieve the patterns, somehow, somewhere, for the attempted reproduction to work, somehow, somewhere.
>
> *(Damasio 2010, p. 140)*

Reasoning deductively, given that we can recognise that certain current experiences are similar to certain past experiences, then our memories of past experiences must follow the same logics and representational forms as our current experiences. Moreover, the representational architecture of memory must be lean enough to be parsed rapidly and subtle enough to distinguish between multiple similarities and differences. It would also seem that matching current experiences and memories to past experiences need only be approximate, it only needs to suggest a match.

When current experience is matched to memories of similar experiences, it would seem that the neural pathways involved are amplified while others are inhibited, something Kurzweil (2012) described in terms of probability matching and computing. From one moment to the next, cycles of pattern matching and inhibiting and amplifying different patterns happens across much of the cortex, reaching out to wherever the relevant memories are stored and 'zeroing in' on the most likely candidates:

> The brain quickly tries to find a part of its world model that is consistent with any unexpected input. Only then will it understand its input and know what to expect next.
>
> *(Hawkins 2004, p. 159)*

Similarly, Kurzweil observed that:

> Signals go up and down the conceptual hierarchy. The signal going up means, 'I've detected a pattern'. A signal going down means, 'I'm expecting your pattern to occur', and is essentially a prediction. Both upward and downward signals can be either excitatory or inhibitory.
>
> *(Kurzweil 2012, p. 91)*

Identifying and sorting candidate patterns that match sensory inputs is of necessity an abductive process, one that tests multiple possible solutions and that selects and promotes better fitting candidates, in part through the associations one transient candidate pattern has with other patterns – see Figure 4.1.

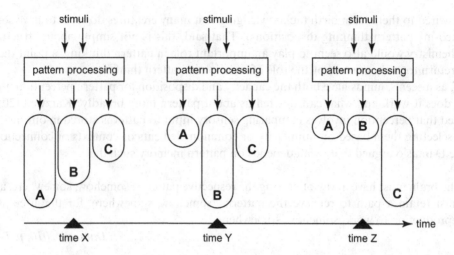

FIGURE 4.1 Different pattern candidates can be promoted or relegated over time. At time X, pattern candidate A seems the most likely fit. A little later at time Y feedback on the fit suggests A is not a good fit after all but B now seems the better fit. Later still at time Z, both A and B have been relegated and pattern C is the perceived better-fit pattern.

Diagram prepared by the author.

Not only does this suggest that pattern perception is transient and adaptive rather than fixed, given the processing required, the searching, sorting, and prioritising of pattern memories takes time:

> Even though every pattern recognizer is working simultaneously, it does take time for recognitions to move upward in this conceptual hierarchy. Traversing each level takes between a few hundredths to a few tenths of a second to process. Experiments have shown that a moderately high-level pattern such as a face takes at least a tenth of a second.
>
> *(Kurzweil 2012, p. 51.)*

This progressive and time-dependent nature of matching sensory data to pattern memory is echoed in the cascades of processes our minds go through in processing incoming data in recognising faces. For instance, Shermer described:

> It appears that the brain first processes the global shape of a face, such as the general outline with two eyes and a mouth, and then processes the details of facial features, such as the eyes, nose, and mouth.
>
> *(Shermer 2011, p. 70)*

Even as provisional pattern matches are identified and promoted, new sensory data can create a dynamic cascade of selection and deselection of candidate patterns. A perceived pattern is not 'correct' in any absolute sense, partly because it is only the best fit that a particular mind can come up with given the time and resources available, and partly because it can only draw on the extant memories of an individual mind.

Even though a pattern may be perceived instantaneously (or at least it seems instantaneous), there must still be processing involved (Litman & Reber 2005). We might call this *preconscious pattern recognition*, and the patterns it invokes will tend to be more familiar, less ambiguous, and are to some extent anticipated. Sometimes finding a matching pattern may take longer and, when it does, we may become aware of the search. We might call this *conscious pattern recognition*, which is delineated by our becoming aware of the time and effort needed to find a pattern match. The patterns that conscious pattern recognition eventually pulls up are likely to be less anticipated, less familiar, and more ambiguous than those invoked by preconscious pattern recognition.

Preconscious and conscious pattern recognition are not distinct phenomena; they both involve searching for appropriate patterns that correspond to sensory input, but they differ in terms of the time and effort required, the extent to which we are aware of the process, and the extent to which we consciously intervene in it. Sometimes a pattern cannot be found to match current experience. Assuming we do not simply lose interest, our minds then have to devise a suitable pattern. We might call this *pattern development*, something that tends to be much more deliberate and analytical than either mode of pattern recognition as well as relatively slow and effortful.

In writing this chapter, I had originally called pattern development *pattern discovery*, in great part because of the extant literature from the computational sciences on pattern discovery that distinguishes it from pattern recognition. However, discovery implies that patterns exist out in the world and can, therefore, be simply found and acquired. That said, patterns can in one sense be acquired, for instance, through instruction, reading, or listening to the radio, but this is about exposure to ideas, not the transfer of pattern memories between minds. All patterns are necessarily unique to the minds in which they occur. They are woven from that individual's idiosyncratic memories, and this means that patterns cannot be discovered but they can be developed in response to experiences (including instruction and reasoning).

There are parallels with dual process theories of mind here, in that system 1 thinking has been described as being fast, largely preconscious, and heuristic in nature, while system 2 thinking has been described as being slower, more deliberate, more effortful, and more analytical in nature (Kahneman 2011). Bringing the two theories together, if we immediately recognise a pattern then we are using system 1 thinking, if not then we default to using system 2 thinking (Croskerry 2009). These kinds of connections between dual processing theory and pattern thinking were also noted by Kurzweil (2012) and by Bor (2012) who suggested that mixing these modes (such as trying to deliberately access preconscious thought) exerts an even greater load. While there are parallels between pattern theory and dual process theory, they are not exact – see Figure 4.2.

I marked the point of inflection between preconscious and conscious pattern recognition in Figure 4.2 as the point at which we become aware we are searching for a pattern. Quite when this happens is likely dependent on many contingent factors, such as the configuration of individual minds, their levels of stress or fatigue, and the contextual realities of what seem to be more or less likely pattern candidates at any one point in time. Similarly, the point of inflection between recognition and development is the point at which one switches from hoping the pattern will simply appear to actively and deliberately searching for it, and that inflection point will also be subject to many contingent factors. To that end, I would draw on De Neys' (2022) suggestions that we switch between system 1 and system

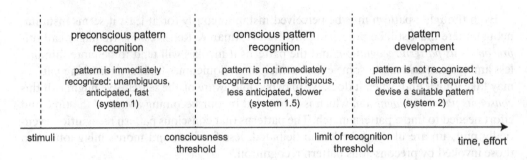

FIGURE 4.2   Stages of pattern perception in terms of the time and effort required.

Diagram prepared by the author.

2 thinking based on competing intuitions in system 1. If one intuition dominates then there is no switch but if there is conflict between them then uncertainty grows, and, at some point, an uncertainty threshold is crossed and there is a switch to system 2 thinking. I reinterpret this as growing uncertainty (and its limbic consequences) over possible matches that contribute to growing awareness and that uncertainty becoming discomfort that contributes to the transition to pattern development. It seems likely too that we progressively move from one modality to the next whenever the current approach fails to provide a good solution, suggesting that there is an underlying heuristic aspect of pattern perception.

Heuristics are the 'rules of thumb' we use for decision-making or problem-solving, particularly in fast moving and/or low resource settings (Hutchinson & Gigerenzer 2005). In ways that seem to echo matching in pattern perception, we seem to select a heuristic strategy from our repertoire because it seems to be the best fit to the problem at hand. As we learn more about the problem, we may abandon one heuristic strategy in favour of another that seems to be a better fit. In assessing that fit, we draw on what we know has worked for us previously and what has worked in the contexts to hand (Gigerenzer & Brighton 2009). A heuristic does not have to provide a perfect solution; it simply needs to be "satisficing" (Simon 1956).

There are many parallels between pattern perception and heuristic thinking. They both work well with missing or ambiguous information, they are both adaptive, satisficing, lean, fast, and based on experience and emerging understanding, they both focus on sufficiency rather than perfection, and we improve our pattern thinking and heuristic thinking through practice. I would argue then, that, rather than being distinct but similar phenomena, heuristic thinking is a particular kind of pattern thinking and individual heuristics are particular kinds of problem-solving patterns.

How does this proposed model relate to existing pattern recognition theories such as template matching, prototype-matching, recognition-by-components theory, and top-down processing (Rookes & Willson 2007)? Template matching theorists proposed that we save templates of everything we perceive in long-term memory, which we subsequently match to new experiences (Shugen 2002). Prototype-matching theorists proposed that our experiences of similar things (dogs, cars, spoons) contribute to a common prototype in memory, with the match (allowing for variance) between sensory input and the prototype that afforded pattern recognition. Recognition-by-components theorists proposed that we recognise objects by breaking them down into their basic visual components and then reassembling them in our minds and matching them to our memories. Top-down processing theorists proposed that patterns are recognised through associations with and cues from context and past experiences.

Rather than favouring any one of these pattern recognition theories, they all seem to reflect aspects of pattern perception. Templates and prototypes both reflect the qualities of patterns as I have defined them, although, as I will argue, pattern is much more than either. Indeed, the erstwhile distinctions between bottom-up and top-down processing seem rather artificial. Rather, we should consider perceptions to be the product of bottom-up and top-down processes (Gregory 1970) that depend on the stimuli and the disposition of the individual (Fang et al. 2008). They are bottom-up in terms of pattern elements being promoted when they match aspects of sensory input, and they are top-down when pattern associations and cues support the sorting and selection process and when deliberate pattern development is involved. Moreover, these selection and promotion processes would seem to be both dynamic and ongoing in that they involve finding a sufficiency of alignment between experience and memory. A solution need not be perfect, but it should be identified quickly, which might reflect an evolutionary capability to interpret and respond to potentially hostile and fast-moving situations (Foster & Kokko 2009). Shermer (2011) described the evolutionary advantage of making Type I errors (false positives) and avoiding Type II errors (false negatives) as the evolutionary basis for developing pattern thinking. If this is true, then all cognitively endowed species alive today have been more or less optimised for pattern thinking according to their various evolved capabilities and opportunities. For humans, the difference seems to have been particularly invested in the evolution of the neocortex and its abstracting capabilities. These evolutionary developments might make us the most adept pattern thinking species we know of, but by no means the only species with pattern-thinking capacities.

Encountering something that we do not recognise does not necessarily lead to pattern development. Rather than trying to understand something unfamiliar we may just ignore it, or we may be satisfied with a poor pattern match if the effort of improving it or engaging in pattern development is beyond our means or level of commitment. This reflects an intrinsic selectivity in what our minds perceive, what we notice, and what we remember. I explore aspects of this selectivity in the next section on pattern memory and in later material on noticing and significance.

Abductive, Integrative, Efficient

Not only does pattern perception depend on your mind's ability to evaluate similarities and differences between your current experiences and your pattern memories of past experiences, assessments of similarity and difference (and the results thereof) change according to the characteristics you are interested in. For instance, peas and cabbages are similar in terms of their being green, parts of plants, and edible, but they differ in terms of size, species, and flavour. You might just focus on one characteristic (such as colour or size), you might evaluate characteristics in a particular sequence (colour then size) or you might consider multiple characteristics at the same time (colour, size, shape, flavour, etc.). As an example, consider birdwatchers' concept of "jizz" or GISS (general impression of size and shape) where rapid and ambiguous impressions are all that are available to a watcher to identify an elusive animal. Perceiving a bird's jizz requires quick thinking and typically works with ambiguous information. As such, it is a heuristic response, and like any heuristic it can be refined through deliberate practice. If you are not a birdwatcher, then maybe you experience a similar capacity in recognising a song from just a few notes or a face from a small portion of an image. Assessments of similarity and difference also depend on the point of reference, which may be either experience or memory. On one hand, if you are searching for a mislaid object, then your pattern memory is your reference as you seek instances thereof. On the other hand, if you are trying to make sense of what your senses tell you then they are the point of reference to which possible pattern matches are compared.

There would seem to be a general consensus among psychologists that assessing similarity and difference is a matter of judgement in which information is considered (Medin et al. 1990; Richie & Bhatia 2021). Different models of judgements have focused on comparisons (inductive pattern recognition), rules (deductive pattern recognition) (Wirebring et al. 2018), or probability and risk (Tenenbaum & Griffiths 2001). Rips (2009) described three modalities in assessing similarity: categorisation ("what class does this belong to?"), typicality ("how prototypical is this?"), and resemblance to previous encounters ("have I seen anything like this before?"), which may all plausibly play a role in pattern matching and sorting. Different similarity assessments seem to draw on different pattern dimensions (elements, associations, variances, decompositions, etc.) and they behave heuristically (they are lean, adaptive, satisficing). However, as with other phenomena discussed in this chapter, our understanding of how we assess similarity and difference, particularly in pattern matching and sorting, is far from complete.

Given the inescapable imprecision, incompleteness, fuzziness, and transitory nature of experience and memory, and the dependence of perception on searching, selection, and matching patterns, pattern perception is essentially *abductive*. Peirce described abduction as "inference to the best explanation" (Douven 2017) based on iteratively building, testing, and revising putative explanations until a satisfactory solution is found. The iterative nature of abduction allows for new and changing information. We might look at something from different angles, we might move towards or away from it, we might touch it, or pick it up, or otherwise interact with it. From one moment to the next there may be new insights or new perspectives suggested by this information, which may confirm or counter various pattern perception suggestions our minds are considering. As Dennett (1991) observed: "differences in knowledge yield striking differences in the capacity to pick up patterns".

Even if we had instant access to everything our senses could tell us about a thing, there would always be much more that our senses could not provide. We typically cannot sense the interior of a thing, nor can we sense its past or its future. Moreover, we are limited to our present circumstances, we cannot be everywhere at once. Perception might be extended through remote surveillance, telescopes, or microscopes, but, even then, no human mind can perceive everything, everywhere, and at every point in time. That the information we draw on in pattern perception is intrinsically incomplete has been a recurring issue in perception research, and whether or to what degree we draw on memory to resolve ambiguity has long been debated (Roger 2017). I would argue that pattern perception responds to, and to a degree interprets, dynamic changes in sensory information over time, and it draws on the apparent fit between putative pattern solutions built from our memories of past experiences and the feedback such evaluation generates. This helps to explain why it is that different observers of the same event may perceive quite different patterns, and the same observer may perceive different patterns in the same thing over time. Clearly, we work with imprecision at both ends of pattern perception:

> You don't have to have the entire pattern you want to retrieve in order to retrieve it. You might have only part of the pattern, or you might have a somewhat messed up pattern. The auto-associative memory can retrieve the correct pattern, as it was originally stored, even though you start with a messy version of it.
>
> *(Hawkins 2004, p. 30)*

Although I agree with Hawkins' assertion that we can perceive or recognise patterns on the basis of attenuated and ambiguous inputs, I would disagree with his suggestion that we necessarily identify "correct" patterns. Indeed, what is a "right" or a "correct" pattern? One might mean correct in the sense that the pattern I perceive reflects the reality of the thing in front of me (perceiving an actual chair as a chair-like thing), or it might mean that the pattern I perceive is more or less what others perceive (although those others might also be wrong in the first sense). Alternatively, a correct pattern might mean that the perception led to a useful response (seeing signs of a predator triggering my getting out of danger). In the latter case, the cue may be minimal; a rustle of grass or a momentary shadow may be enough to trigger a danger response. Perceiving what might be a tiger when the lurking animal is in fact a leopard does not invalidate the threat response, but neither is the perceived pattern strictly correct. This leads to Gettier problems and other logical paradoxes where justification, perception, and response may not logically align but they nevertheless prove useful (Gettier 1963). However, this would distract from the point that pattern perception need not be correct in any absolute sense, it only needs to be useful from one moment to the next.

The pattern memories that we call on in support of perception are also imprecise and partial. Patterns are specific to the mind in which they formed, they are open to change, and they reflect the idiosyncrasies of individual experience and thought. Our minds respond to this imprecision by 'filling in the blanks' and by evaluating possibilities and probabilities associated with the impressionistic perceptions we form. It would seem then that some aspect of current experience is matched as a part of a pattern memory. This partial pattern match would seem to *pull* on the rest of the pattern, and variations thereon. This echoes Grenander's argument that understanding patterns requires some capacity to work with their transforms and variations. All we need is some sense of similarity:

> We can recognise a pattern even if only part of it is perceived (seen, heard, felt) and even if it contains alterations. Our recognition ability is apparently able to detect invariant features of a pattern – characteristics that survive real-world variations.
>
> *(Kurzweil 2012, p. 30)*

Even a relatively limited range of memories can be partially connected to new experiences. Not only does this allow for rapid and adaptive pattern perception and recognition, it also allows us to accommodate variations and transforms in perceived phenomena so as to accommodate any ambiguities and novelties that we might encounter. In this way, we can perceive things we have never encountered before by drawing on our memories of those things that we have previously encountered. This would seem to afford significant redundancy, which allows us to create new memories based on rich associations to our existing memories:

> Our conscious experience of our perceptions is actually changed by our interpretations ... we are constantly predicting the future and hypothesising what we will experience. This expectation influences what we actually perceive.
>
> *(Kurzweil 2012, p. 31)*

It is important to note that we do not typically match memories to sensory inputs in discrete paired items (like a game of Snap!), not least because this would be a somewhat arduous process of parsing our entire memories for every match we sought. Rather, the connections between patterns assist in the matching process such that one pattern can 'suggest' its connected patterns as alternatives:

> Patterns triggered in the neocortex trigger other patterns. Partially complete patterns send signals down the conceptual hierarchy; completed patterns send signals up the conceptual hierarchy. These neocortical patterns are the language of thought.
>
> *(Kurzweil 2012, p. 68)*

It would also seem that one pattern may serve as a part of another pattern, such that we seem to form continua of associative memory that can come together quickly and efficiently to create solutions to novel experiences. One implication of this is that the distinctions between patterns would seem to be somewhat arbitrary and dynamic. We may have a particular pattern (or an idea of a pattern) as the focus of our perception or attention, but there are always peripheral connected patterns clustering about it. Moreover, we need to consider the parts of a pattern that represent the structure and boundaries of what it refers to as being distinct from its cognitive boundaries. Put another way, a pattern may contain spatial, temporal, or other structural components that refer to the nature of expression of a pattern, but these do not necessarily define the boundaries of a pattern as a cognitive phenomenon. Structure that is encoded in a pattern is unlikely to be the same as the cognitive structure of a pattern.

Attention, Cueing, Anticipation

Although I have described individual cycles of stimulus and pattern response, our experiences are rarely so linear or discrete. Much of the time we are immersed in dynamic, plural, and entangled cycles of stimulus and response. Even when we are asleep or when things around us are quiet and subdued, our minds continue to revise and consolidate our memories, feelings, and thoughts. How can our minds cope with this tsunami of thought and experience? In the previous section I described how one recalled pattern can *pull* on its associated patterns to add depth and nuance to perception. This clustering of patterns means that we can experience a bolus of memory that adds to what our senses are telling us. That we often recall emotions, images, sounds, even odours that our minds associate with some triggering stimulus is exemplified by Proust's description of how the scent of baking flooded his protagonist's mind with entangled memories, feelings, and meanings. Indeed, it has been noted that this kind of memory is often rich in sentiment and affect such that we recall an edited or even revised version of the original experience (Herz & Schooler 2002). More prosaically, contextual pattern recall can help us with filling in the ambiguities and unknowns in perception; it is what constitutes the sense of familiarity in places, people, and activities we have encountered before.

A series of interlinked cues can, therefore, precipitate a state of *anticipation* if the cues correspond and cumulatively suggest patterns that contain deeper understanding of a phenomenon. Anticipation is not simply associative; it is predictive and preparative. Predictive in the sense that it suggests that something is likely to happen (suggesting degrees of certainty) and preparative in the sense that relevant patterns needed to respond can be activated in advance of their being needed. Perceiving pattern A may lead to an anticipation of pattern B because B typically follows or accompanies A – see Figure 4.3.

Patterns can include emotional memories such that recalling a pattern's emotional aspects can tell us how we feel (or should feel) about it. Indeed, not only can contextual pattern associations allow us to quickly assemble a sense of what is around us, they can help us to anticipate what we might encounter. Again, there can be efficiency in this in that we need only attend to dissonances in our anticipated contextual patterning rather than having to parse every single percept. Contextual cueing may reflect memories of concurrence, but it can also reflect memories of causal relationships and of symbolic and affective connections between things. Indeed, any relationality we can experience we can cast into memory, and any pattern memory can be recalled to help us interpret new experiences. Although contextual cueing can help to make pattern perception more efficient, it cannot resolve the many possible foci of attention available to us. How then do we distinguish, select, and attend to certain things while disregarding others? Clearly, we are reasonably adept at focusing our attention. For instance, in walking in the wild we may recognise trees and grass, but we pay them little attention, but if we recognise a predator (or even the suggestion of one) then that will likely grab and hold our attention. It would seem, therefore, that recalling the emotional elements of pattern memory guides our attention: the greater the emotional response the more attention we pay.

Whether it is co-occurring, or correlated, or even causally linked, anticipation is both enabled by pattern thinking and is a powerful contributor to efficiency and even expertise in pattern thinking. Indeed, we might understand expertise as matching and perceiving a pattern from a few tentative initial indications and then adjusting and focusing clusters of pattern matches to identify and respond to novel situations with precision and accuracy. Of course, although we may anticipate many things that do not happen, being prepared for their possibility can be a reasonable commitment of effort. For instance, most of us are

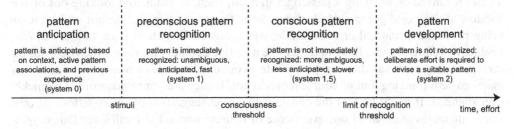

FIGURE 4.3 Extending Figure 4.2 to include the precondition of pattern anticipation.

Diagram prepared by the author.

taught to drive defensively in anticipation of other drivers acting in unsafe ways. That most of us do not drive unsafely does not mean that driving in a state of anticipation is misguided or wasted effort. Indeed, returning to the evolutionary case of being vigilant for signs of predators even if they are not there, this might be better understood as anticipation of the possibility of predators given the anticipatory activation of patterns associated with previous encounters with predators either directly or through the stories we are told. That much of this can occur without deliberate thought or even awareness is reflected in the placebo effect where the anticipation of something happening can unconsciously alter the body's physiological and perceptual states even when the anticipated event or interaction does not actually occur (Colloca & Miller 2011). Anticipation can also be mistaken and misdirected, for instance, reflecting the biases that lead us to anticipate unlikely events, as well as our tendency towards paranoia, hypervigilance, and hypochondria.

We can engage this anticipation effect deliberately. For instance, creativity would seem to be based on deliberate anticipation in directing the creation of crafted or engineered artefacts, structures, or philosophies. Anticipation can also be actively engaged by learners such as in the "three cueing" model used in early literacy training where children are encouraged to work out an unfamiliar word by thinking about the sentence context, the grammar, and the letters and sounds involved (Adams 1998). Setting aside the efficacy and variance in such practices, clearly anticipation is used in many aspects of human thought and interaction. Indeed, however we look at it, cueing and cumulatively anticipation seem to be basic functions of patterned minds.

Micropattern, Mesopattern, Macropattern

Earlier I noted the aggregating and abstracting tendency of pattern thinking. In this section, I develop this idea to outline a continuum of pattern memory and thought that builds from impressionistic micropatterns through perceptual mesopatterns to abstract macropatterns.

Micropatterns

The moment-to-moment matching of patterns to experiences are not particularly coherent, logical, or complete, as they are rather impressionistic reflections of past experiences. These are what I call *micropatterns*. We experience micropatterns as impressions, feelings, or suggestions, that are every bit as fleeting as the cognitive events that stimulated them. One micropattern may pull on other micropatterns but the rapidity and fluidity of micropattern perception will not allow for much of this before other micropattern impressions crowd in. Think, for instance, of being a passenger in a car, train, or plane and looking out of the window at the world going past. As long as the things you perceive do not strike you as being particularly unusual or remarkable, you barely register them (tree, house, car, hill, field, cow) before the next thing and then the next thing comes into view. Much depends on certain micropatterns as they can precipitate a cognitive cascade of mesopattern and even macropattern thinking, but of course most do not. If they did our minds would be quickly overwhelmed. It might be that the constant low-level suggestion of micropattern perception contributes to much of our experience of consciousness, but I will leave that suggestion for another time. I will say though that those micropatterns that do grab our attention are what can become mesopatterns.

Mesopatterns

A micropattern perception might lead to a mesopattern perception if it grabs our attention or if we focus our attention on it. *Mesopatterns* allow us to answer questions such as "what is this?", "what is going on?", and "what will happen next?". Impressionistic pattern perception can become coherent pattern perception if we pull a coherent pattern (rather than an impressionistic part pattern) into working memory. This takes more time than micropattern perception and as such we may only work with a few mesopatterns at a time. I call these mesopatterns because they are more fully formed than micropatterns, certainly they are sufficiently well-formed to focus attention on something. That said, mesopatterns need not be particularly logical or objective, they only need to be sufficient to our current needs. Given that mesopattern perception is more active and longer lasting than micropattern perception, there is time for conscious thought and emotional reaction, both of which might be woven back into pattern memory.

Although I differentiate between micropatterns and mesopatterns, there would seem to be no one definitive inversion point or threshold that separates them. Rather, we can think of a continuum of pattern association with longer-lasting and more coherent instances of pattern recall being more mesopattern-like and more ephemeral and impressionistic pattern recall being more micropattern-like. Much of this depends on the idea of *pulling* on memories. We *pull* on one or more pattern nexuses when they match sensory or other information and that in turn *pulls* on the pattern's connections and associations. We might, therefore, think of pattern recall in terms of pulling (or perhaps *plucking*) at a multidimensional mesh or membrane of interconnected pattern memories that have no distinct edges. We need not match a coherent memory element for element to establish a match, nor would this be an efficient way for things to work. Rather we need only match current experience to a smattering of pattern elements for a pattern to begin to be activated or pulled into active thought. If our experience or attention changes, then this pulling process stops or fades and all we have are fleeting micropattern impressions. If we sustain experience or attention then more and more of a matching pattern is pulled into working memory, establishing an ever more mesopatterned response. The same underlying pattern may be involved in both kinds of recall; it is the extent of pulling and the consequences thereof that matter in distinguishing between micropatterns and mesopatterns. *Pulling* is of course a metaphor, one that reflects a sense of the web- or fabric-like nature of pattern memory. It is precisely because memory elements are connected that we can match one element but then pull in the rest of the pattern.

Pulling (or plucking) at the metaphorical fabric of pattern memory suggests that any pattern that may seem to us to be relatively discreet (such as a word or concept) is no more (or less) than a relatively localised and coherent part of a pattern continuity, such that what seems like distinctiveness is more akin to useful distinctions and categorisations than intrinsic structure. Indeed, our will to find structure is itself a pattern-like attempt to understand our own thoughts. If what we perceive to be a pattern is no more than a localised nexus, then we may pull on or invoke it from many directions. The same pattern nexus may therefore seem different depending on the way it was approached or invoked. This means that, rather a pattern having a fixed meaning or point of access, its meaning may change or take on a different character depending on how the pattern was invoked. It is also important to note that the numbers of connections and associations within and between patterns, and the relative strength of those connections are, therefore, not simply bonds, they are the very substance of pattern.

Macropatterns

Macropatterns continue the trend in terms of size and the time it takes to pull them into working memory, but they differ from micropatterns and mesopatterns in a couple of important ways. When we think about (reflect, ponder, analyse, imagine) our mesopattern perceptions, we may notice or make broader and more abstract connections between them, and in so doing we create larger and more abstract patterns that can help to answer questions such as "what kind of thing is this?", "why is this the way it is?", and "what do these things have in common?". As before, there is no distinct threshold where we can say a mesopattern switches to being a macropattern. Rather, there is a continuum where a pattern becomes more macro the larger it gets, the more abstract it gets, the more aggregated it gets (weaving together otherwise unassociated patterns), and the more long-lasting and deliberate it is. We might suggest other qualities of macropatterns, such as increasing levels of logic or coherence, but while this might be true of some macropatterns they do not seem to be intrinsic qualities as macropatterns can also include aspects of ideology, culture, and belief. We might say, therefore, that macropatterns tend to be the product of our explicit, adaptive, and deliberate uses of pattern thinking. We may develop our own macropatterns (such as our theories and beliefs about how the world works) or we may draw on macropatterns that others have developed and shared with us (such as linguistic and cultural norms).

Macropatterns include theories, methods, and abstractions, procedural patterns of language and activities, classifications and typologies, and the structures, rules, and procedures that constitute our social and technical systems. Indeed, most if not all of the patterns in the work of the authors I reviewed in Chapter 2 were, from this perspective, macropatterns. Macropatterns are patterns in the sense that they consist of webs and nexuses of associated and connected pattern elements, but these elements are likely to themselves be patterns of other patterns.

It is interesting (although not essential) to consider the possible differential role that the left and right hemispheres of the brain might have in supporting macropattern thinking. It has long been noted that there is asymmetrical specialisation between the two sides; the left hemisphere focuses on language, analysis, categorisation, and abstract thinking, while the right hemisphere focuses on image and spatial processing (Toga & Thompson 2003). Neither hemisphere dominates overall, and each hemisphere behaves according to what it is trying to do. There is much variation between individual brains; left- and right-handed people's brains tend to differ in how functions are distributed between the hemispheres, and some brains are highly differentiated in the distribution of their left and right hemispherical functions while others are more functionally symmetrical (Geschwind 1979). Hemispheric topology notwithstanding, as McGilchrist (2019) observed, the two hemispheres tend to process things differently such that our thinking is the product both of their cooperation and of their conflicts. It might be that macropattern thought tends to be left hemisphere-focused in that it is made up of abstractions and aggregations, but it might make more sense to say that macropatterns can also involve emotion, imagination, and morality and as such different hemispheres may be more or less active in working with different kinds of macropatterns.

Macropatterns reflect Damasio's (2021) concept of brain maps in that it is macropatterns that give us our capacity for "reflection, planning, reasoning, and, ultimately, the generation of symbols and the creation of novel responses, artefacts, and ideas". I would argue that macropatterns also reflect Kaplan's concept of a "pattern model" (although I disagree with his argument that a "pattern model" can only consist of objective relations and that it axiomatically excludes feelings and subjective perceptions):

> We understand something by identifying it as a specific part in an organized whole ... as we obtain more and more knowledge it continues to fall into place in this pattern, and the pattern itself has a place in a larger whole.
>
> *(Kaplan 1963)*

Returning to the distinction I made between pattern recognition and pattern development as two modes of pattern perception, macropatterns are very unlikely to be developed without deliberate thought, discipline, and focus. This is one way in which macropatterns differ from micropatterns and mesopatterns. Macropatterns also allow us to collaborate with others and to participate in domains and disciplines of action, such as professions, scientific disciplines and paradigms, cultures, and societies. Indeed, macropatterns may become so codified as to constitute laws or philosophies of identity and of action (Baggini 2018).

A Pattern Continuum

We can consider micropatterns, mesopatterns, and macropatterns as existing within a continuum of pattern perception that ranges from fleeting pattern memory matches to in-depth and extended deliberate pattern development. The distinctions between micropatterns, mesopatterns, and macropatterns are relative and constitutive rather than absolute. For instance, if pattern A is used as a part of pattern B then A functions as a micropattern relative to B and B functions as a macropattern relative to A.

I have proposed the terms *micropattern*, *mesopattern*, and *macropattern* but have differentiated them from the basic fabric of pattern memory. They might be considered to be aspects of pattern coherence, recall, and aggregation rather than pattern elements (engrams), so should I call them patterns at all? I do so because I think it is problematic to treat pattern memory as something entirely different from pattern thinking. Micropatterns are fleeting associations and impressions, they draw on memory more than they trigger memory, mesopatterns may be significant enough to be committed to memory, at least in ways to serve future pattern perception, while macropatterns are much more likely to be woven into memory with increasing abstraction. That is not to suggest that we do not also sometimes externalise mesopatterns or even micropatterns, for instance, in the form of poetry or painting, only that so much of science and culture would seem to be based on macropattern thought.

If, to paraphrase Feynman, it is patterns "all the way down", then we can understand there being chains, webs, and clusters of associations (whether or not they are apparent) reaching from pattern elements to patterns to macropatterns, with multiple interactions and adaptive cycles (which we might call *learning*) occurring between them. We might draw on ecological systems theory here to consider these adaptive cycles both within and between pattern strata or pattern clusters (Holling et al. 2002). An adaptive cycle within a web or cluster might be considered a routine pattern elaboration, responding say to another example of a familiar pattern (a dog, a bus, a bird). An adaptive cycle in one pattern or pattern web may also have propagating adaptive impacts in any patterns it is associated with. For instance, changes to the pattern "how I travel to work" may impact patterns associated with "getting home from work" and "getting ready to go to work". There may be deeper pattern elaborations where basic pattern assumptions are found wanting. For instance, learning that dolphins are mammals rather than fish (assuming you had previously thought that dolphins were fish) may impact patterns associated with other animals such as whales and seals, and it may also impact broader patterns associated with fish and with mammals. In this way, these adaptive cycles can be panarchic (Gunderson & Holling 2002) in that changes at one stratum or level have adaptive and even disruptive propagating effects on other levels.

Another way of looking at this continuum is to adapt a model advanced by Dennett (2017) based on three continua or axes to define a three-dimensional space within which different kinds of patterns can be mapped – see Figure 4.4. Rather than suggesting that perception is based on some kind of threshold of awareness, we might use this model to consider different kinds of pattern perception: gestalt (bottom-up, unconscious, understanding), expertise (bottom-up, deliberate, understanding), curiosity (bottom-up, deliberate, impression), aesthetic (top-down, deliberate, impression), belief (top-down, unconscious, understanding), and bias (top-down, unconscious, impression). I do not claim this as authoritative or central to a pattern theoretical approach, only that it is one way of building from pattern theory to develop models and hypotheses for empirical exploration.

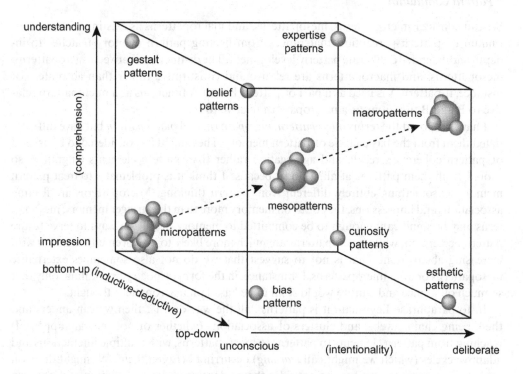

FIGURE 4.4 A model of a pattern perception continuum based on three orthogonal dimensions (level of comprehension, level of intentionality, and inductive-deductive – after Dennett). The main sequence of micro-meso-macro pattern development reflects different stages and kinds of perception.

Diagram prepared by the author.

Pause for Thought

As much as it is apparent that cognitive phenomena emerge from neural phenomena, the connections between the two are as much built of hypotheses, syllogisms, and theories as they are built on empirical evidence. For instance, as Mattson observed:

> While some principles by which the brain uses pattern recognition and encoding to represent the past and the future have been established, a clear understanding of the underlying molecular and cellular mechanisms is lacking. How the encoded patterns are recalled and processed to generate enduring memories of the different patterns and their association with other encoded patterns ... is also not well understood.
>
> *(Mattson 2014)*

Like many cognitive theories (dual processing, cognitive load, etc.), the pattern theory I have outlined aligns with our experiences of our own minds and with neurological findings and theory. It also reflects the work of other scholars who have linked pattern and mind. For instance, Feldman Barrett argued that perception is constructed primarily around the memories (patterns) we activate in response to sensory input. This raises issues of selectivity and efficiency, and of perception arising from prediction (pattern anticipation) and pattern understanding and interpretation, not least because this suggests that we can only perceive those things that we can understand or make sense of:

> Your experiences are not a window into reality. Rather, your brain is wired to model your world, driven by what is relevant for your body budget, and then you experience that model as reality ... your experiences might seem to be triggered by the world outside the skull, but they're formed in a storm of prediction and correction.
>
> *(Feldman Barrett 2017, p. 289)*

As much as there seems to be a great deal of congruence, caution is needed in tying all these phenomena into a single coherent pattern theory. As with all such work, more research is needed, and, to that end, I note a few weak spots in my arguments that will need further consideration.

First, I have rather ambiguously used the concepts of sensory data and sensory information in describing pattern perception. When Ackoff differentiated between data (signals, symbols), information (data with formatting, structure, and meaning), knowledge (information with purpose), and wisdom (knowledge with judgement), he was clear that each builds upon the other such that:

> An ounce of information is worth a pound of data. An ounce of knowledge is worth a pound of information. An ounce of understanding is worth a pound of knowledge.
>
> *(Ackoff 1989)*

So, is it data, information, or knowledge, or perhaps even wisdom that informs or triggers pattern perception? We might say that sensory data are what flow from our senses to our brains, except that this data has been generated by the radiation (light, heat), kinetics, or chemicals interacting with our senses, which means there is already some structure added (temporal, spatial) before getting to the brain. Then there is the in-brain processing in, say, the visual or auditory cortexes, which appear to do much of the work of translating signals into a sense of an external reality before any semantic interpretation is attempted (Pautz 2021). This may happen before pattern perception or in parallel with it, or it might be considered a part of pattern perception, such that depth, substance, and space are also patterns (depending on where our pattern definitions end up of course). As an example of this, Milner & Goodale argued that our brains have two semi-independent perception processes, one focused on spatial reasoning that guides our actions and reactions as physical mobile beings, and the other that generates conscious perception. The former reflects this idea of pre-processing sensory data, the latter is roughly equivalent to the pattern selection model I have advanced here (Milner & Goodale 2006). However, while experience suggests that this may indeed be the case, the evidence to back this up is at best provisional:

> The debate about the existence of an unconscious mental life is as old as psychology itself. An overview of the contemporary opinions about the nature of the unconscious shows that, despite countless studies, the opinions expressed by the researchers working in this field are as diverse today as they were when the debates began.
>
> *(Litman & Reber 2005, p. 431)*

My point is that our brains seem to convert data into information before they enter the cortical matching and sorting process – see Figure 4.5. Even our eyes seem to be selective in what they send along our optic nerves, such that sensory data is augmented and filtered long before it reaches the theorised pattern matching and selecting stages. We might argue, therefore, that pattern perception works with information rather than data, and often it works with knowledge rather than information. Indeed, how could pattern matching ever work based on raw sensory data? Our ability to perceive pattern depends on pattern memory with all its filtering, structuring, and associative dimensions. Pattern perception would therefore seem to occur relatively late in the cognitive pathway and as such it depends on earlier more fundamental pattern processes of the cortexes that turn sensory data into

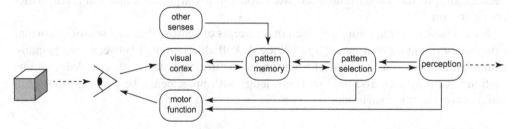

FIGURE 4.5 Hypothetical set of pathways in perceiving pattern from visual data. Light from an object falls on our eyes which triggers nerve activity sending signals to the brain. These are initially processed by the visual cortex where they are translated into a sense of depth, form, etc. and then these augmented signals are entered into the pattern perception process along with other data, and that process and/or the resulting perception may in turn influence our bodies to move our eyes to see the object in different ways to help resolve or improve on the pattern perception.

Diagram prepared by the author.

sensory information. From this, and despite my somewhat reductionist models, it is hard to say definitively quite where pattern perception begins and ends. Rather, there would seem to be many pattern processes with feedback loops that form an emergent, dynamic, and adaptive pattern perception system.

Another concern is where the pattern selection or matching 'engines' are in the brain. Are these metaphorical mechanisms distinct from pattern memory recall or tightly integrated with it? Is pattern perception a top-down executive function or a bottom-up emergent phenomenon? The answer from a process perspective seems to be both and neither in that they seem to be more abductive than inductive or deductive. From a neurological perspective, there is research that suggests that these engines or mechanisms might be relatively region-specific in the brain, albeit with some degree of uncertainty and fluidity (Konovalov & Krajbich 2018). Beyond that, we only have theory and conjecture.

Another issue is the nature of the matching and selection process (or processes). We might consider this sorting as a kind of natural selection (only the fittest survives), albeit one that happens in milliseconds rather than across many lifetimes. However, as we are voluntary and enculturated beings, the pattern selection process cannot be assumed to be a direct analogue of natural selection. We curate our memories, we interpret and create identities, values, and preferences, and we acquire and articulate biases, all of which can shape our pattern thinking. Moreover, a 'losing' candidate match is not erased; the pattern memory is still there and may indeed be elaborated as a result of a failed match. Indeed, there is nothing to suggest the selection process is particularly algorithmic or logical, so a rapid evolutionary selection of better fitting patterns seems more likely.

Yet another issue I wrestled with was the regionality of the brain in pattern perception. Research has shown time and again that much of the cortex fulfils specialised functions, albeit with a degree of variation and plasticity. Likewise, many studies have suggested that the left and right hemispheres engage in different kinds of reasoning, the left side tending to more inductive probabilistic reasoning ("how likely is this?") and the right tending to more deductive reasoning ("does this comply with the rules?") (Parsons & Osherson 2001). This may have implications for different kinds of pattern reasoning, and it potentially contradicts or at least complicates the Hawkins-Kurzweil hypothesis that the cortex is a pattern memory and association machine. If different kinds of pattern logics are not evaluated across the cortex (where memory might be expected to be accessed and processed) but regionally then perhaps different regions and, therefore, different pattern logics are adjacent to different memories. This is well beyond my area of expertise and not essential to my general arguments regarding pattern thinking, but there are implications, for which, naturally, yet more research is required.

There are many issues that I have not (yet) explored, largely to avoid going down too many rabbit holes. For instance, I decided not to explore issues of neuroplasticity and how pattern thinking changes the brain as well as the mind, even though I acknowledge that:

> The various regions of the brain develop at different rates in different people. No two people's brains store the same information in the same way in the same place.
>
> *(Medina 2008, p. 70)*

I also sidestepped implications of embodiment, such that mind and body are a single cognitive entity. We might seek to explore to what extent patterns or pattern thinking extend into our distributed nervous systems. I did not consider the role of brain chemistry in patterns and pattern thinking (such as the effects of blood oxygenation or the varying levels of different hormones such as adrenaline and cortisol) even though these are likely to play an important role, not least in terms of forming and drawing on affective pattern elements. So many questions, so little time.

Chapter Summary

Philosophers have often sought to connect concepts of pattern with concepts of mind and thought. Descartes explored the ideas of the patterns we experience (Morris 1969), Kant's "constitutive rules" (and Searle's later reworking of them) were also patternlike (Ludwig 2017), as were Hume's "regularities". Even Plato's interest in causality reflected a sense that the patterns humans perceived in phenomena indicated underlying causalities. Central to all these perspectives is the idea of dependence on perception, a point that was captured in Whitehead's argument that:

> Pattern is the 'manner' of a complex contrast abstracted from the specific eternal objects which constitute the 'matter' of the contrast … the realisation of a pattern necessarily involves the concurrent realisation of a group of internal objects capable of contrast in that pattern.

(Whitehead 1978, p. 115)

Pattern is not *in* the phenomenon in which it appears, it is the basis for our perceptions of it as a phenomenon.

Having argued that pattern is an artefact of memory and therefore mind, I outlined concepts of pattern perception, pattern recognition, and pattern development. I described pattern as nexuses of connected memories that can be searched, matched, and pulled into working memory to form our perceptions. I considered the temporal and adaptive nature of pattern perception and its ability to work with ambiguous, incomplete, and changing information, and how patterns are an efficient basis for long-term memory and dynamic pattern perception through their associations and accommodation of redundancy and heuristic reasoning. I distinguished between micropatterns as impressions, mesopatterns as conscious perceptions, and macropatterns as deliberate, formal, and long-lasting abstractions. I closed with a consideration of the limitations of the arguments and lines of reasoning I pursued in this chapter.

I should note that the neuroscience literature I drew on had a recurring focus on "patterns in the data" and "patterns of behaviour" in terms of what investigators notice, interpret, emphasise, or extract from the phenomena they observe. Indeed, much of the neuroscience literature has been focused on computational pattern recognition applied to data acquired from brain function research (Churchland & Sejnowski 1992; Schrouff et al. 2013). There is, therefore, some confusion in talking about neurological or cognitive patterns as to whether this refers to patterns that are perceived in phenomena or to patterns that constitute or generate the phenomena. My simple guiding response is to reassert that, like the ghost in the machine, pattern is not in the data, pattern is in our minds, our perceptions, and in our interpretations of what we perceive.

References

Ackoff RL. From Data to Wisdom. *Journal of Applied Systems Analysis* 1989; 16: 3–9.

Adams MJ. The Three-cueing System. In Osborn J & Lehr F (Eds.), *Literacy for All: Issues in Teaching and Learning.* New York, NY: The Guilford Press: 1998.

Ashwell K. *The Brain Book.* Buffalo, NY: Firefly Books: 2019.

Baddeley AD, Warrington EK. Amnesia and the Distinction between Long- and Short-term Memory. *Journal of Verbal Learning and Verbal Behavior* 1970; 9(2): 176–189.

Baggini J. *How the World Thinks: A Global History of Philosophy.* London, UK: Granta: 2018.

Bisaz R, Travaglia A, Alberini CM. The Neurobiological Bases of Memory Formation: From Physiological Conditions to Psychopathology. *Psychopathology* 2014; 47(6): 347–356.

Bor D. *The Ravenous Brain: How the New Science of Consciousness Explains Our Insatiable Search for Meaning.* New York, NY: Basic Books: 2012.

Brewin CR. Coherence, Disorganization, and Fragmentation in Traumatic Memory Reconsidered: A Response to Rubin et al. (2016). *Journal of Abnormal Psychology* 2016; 125(7): 1011–1017.

Colloca L, Miller FG. How Placebo Responses Are Formed: A Learning Perspective. *Philosophical Transactions of the Royal Society of London. Series B, Biological Sciences* 2011; 366(1572): 1859–1869.

Croskerry P. Clinical Cognition and Diagnostic Error: Applications of a Dual Process Model of Reasoning. *Advances in Health Sciences Education* 2009; 14, 27–35.

Damasio A. *Feeling and Knowing: Making Minds Conscious.* New York, NY: Pantheon: 2021.

Damasio A. *Self Comes to Mind: Constructing the Conscious Brain.* New York, NY: Pantheon: 2010.

De Neys W. Advancing Theorizing About Fast-and-slow Thinking. *The Behavioral and Brain Sciences* 2022; 1; 68.

Dennett DC. Real Patterns. *The Journal of Philosophy* 1991; 88: 27–51.

Dennett DC. *From Bacteria to Bach and Back: The Evolution of Minds.* New York, NY: WW Norton: 2017.

Douven I. (2017). *Peirce on Abduction.* In: Zalta EN. Stanford Encyclopedia of Philosophy. https://plato.stanford.edu/entries/abduction/peirce.html

Edelman GM. *Remembered Present: A Biological Theory Of Consciousness.* New York, NY: Basic Books: 1990.

Fang F, Kersten D, Murray SO. Perceptual Grouping and Inverse fMRI Activity Patterns in Human Visual Cortex. *Journal of Vision* 2008; 8: 2–9.

Feldman BL. *How Emotions Are Made: The Secret Life of the Brain.* New York, NY. Houghton Mifflin Harcourt: 2017.

Foster KR, Kokko H. The Evolution of Superstitious and Superstition-like Behaviour. *Proceedings of the Royal Society B* 2009; 276: 31–37.

Frankland P, Bontempi B. The Organization of Recent and Remote Memories. *Nature Reviews. Neuroscience* 2005; 6: 119–130.

Geschwind N. Specializations of the Human Brain. *Scientific American* 1979; 241(3): 180–201.

Gettier EL. Is Justified True Belief Knowledge? *Analysis* 1963; 23(6): 121–123.

Gigerenzer G, Brighton, H. Homo Heuristicus: Why Biased Minds Make Better Inferences. *Topics in Cognitive Science* 2009; 1: 107–143.

Gregory R. *The intelligent eye.* London: Weidenfeld and Nicolson: 1970.

Gunderson LH, Holling CS. *Panarchy: Understanding Transformations in Human and Natural Systems.* Washington, DC; Island Press: 2002.

Hawkins J. *On Intelligence.* New York, NY: St Martins Griffin: 2004.

Hebb DO. *The Organization of Behavior; A Neuropsychological Theory.* New York, NY: Wiley: 1949.

Herz RS, Schooler JW. A Naturalistic Study of Autobiographical Memories Evoked By Olfactory and Visual Cues: Testing the Proustian Hypothesis. *The American Journal of Psychology* 2002; 115(1): 21–32.

Holling CS, Gunderson LH, Ludwig D. In Quest of a Theory of Adaptive Change. In: Gunderson LH, Holling CS (eds). *Panarchy: Understanding Transformations in Human and Natural Systems.* Washington, DC: Island Press: 2002.

Hutchinson JMC, Gigerenzer G. Simple Heuristics and Rules of Thumb: Where Psychologists and Behavioural Biologists Might Meet. *Behavioural Processes* 2005: 69(2); 97–124.

Johns G, Saks AM. *Perception, Attribution, and Judgment of Others. Organizational Behaviour: Understanding and Managing Life at Work*. North York, ON: Pearson: 2011.

Kahneman D. *Thinking, Fast and Slow*. New York, NY: Farrar, Straus and Giroux: 2011.

Kaplan A. *The Conduct of Inquiry: Methodology for Behavioral Science*. Chandler: 1963.

Konovalov A, Krajbich I. Neurocomputational Dynamics of Sequence Learning. *Neuron* 2018; 98(6): 1282–1293.

Kurzweil R. *How to Create a Mind: The Secret of Human Thought Revealed*. New York: Viking Books: 2012.

Litman L, Reber AS. Implicit Cognition and Thought. In: Holyoak KJ, Morrison RG (eds.). *The Cambridge Handbook of Thinking and Reasoning*. Cambridge UK: Cambridge University Press: 2005.

Ludwig K. Constitutive Rules and Agency. *From Plural to Institutional Agency: Collective Action II*. Oxford, UK. Oxford University Press: 2017.

M'Closkey K, VanDerSys K. *Dynamic Patterns: Visualizing Landscapes in a Digital Age*. London, UK: Routledge: 2017.

Mattson MP. Superior Pattern Processing is the Essence of the Evolved Human Brain. *Frontiers in Neuroscience* 2014; 8: 265.

Mayes AR. Memory and Amnesia. Behavioural Brain Research 1995; 66(1–2): 29–36.

McGilchrist I. *Ways of Attending: How Our Divided Brain Constructs the World*. Abingdon, UK: Routledge: 2019.

Medin DL, Goldstone RL, Gentner D. Similarity Involving Attributes and Relations: Judgments of Similarity and Difference Are Not Inverses. *Psychological Science* 1990; 1(1): 64–69.

Medina J. *Brain Rules: 12 Principles for Surviving and Thriving At Work, Home, and School*. New York, NY: Pear Press: 2008.

Milner D, Goodale M. *The Visual Brain in Action* (2nd ed.). Oxford, UK: Oxford University Press: 2006.

Morris J. Pattern Recognition in Descartes Automata. *Isis*1969; 60(4); 451–460.

Ortega-de San Luis C, Ryan TJ. United States of Amnesia: Rescuing Memory Loss from Diverse Conditions. *Disease Models & Mechanisms*. 2018; 11(5): dmm035055.

Parsons LM, Osherson D. New Evidence for Distinct Right and Left Brain Systems for Deductive versus Probabilistic Reasoning. *Cerebral Cortex* 2001; 11(10): 954–965.

Pautz A. *Perception*. London, UK: Routledge: 2021.

Richie R, Bhatia S. Similarity Judgment Within and Across Categories: A Comprehensive Model Comparison. *Cognitive Science* 2021; 45: e13030. DOI: 10.1111/cogs.13030

Rips LJ. Similarity, Typicality, and Categorization. In: Vosniadou S, Ortony A. (eds.) *Similarity and Analogical Reasoning*. Cambridge, UK: Cambridge University Press: 2009.

Roger B. *Perception: A Very Short Introduction*. Oxford, UK: Oxford University Press: 2017.

Rookes P, Willson J. *Perception. Theory Development and Organisation*. London, UK: Routledge: 2007.

Schacter DL. The Seven Sins of Memory: An uPDATE. *Memory* 2022; 30(1): 37–42.

Schrouff J, Rosa, MJ, Rondina JM et al. PRoNTo: Pattern Recognition for Neuroimaging Toolbox. *Neuroinformatics* 2013; 11, 319–337.

Churchland PS, Sejnowski TJ. *The Computational Brain*. Cambridge, MA: MIT Press: 1992.

Shermer M. *The Believing Brain*. New York, NY: St Martin"s Press: 2011.

Shugen W. Framework of Pattern Recognition Model Based on the Cognitive Psychology. *Geo-spatial Information Science* 2002; 5(2):74–78.

Simon HA. Rational Choice and the Structure of the Environment. *Psychological Review* 1956; 63(2): 129–138.

Tenenbaum JB, Griffiths TL. Generalization, Similarity, and Bayesian inference. *The Behavioral and Brain Sciences* 2001; 24, 629–640.

Toga AW, Thompson PM. Mapping Brain Asymmetry. *Nature Reviews. Neuroscience* 2003, 4(1): 37–48.

Voss JL, Bridge DJ, Cohen NJ, Walker JA. A Closer Look at the Hippocampus and Memory. *Trends in Cognitive Sciences* 2017; 21(8): 577–588.

Wason PC, Evans J. Dual Processes in Reasoning? *Cognition* 1974; 3(2): 141–154.

Whitehead AN. *Process and Reality*. New York, NY: The Free Press: 1978.

Wirebring LK, Stillesjö S, Eriksson J, Juslin P, Nyberg L. A Similarity-Based Process for Human Judgment in the Parietal Cortex. *Frontiers in Human Neuroscience* 2018; 13. DOI: 10.3389/fnhum.2018.00481

Yadav N, Noble C, Niemeyer JE, Terceros A, Victor J, Liston C, Rajasethupathy P. Prefrontal Feature Representations Drive Memory Recall. *Nature*. 2002; 608: 153–160.

5

PATTERN THINKING

In which I consider different ways in which our minds use patterns.

Pattern thinking is an umbrella term for any mental process that engages pattern in some way. This contrasts with theories that have framed pattern as a more specific form of thought. As an example, Baron-Cohen focused on pattern thinking as understanding of "if-and-then" causal relationships between things, arguing that, as a species, one third of us are systemising (pattern) thinkers, one-third are empathic (non-pattern) thinkers, and one-third are adept at both (Baron-Cohen 2020). While I agree that some people seem to be more adept at some kinds of pattern thinking than others, I do not agree that patterns are only causal and logical, that patterns are primarily the domain of science and technology, or that patterns are solely about systems of rules and relationships. It might well be that there are individuals who tend to be systemising in their thinking and others who are more empathic in their thinking, but I would argue that these are simply different kinds of pattern thinking.

Although I earlier focused on sensory information as stimuli for pattern perception, I should be clear that pattern thinking can have many different triggers including emotionality, curiosity, identity, creativity, and even our dreams in which it seems we sort and tidy our pattern memories (Crick & Mitchison 1983). Indeed, pattern thinking would need, deductively speaking, to be somewhat continuous and integrated to build and sustain any sense of a life of the mind.

The depth and primacy of pattern thinking is reflected in our use of language. Chomsky's (1980) argument that all human minds come with a "universal grammar" has been robustly challenged, reflecting wider debates between those who believe that language abilities are hard-wired into our brains and those who believe language abilities are learned (Sampson 2005; Harnad 2017). Clearly, specific languages are not inherited; Japanese babies are no more pre-programed to speak Japanese than Spanish babies are pre-programed to speak Spanish (or Japanese). The languages a child learns are the ones they are immersed in.

DOI: 10.4324/9781003543565-5

That said, humans as a species seem to have a remarkable capacity for acquiring language, whatever their context or culture. However, rather than having a specific "language instinct", I would argue that we have a disposition and capacity for pattern thinking, and it is that which forms a ready substrate for developing language. Rather than being intrinsically adept linguists, I would argue that we are particularly adept abstract pattern thinkers and our most fundamental abstract pattern thinking modality is language. Understanding language development in terms of kinds and capacities of pattern thinking affords some interesting perspectives, for instance on the exploration of non-human languages (Shah 2023). Indeed, although Pinker broadly supported Chomsky's ideas of a universal grammar, he did describe patterns as being central to how our brains work:

The brain has around 100 billion neurons connected by 100 trillion synapses, and by the time we are 18 we have been absorbing examples from our environments for more than 300 million waking seconds. So we are prepared to do a lot of pattern matching and associating.

(Pinker 2021, p. 107)

Human minds seem particularly adept at associating symbols with patterns such that a symbol can both invoke and stand in for the more complex pattern construct it refers to: words can represent things, feelings, or concepts, while numbers can represent quantities, relationships, and probabilities. Indeed, over millennia, human societies have learned to use this capacity to build ever more elaborate systems of interacting concepts and symbols. We might, therefore, consider patterns as consisting of a web of memories of percepts (impressions, experiences, etc.), abstractions (characteristics, similarities, variances etc.), and symbols (spoken, written, graphic, etc.) – see Figure 5.1.

The symbolic parts of a pattern would seem to connect with the symbolic parts of other patterns by using common symbols. These connections between symbols are themselves abstractions, which can also drive abstractions within and across patterns, thereby creating the basis for language, mathematics, music, and other macropattern systems. Patterns can also reflect different kinds of relational symbolism, such as allegory (a mundane symbol represents bigger issues), archetype (recurring typical symbols), irony (symbols that reflect their opposites), metaphor (a symbol that suggests a paradoxical connection), and simile (a symbol that suggests connections between two dissimilar things).

Symbolic pattern relationships can be highly idiosyncratic. Indeed, there are various psychoanalytic techniques of word and image association that have been developed as a way of exposing and analysing the idiosyncratic pattern connections in individual minds. As Jung observed:

What we call a symbol is a term, a name, or even a picture that may be familiar in daily life, yet that possesses specific connotations, in addition to its conventional and obvious meaning. It implies something vague, unknown, or hidden from us.

(Jung 1964, p. 3)

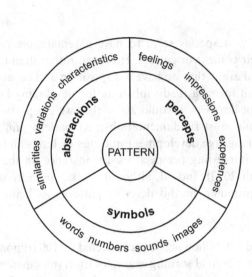

FIGURE 5.1 An abstract model of a pattern as a web or nexus of memories with distinct memory elements of percepts, abstractions, and symbols.

Diagram prepared by the author.

We each have our own somewhat partial and fragmentary symbol systems, and we also participate in the symbol systems of our societies, cultures, religions, and professions. It has been reasonably suggested that developing the capacity for collective thinking with and through symbolic thought expressed as language must have conferred evolutionary advantages to our ancestors (Medina 2008). We do not know whether pattern preceded language in our evolutionary past or whether language preceded pattern, or whether (as seems more likely) they co-evolved. Pattern thinking is perhaps the more abstract and the more foundational of the two as it underpins many other cognitive capabilities. On the other hand, the evolutionary advantages of developing language as a way of supporting more sophisticated and advantageous forms of social communication and collaboration seem to be so great that language may have been the first (or among the first) abstract pattern thinking capabilities we developed.

Affordances

Recognising something as an instance of a pattern memory does more than provide the identity of something (e.g. "that is a dog"), it also pulls on associated thoughts and feelings that can tell (or remind) us what that identity means to us ("friend", "danger", "smelly", "hairy", etc.). This brings me to the knotty problem of *affordances*, a term defined by Gibson in the context of animals (including humans) reading the environments through which they move:

> The affordances of the environment are what it offers the animal, what it provides or furnishes, either for good or ill
>
> *(Gibson 1979, p. 119)*

Gibson argued that animals are somehow resonant with or attuned to their natural environments such that they do not need to think about the things they encounter, they simply know what they are and what they mean. I think this is mistaken; thought is always present. After all, only some of our pattern thinking is conscious and deliberate even though our minds are constantly drawing on pattern memory both in anticipation of events and in response to them. I would argue that Gibson's misstep was a failure to notice or acknowledge the anticipatory pattern activation that allows for the rapidity and fluidity of thinking in familiar contexts. Some affordance theorists have tried to work around the elision of mind by adding concepts such as "solicitations" that reflect selectivity and intentionality (that do have cognitive correlates):

> ... those affordances that are experienced as relevant by a situated individual ... It is those affordances that allow us to improve our grip on the situation that are relevant ... When an individual acts or when the situation itself develops, the individual–environment relation is changed and other solicitations arise ... What is foreground and what is background shifts continuously, the field is in flux over shorter timescales.
>
> *(Kolvoort & Rietveld 2022, p. 17)*

I would argue that pattern thinking (and pattern anticipation in particular) can fully account for the speed of response and the parsimony of cognitive load that Gibson described. We do not need to directly perceive and understand a mug as a cylinder with a loop on the side, and from that unconsciously intuit what to do with it. If we have ever seen a mug, then we just seem to know that this thing is another instance of the pattern of a mug. If we had never seen anything like a mug, we might engage in something like deliberate affordance reasoning (mirroring Kahneman's sense of System 2 thinking), but even then our past experience and understanding informs current perceptions. Pattern thinking is simpler, faster, and quite sufficient for most day-to-day experience, and it can also account for our misperceptions (both Type I and Type II errors).

Some affordance theorists, like Heras-Escribano (2022), came close to suggesting that pattern and affordance are linked, but his rejection of a cognitive dimension to affordance led him to argue that pattern is still in the world rather than in the mind. This argument must be wrong because we know that perception varies between perceivers, that it can easily be mistaken or confused, and that illusions and delusions can further complicate matters. That said, not all affordance theorists have been so determined to excise the role of mind in affordance (Gregorians & Spiers 2022). Borghi reflected on the role of language and other symbolic (pattern) structures as part of our perception of affordances:

> The fact that participants are sensitive to object affordances during language processing confirms that language is grounded in perception and action systems; nouns evoke a simulation of their referents, and verbs a simulation of possible actions. Second, evidence indicates that language constrains the activation of affordances. It activates stable rather than variable affordances.
>
> *(Borghi 2022, p. 114)*

If language (as a pattern system) shapes affordance then not only is there cognitive processing going on, but it is of necessity a form of pattern thinking. That some patterns reflect experiences of concrete things (such as apples and doors) with clear external referents, while less concrete patterns reflect experiences of practices, concepts, and contexts with more interpretive and subjective referents, still suggests that their commonality is not in the world but in the fabric and function of our minds. I see this reflected in the work of those affordance theorists who have proposed a more dyadic relationship between mind and environment and who acknowledge the subjectivity and positionality of such readings (Krueger 2022). All of this would seem to support my proposition that different people have divergent patterns even when they are related to the same external phenomena, that patterns involve memories, feelings, and other constructive and subjective elements, and that the patterns we use at any given time are cued not only by context but also by experience, need, attention, capacity, and capability. This is further supported by Krueger's argument that:

> … a question Gibson doesn't explicitly consider is what happens to bodies when this access is ruptured or impeded? … Autistic people, for example, or people living with schizophrenia, clinical depression, obsessive-compulsive disorder, or anorexia nervosa often describe feeling as though they've lost access to bits of the world, to different affordances, that others take for granted. Some even describe feeling as though they inhabit a different world altogether.
>
> *(Krueger 2022, p. 141)*

When an organism 'reads' its environment familiar and anticipated patterns can be rapidly recalled and interrogated to provide the organism with a perception of what is going on (least provisionally) and what it might mean to us or to others. We need not formulate a definitive macropattern solution, a mesopattern emerging from the micropatterns that our senses trigger typically satisfices. It might be better to say, therefore, that our perceived patterns cue affordances in terms of what we perceive as possible actions based on similar previous encounters, similar contexts, and/or similar feelings.

Returning to Gibson (1979), one of his original arguments for rejecting a cognitive component in his theories of affordance was to challenge the then orthodox belief that perception and memory were distinct phenomena. In this issue, Gibson is perhaps more aligned with a pattern theoretical approach than might at first have seemed the case given that pattern (memory) and pattern thinking (perception) are functionally inseparable, particularly given contextual association and activation that leads to cueing and anticipation. Indeed, the noncognitive aspect of Gibsonian affordance may be less of an issue if it is read as a challenge to then common computational and modular models of cognition and a pursuit of a more organic and adaptive model of perception such as that afforded (!) by pattern theory, particularly the role of cueing and anticipation. Clearly, many questions remain. Is the experience of affordance a consequence of pattern perception, or is the affordance a part of the pattern? Does the affordance become part of the pattern instance in that moment?

Interpretations

Interpretation plays a central role in pattern thinking, not least because it depends on the constant assessment of similarities and differences rather than on establishing binary identities. This interpretive basis brings me to consider hermeneutic perspectives on pattern thinking. I should be clear that I refer here to the philosophical hermeneutics of Heidegger,

Gadamer, and Vattimo with their broad focus on experience as interpretation, rather than the hermeneutics of religious scholars (although I would argue that their work is also intimately and inevitably tied up with pattern thinking, both individual and collective). I should also note that hermeneutics is the study of interpretation rather than the interpretation itself, and as such we can take a hermeneutic perspective on pattern thinking rather than saying that pattern thinking is itself hermeneutic. To that end, we can adopt a "hermeneutic gaze" to consider meaning making and interpretation in the context of individual pattern thinkers (Caputo 2018).

To what extent meaning-making is limbic and intuitive (micro-meso) or abstract and analytical (meso-macro) would seem to depend on the associated memories of whatever pattern is pulled upon. Indeed, given the multitude of ways in which different pattern memories (thoughts, experiences, feelings, etc.) can be connected or triggered, the ways in which pattern thinking can contribute to interpretation and meaning making might take many forms. That said, there would seem to be some commonalities. One is that there must be some kind of interpretation or meaning made before it can be cast into memory and then subsequently recalled. This is, I would propose, an aspect of pattern thinking in that it can both draw on pattern memory and lead to pattern memory, but it may not happen in pattern memory. Put another way, interpretation and meaning making would seem to function, metaphorically at least, like DNA being spliced. Memories are not wiped and replaced with new versions. Rather, some connections may be weakened or removed, while others are strengthened or added, and new memories can also be woven into existing ones. Indeed, it would seem that we need not devise complete interpretive solutions in one go, we need only abductively associate and elaborate pattern memories based on our thoughts and reflections such that interpretation gradually emerges from such associations. I would also suggest that emotion can play an important role in establishing and consolidating interpretation by adding a sense of significance. This is not to downplay the importance of other processes (such as deliberate analysis and macropattern development) in establishing meaning and significance, only that there are likely to be many contributing pattern processes involved.

Gadamer argued that interpretive thought is anchored in our past experience and understanding such that we are always in a subjective state of anticipating that our experiences are interpretable. He described this as the "anticipation of completeness", a sense of expectation that things can be understood that compliments the specific anticipations and prejudgments that shape our approach to meaning making. And yet, how aware are we of our pattern thinking abilities and dependencies? To what extent do we construct identity and purpose around these understandings, and what impact do they have on interpretation and meaning making? It may be that pattern theory has much to add to meta-hermeneutic debate.

Another Gadamerian concept to consider in the context of pattern thinking is that of horizons of interpretation. Gadamer argued that everyone who engages in interpretation has an interpretive horizon of their own expectations and enculturation (their specific pattern memories and pattern thinking abilities and dispositions):

> A person who is trying to understand a text is always projecting. He projects a meaning for the text as a whole as soon as some initial meaning emerges in the text … the initial meaning emerges only because he is reading the text with particular expectations in regard to a certain meaning.
>
> *(Gadamer 1994, p. 267)*

This reflects pattern theoretical perspectives on perception, cueing, and anticipation, as well as on bias, error, and illusion. I would go further though to consider cultural and para- digmatic pattern dispositions that we develop through our participation in common social activities and collaborations such as science, art, and culture. Although it is true that pat- tern memories only occur in individual minds, there will necessarily be some degree of convergence between the patterns in the minds of individuals who collaborate with each other. We should, therefore, explore the implied social patterns and pattern thinking of a social group (family, tribe, community, culture, profession, class, society).

Gadamer was explicit in arguing that our hermeneutic horizons are not fixed. Rather, they change with experience, context, etc. and they can interact with the horizons of others. This can be seen in the ways in which one person seeks to interpret the ideas of another person, or in the ways that a group of people strives towards consensus in interpretation. This suggests that much as pattern memory and pattern thinking are situated in individual minds and their accumulated thoughts and experiences, we develop much of our meaning making collaboratively and socially.

Gadamer also argued that language serves as a universal horizon of interpretation and meaning making among those who participate in it. We might, therefore, consider conver- sations, debates, and arguments over meanings and interpretations to be extensions of our pattern thinking into social spaces. This in turn raises the point that a hermeneutic gaze is not solely or even particularly focused on individual minds. Indeed, Gadamer argued that meanings do not solely spring from us as individuals, rather we are steeped in the many meanings of culture, language, society, and family that provide a counterpoint to our indi- vidual experiences (Gadamer 1977).

The utility of hermeneutics notwithstanding, there are other perspectives on meaning making that also help to explain pattern thinking. For instance, Taylor (2016) described language in terms of its shaping our thinking such that language directs rather than responds to meaning making. Indeed, Taylor argued that our sense of self is generated out of the habits and practices of speech and language. I would argue that any study of under- standing and interpretation requires us to consider patterns and pattern thinking (whether those involved are aware of them or not). On that basis, not only can hermeneutics lend pattern theory depth and sophistication, pattern theory can provide hermeneutics with a cognitive and possibly even a neurological foundation.

Pattern Reasoning

Reasoning is another broad and polysemic concept in the English language that can refer to thinking in a logical way, influencing others through logic and persuasion, or discussing, arguing, or questioning things. Some accounts of reasoning simply contrast deductive and inductive reasoning, others consider the abductive nature of reasoning, while others con- sider more expansive typologies. Rather than engaging in yet another word game where reasoning means whatever I or anyone else wants it to, it might be more productive to con- sider *reasoning*, whatever its cognitive bases, to be another pattern concept that has many parts (such as memory, reflection, logic, inference, implication, anticipation, and debate), modes (such as inductive, deductive, abductive, and counterfactual), applications (such as problem-solving, similarity-difference, and priority), and associations (such as decision- making, prediction, and risk assessment). I would argue that this aligns well with Edelman's

(1990) observation that concepts (patterns) do not only refer to things (objects, feelings, thoughts, etc.) but also to relationships between things, and more abstractly ways of thinking about other concepts (patterns again). I would also suggest that reasoning is heuristic in that it selects from a range of approaches, switches between them as necessary, it works with incomplete information, and it provides satisficing rather than perfect solutions.

In "The Speckled Band", Sherlock Holmes pronounced: " 'I had,' said he, 'come to an entirely erroneous conclusion which shows, my dear Watson, how dangerous it always is to reason from insufficient data.'" As much as I enjoy the Holmes canon, I think that Conan Doyle missed a central fact of reasoning, that it is *always* conducted with insufficient data (or information or knowledge). After all, if you had enough data to be sure of something, you would not need to reason your way to understanding it. Reasoning and insufficiency therefore go hand-in-hand; reasoning is employed because we lack knowledge. We do not always reason in such situations, we may just guess or trust to chance, but any attempt to account for missing information logically or systematically is a way of defining reasoning. If so, then whenever we seek to complete a pattern or invoke a pattern from its parts (which are also based on incomplete, imperfect, and ambiguous information), we must engage in some kind of reasoning, however imperfect and whether or not we are aware of doing so.

Reasoning to Pattern

In the previous chapter I contrasted *pattern recognition* (rapid matching stimuli to patterns from memory) with *pattern development* (deliberate and analytical approaches to identifying or developing pattern understanding). I suggested that we usually start with pattern recognition, and only when this fails to produce a satisfactory result do we shift to pattern development. A pattern need not be a perfect fit to be a candidate; a putative match only needs to be within the range of acceptable variations to be recognised as an instance of that pattern. Matching parts might include pattern elements, their variations, the links and associations between patterns (such as which ones commonly occur together or what patterns are typically found in a particular kind of context), how the pattern works (its processes, logics, and dependencies), how the pattern behaves (what it does and how it does it), or some combination of these characteristics. Similarly, Rips (2009) described modalities of categorisation (what class something belongs to), typicality (how prototypical something is), and resemblance (characteristics they have in common), while Collins & Burstein described different kinds of correspondences (between one or more components or properties), and from this they outlined a typology of similarity assessment based on comparative judgments (similarity typicality, categorisation, identity, overlap, difference), mappings (property, component, conceptual), and topologies of relationship (that differ according to what is being compared to what) (Collins & Burstein 2009). All of these might be considered elements of pattern reasoning, to which we might also add comparisons of values, biases, and cultural orientations. There are also many different *kinds* of reasoning that may play a part in pattern acquisition – see Table 5.1.

Quite when and how different kinds of reasoning might be used in pattern development would seem to depend on many factors, including familiarity and experience in using different reasoning approaches, the time available or that the reasoner is prepared to invest in pattern reasoning, as well as other dispositional qualities of the reasoner. For instance, in the game of "20 questions", a player may start with categorical reasoning based on whether

TABLE 5.1 A variety of types of reasoning that can be applied as a part of pattern acquisition (drawing on Holyoak & Morrison [2005] and Parnafes & Disessa [2004])

Reasoning	Questions
Deductive	What patterns should apply here?
Inductive	What here suggests patterns?
Abductive	What patterns can I combine or adapt here?
Analogical	What can I learn from similar patterns here?
Reductionist	What do the patterns relevant to parts tell us about wholes?
Causative	What patterns are suggested by the relationships between causes and outcomes?
Counterfactual	What if this apparent pattern is wrong?
Evaluative	What patterns are more useful?
Model-based	What patterns does my mental model suggest?
Constraint-based	What patterns apply given the constraints here?
Affordance	What patterns are suggested by my immediate context?

something is animal, vegetable, or mineral, and then switch to analogical reasoning in terms of what the mystery thing is like, and then constraint-based reasoning in terms of where or how it may or may not be found. Woven through these approaches there may also be cycles of deductive, inductive, and abductive reasoning. It is only through the combination of tangents drawn, that the player can hopefully infer what the answer is. Of course, there are other ways in which patterns might be acquired in which reasoning would seem to play a minor role. For instance, we might simply look something up, although this would still require some pattern thinking to know what kind of solution it might be and where and how to find it. Similarly, someone might suggest a pattern to us (through conversation or through mentorship or guidance), or we might serendipitously or randomly stumble across a suitable pattern solution. Indeed, we might reason that it would be faster and more effective to find a pattern solution by asking for someone's help or reading up on the topic. However we do it, reasoning can play a significant role in pattern acquisition.

When we do find or develop a pattern that fits, we have an emotional response, as the significance, understanding, and import of the pattern becomes apparent, particularly if the effort to come up with a pattern is substantial, the need for a pattern is acute, or the significance of the pattern is high. A pattern can make something tractable, it can diffuse its apparent threats, and it can suggest direction, intent, and purpose. This is an inflection point between reasoning toward and reasoning from pattern, the moment when struggle can turn to relief and wonder, or to dismay and dread, depending on the pattern identified. Of course, it may turn out that a pattern we matched on proves less appropriate or useful than we supposed, in which case we would likely revert to pattern acquisition (or simply give up). However, if we have acquired a pattern that satisfices, we can move to reasoning *from* pattern.

Reasoning from Pattern

Once we have recognised (found and activated) or developed a satisfactory pattern, it can tell us (according to the specifics of the pattern in question) what something is, what it might do, what we might do with or to it, what it means, and what its associations are.

Reasoning may depend as much on a pattern's connections and associations as it does on the elements that are connected and associated. Deductively, we might reason that, if this thing is an instance of that pattern, then it should have the qualities of that pattern or of other things also associated with it. Inductively, we might reason that any differences the thing has relative to the pattern might mean that we need to elaborate on the pattern to accommodate them.

Patterns can direct our attention. It would seem to be a useful evolutionary trait that we "notice" things that reflect certain patterns, particularly when they indicate danger or opportunity. Not noticing things does not mean that we are unaware of them, or even that we do not or cannot understand them, it may simply be a failure to note the significance of certain things. Noticing and significance would seem to be closely linked to affective (emotional) pattern elements such as happiness, fear, curiosity, lust, disgust, horror, and anger, in that invoking the pattern also invokes memories of these responses. Moreover, by focusing attention on an instance of a particular pattern we can pull on more and more of the pattern's connections and associations such that we can experience a wellspring of thought and possibility associated with it. Noticing can also be understood in terms of deepening experience, adding nuance and subtlety, and extending understanding.

Pattern reasoning, by connecting memory, feeling, understanding, and prediction, can help us to identify the presence of risk, assess it, and shape our reactions to it. Indeed, I would argue that without pattern thinking we could have no concept of risk, only of danger at an autonomic level. Even an insect that shies at movement around it need not conceive of a human being for it to autonomically match a pattern that triggers a threat response.

Reasoning from pattern may involve many reasoning modalities (see Table 5.2), and, as much as pattern reasoning can help us to respond to immediate challenges and uncertainties, it also plays a role in longer term pattern reflection and elaboration. Given that the more examples or reference points pattern reasoning can draw on (within the cognitive limits of our minds) the stronger its underlying structures and implications are likely to become.

Pattern reasoning is also a key part of our development of macropatterns. I would argue, therefore, that (in its more expansive forms) pattern reasoning is about elaborating and developing an increasingly macropatterned understanding of our experiences and

TABLE 5.2 Types of reasoning applied to pattern articulation

Reasoning	Questions
Deductive	What does the pattern tell us about this instance?
Inductive	What do the idiosyncrasies of this instance tell us relative to the pattern?
Abductive	How might we adapt and refine our pattern understanding of this instance?
Analogical	In what ways is this instance similar to and different from other instances?
Reductionist	What do the pattern's elements tell us about this instance?
Causative	What does the pattern tell us about the causes and origins of this instance?
Counterfactual	What if this were an instance of a different pattern?
Evaluative	What does the pattern tell us about the value of this instance?
Model-based	How does the pattern contribute to a mental model of this instance?
Constraint-based	What do the pattern's constraints tell us about this instance?
Affordance	What does the pattern's contextual associations tell me about this instance?

interpretations, which may include reflection on how far a pattern can be relied on, how much work it can do for us, and how much that pattern can be elaborated while remaining true to the identity of the pattern. Moving away from a reductionist sense of single problem-solution moments allows to understand thought as involving many entangled cascades, loops and networks of pattern thinking and reasoning.

Again, I should be clear that our patterns do not 'do' the reasoning for us, but they do provide the resources that a reasoning mind draws on, the processes and approaches it employs, and the understandings and meanings it generates. This raises a fundamental question that I noted earlier: is pattern thinking purely about patterns as coherent clusters of memories, thoughts, and feelings, or does it also include our capabilities to work with patterns? After all, separating these two aspects does not reflect the reality of our minds where there may not be a meaningful separation between resources and processes. Indeed, can we truly separate process and resource other than as an analytical conceit?

Pattern reasoning also involves working with implications, those nonexplicit meanings, significances, and probabilities that we extrapolate from what we know about a thing. Implications should not be guessed at or imagined (although they may well be), they should follow from what is known and what is likely. Moreover, although they can be tested and perhaps proved, implications typically precede testing and proving; they are possibilities and likelihoods, not established empirical facts. Our ability to develop and understand implications is a direct product of our patterned minds. In part, this is apparent from the basis of implication on evaluating similar differences and different similarities such that the knowledge we have about one thing is extrapolated to another similar thing with our confidence in that extrapolation based on the degrees of similarity and difference.

Pattern is also implied in the ways that we can use patterns that are similar but clearly not a strong match to a current stimulus to make sense (however imperfectly) of what our senses are telling us. On this basis, I would suggest that studying implications is as much about studying our own minds as it is about the phenomenon of interest. That said, implications are not simply a by-product of pattern thinking or reasoning. Thinking about implications contributes to anticipation. Anticipations can lead to physical preparations, a stiffening of the sinews and a summoning up of the blood in response to apparent likelihoods, and as such they are individual pattern perception responses based on conditioned pattern associations. Implications on the other hand tend to be less personal and less embodied and more abstract (macropatterned) and more objective than anticipations (although anticipations and implications are probably best understood as forming a continuum). Indeed, I would suggest that implications form one end of a continuum that is analytical, interpretive, and speculative based on possible and deductive pattern associations while the other end is anchored in the reflexive and limbic pattern memory responses of anticipation. This anticipation-implication continuum would seem to be a significant contributor to the speed and efficiency of pattern thinking if certain pattern memories are activated in advance of their being needed.

The subjective experience of near instant recognition (that confounded Gibson and his followers) is less remarkable if we understand that anticipation and/or implication has already prepared our minds to recognise contextual regularities. The role of anticipation and implication would also explain why we attend to certain things and are less attentive to others. This might be considered a good thing if we were interested in the conditioning of minds, for instance through training, socialisation, and the development of expertise.

Indeed, expertise might be understood in terms of the conditioning of anticipation-implication responses that hold an expert's mind in a state of readiness to apply their trained pattern thinking. Not only are they pre-recognising implied patterns in their rapidly changing and ambiguous streams of experience, cascades of pre-recognition can give a sense of prediction and flow that others may perceive as uncanny mastery.

Anticipation-implication would also seem to open our minds to manipulation and error. Whether it is magic tricks, the invisibility of gorillas, or the psychology of advertising and propaganda, we all seem vulnerable in that our anticipation-implication responses to attend to some things and not to others can be manipulated by others (Chabris & Simons 2010). This vulnerability may even condition pattern memory over time by elaborating and reorganising certain memories to reflect the pattern associations and expectations suggested to them. This in turn suggests that pattern thinking is not simply a matter of our ability to form patterns, to recall them in useful and productive ways, and to iteratively develop and refine pattern memory and the uses thereof. We also need to engage in some degree of filtering and checking of suggested patterns and pattern associations. I could argue, therefore, that highly suggestible people, such as those favoured by advertisers, politicians, and hypnotists, are adept at forming and editing pattern memory but less adept at filtering and checking suggested or implied patterns. The corollary of this is that critical thinkers are perhaps more adept at pattern filtering and, therefore, less prone to pattern suggestion or manipulation. However, critical thinking does not seem to be a global skill but one rather that is specific to a particular domain of activity and thought, which would better reflect the fact that erstwhile experts can still be hoodwinked or scammed in areas in which they lack expertise. The ability to engage in implication is, therefore, both a powerful cognitive capacity and a significant vulnerability without the checks and balances of rational thought.

Affective Pattern Thinking

I earlier noted the role that affect can play in pattern reasoning. Now I expand on this to consider the emotional landscape of pattern thinking more broadly. It is well established that emotions are woven into our memories (Bower 1981) such that when we remember something we also remember how we experienced it and how we felt about it (Feldman Barrett 2017). The emotional elements of the pattern memories we draw on can also *pull* on other patterns and their affective elements. Pattern thinking is not, therefore, simply a matter of organising and drawing on facts and understanding, it is very much tied up with feelings and their abstractions into values, aesthetics, ethics, and other axiological responses. Again, it can be argued that much of this comes from an evolutionary past where:

> Emotions evolved to reinforce memories of patterns of particular significance vis-à-vis survival and reproduction … Pattern processing in its most fundamental manifestation is enhanced by perception of the patterns in an emotional setting.
>
> *(Mattson 2014)*

Affective pattern elements can convey a sense of good and bad, of right and wrong, of harmony and disharmony or of consonance and dissonance. We may feel unhappy about pattern instances that undermine a valued pattern as they may indicate thinking or behaviour that threatens social stability or order. We may feel excited by a pattern whose implications suggest new opportunities for understanding and action. I would argue that it is these axiological pattern elements (memories of beauty, ugliness, rightness, wrongness, and so on) that can help us to transform our understanding into meaning, and once meaning is indicated our attention can be drawn to consider possible future instances. This can include our noticing things when the emotional elements of a pattern call particular attention to something that we perceive as displaying those elements. It can also include a sense of calm and familiarity that "things are as they should be" when pattern instances conform well and do not call our attention to notice them.

I would argue that the affective aspect of pattern thinking is intimately tied up with regularity. We perceive regularity as significant, not in an abstract way but because we have learned to notice it, to attach significance to it. Indeed, we might revisit pareidolia here, not as a cognitive failing but as the triggering of a conditioned signification. We do not just perceive regularity, we look for it as an indicator of "something going on", something important, useful, meaningful. Similarities and differences in pattern affect would also seem to be responsible for our interests and tastes in things such as music, dance, theatre, cinema, and literature. A common quality of artistic and athletic performances that make them enjoyable is a sense of uncertainty within the patterns they embody; a performance that has unexpected consequences, a plot twist or new take on an old idea, anything that both complies with a pattern and yet still innovates and confounds expectations:

> We especially value experience that is surprising, but not too surprising. Routine, super-ficial recognition will not challenge us, and may not be rewarded as active learning. On the other hand, patterns whose meaning we cannot make sense of at all will not offer awarding experience either; they are noise.
>
> *(Wilczek 2015, p. 15)*

The emotional aspects of pattern can also be problematic, even debilitating, at times. For instance, the stronger the emotional content of pattern recall the less cognitive flexibility we are likely to sustain as the emotive pattern dominates our thinking. This is reflected in our tendency to worry about some things to the exclusion of others, and, in more extreme cir-cumstances, in post-traumatic stress when patterns with very strong emotional aspects are easily triggered and can effectively hijack our minds and prevent us from thinking about anything else. The emotional character of pattern can also allow us to be manipulated by others who know how to draw on our patterned fears and paranoias, and they can skew our own judgment when the emotional aspects of our patterns are not proportionate to the realities they represent or refer to. I would also note that strong emotions that do not have their origins in pattern recall can also disrupt our pattern thinking as they likely alter our abilities to search, match, and extrapolate our patterns to guide our understanding and actions. There are also the issues raised by Feldman Barrett in the context of what she called "affective realism", the tendency of our minds to "experience supposed facts about the world that are created in part by our feelings" (Feldman Barrett 2017, p. 75).

Pattern thinking about emotion can allow us to better understand and to some extent regulate our own emotional states and experiences, and it can help us to understand the emotional states and reactions of others. Indeed, there are so many sides to affective pat-tern thinking, that this could easily be the basis of another book altogether with themes of pattern entertainment, pattern satisfaction, and pattern comfort, along with pattern discord, pattern anger, and pattern distress. However, having lingered a little longer than intended on pattern reasoning, I will end this section with this observation from Bor that I think sums up much of the potential for further investigation of affective pattern thinking:

> We really are a decidedly strange species for actively seeking out games with patterns in them, when such activities seem to serve no biological function whatsoever, at least not in any direct way. It's as if we were addicted to searching for and spotting structures of information, and if we do not exercise this yearning in our normal daily lives, we then experience a deep pleasure in artificially finding them.
>
> *(Bor 2012, p. 148)*

Pattern and Learning

If we accept that humans are adept pattern thinkers, then we should ask how it is that we acquire or develop these capabilities. I have already considered this issue in outlining pattern acquisition through individual thought and through participation in social pattern thinking, pattern elaboration, and macropattern development. Although most mammals and birds seem to be pattern thinkers within the limits of their cognitive capabilities, human brains have evolved to be distinctively adept at pattern thinking, particularly in its more abstract forms (Shermer 2011).

Arguing that our brains are naturally disposed to pattern thinking still has echoes of Chomsky's innate universal language capacity. It might be argued that pattern thinking requires some *a priori* reference or foundation from which to build, some kind of inherited basic pattern 'content' and pattern thinking ability. Indeed, it would seem to be a paradox that we cannot engage in pattern thinking without patterns to think with and we cannot develop a basic repertoire of patterns without some capacity for pattern thinking. We are born with basic pain and pleasure responses, aversion to extreme heat and cold, and so on. This basic autonomic pattern wiring may be all we need to initialise pattern thinking in our infant minds, but as these are not memories then they cannot be patterns as I have defined them. Clearly, new-borns seem to have very little cognitive ability to recognise, respond, or act beyond their autonomic triggers. They have brains but their minds have yet to develop substantive pattern thinking abilities.

That said, once a child's mind does begin to develop, it happens at a very rapid pace. Infants quickly learn to distinguish between happy and sad things, between themselves and others, between animate and inanimate things, and so on. Indeed, not only can very young children begin to recognise and classify objects and actions, to identify situations and possibilities, and to work with symbols and other abstractions, their minds seem to yearn for classification, order, and regularity in making sense of the world (Feldman Barrett 2017).

I would suggest, therefore, that much of child development involves developing patterns and learning how to use them, not through reasoning but by noticing, recalling, and reapplying certain similarities and differences in these fundamental responses. These early pattern memories are not the crafted macropatterns of more mature minds, they are almost exclusively impressionistic micropatterns. Developing mesopatterned and then macropatterned thought occur along a developmental arc, rather like matter coalescing around a new star and forming planets. If this is true, then these earliest patterns need only be provisional, satisficing, enough to keep the child moving forward to the next encounter and its associated pattern recognition or development.

Recurrence (regularity) and memory would seem to play a key role in this; we start to notice that some things are very changeable while others are relatively consistent and, in developing memory, we can recall these consistencies and inconsistencies, examine them, and in a very basic way develop theories regarding them. Emotion would seem to an important part of this as, if the individual finds this useful, intriguing, and even enjoyable, then this precipitates cascades of pattern learning about ourselves and about the world around us. Emotion would seem to be a key catalyst in initialising and sustaining pattern learning.

As we learn how to work with similarity and difference we also learn how to categorise, and how to infer and reason. We develop pattern thinking skills that allow us to think with pattern deliberately and not just reactively. An infant developing these skills need not start anywhere in particular, although toys, colours, and other instructional aids are predicated on helping young minds to notice and work with patterns, to make external pattern instances, and to associate patterns to make larger patterns. Indeed, much early learning more or less explicitly encourages this pattern thinking and learning through its use of repetition and reducing the world to simple pattern structures (red versus blue, cat versus dog, etc.).

I would differentiate between patterns *in* child development (the patterns developed in children's minds) and patterns *of* child development (patterns *qua* regularities in how children learn and develop). It would seem far more research has explored the latter than the former. That said, others have noted the role of pattern or pattern-like characteristics in early learning and child development (Inchaustegui & Alsina 2020). As humans we are (with very few exceptions) born into cultural and social contexts rich with shared patterns and pattern thinking. Indeed, societies and cultures would likely never have come into being without our capacity for individual and collaborative pattern thinking:

> Human consciousness is unlike all other varieties of animal consciousness and that it is a product in large part of cultural evolution, which installs a bounty of words and many other thinking tools in our brains, creating thereby a cognitive architecture unlike the "bottom-up" minds of animals.
>
> *(Dennett 2017, p. 370)*

This immersion in pattern thinking is also likely to play a role in stimulating and guiding the development of pattern thinking in young minds, particularly through immersion in language. Given that it is unusual for human minds not to develop pattern thinking abilities, we might look to our development differences to better understand the development of pattern thinking in humans. Some forms of neurodivergence, such as dyslexia and dyscalculia, might reflect variances in the development of pattern thinking. I should be clear though that there is no one 'right' way of pattern thinking, we all differ in the patterns our minds develop and the ways in which those patterns are used. I also think that we should try to understand the interplay between an individual's intrinsic pattern thinking abilities and that individual's opportunity to develop or express them (Nussbaum 2011).

Although we start learning using patterns as soon as we have the capacity to do so, and we continue our pattern learning until we die, much of our time as children is focused on formal learning in preparation for being an adult. Indeed, childhood and learning have become so entwined in the human experience that we might consider the social construction of childhood to be primarily about learning, while much of the role of adults (beyond shelter, food, safety, and affection) is to guide this learning. We have systematised a significant part of this learning in the form of "education" with all its structures, processes, and identities of schools, teachers, curricula, and exams.

Children are taught the many shared patterns of language, mathematics, science, and culture, the procedural patterns of critical and creative thinking and culture, and the embodied patterns of physical exertion (sports, games, and dance). Children learn to recognise and develop patterns, they learn to think using patterns as lenses and tools, they learn to create, elaborate, and adapt patterns, and they learn the basis of pattern thinking. Indeed, as Feldman Barrett described, social reality is to a great extent shaped by our participation in shared pattern thinking:

> When you are born, you can't regulate your body budget by yourself – somebody else has to do it. In the process, your brain learns statistically, creates concepts, and wires itself to its environment, which is filled with other people who have structured their social world in particular ways. That social world becomes real to you as well ... No particular social reality is inevitable, just one that works for the group (and is constrained by physical reality).
>
> *(Feldman Barrett 2017, p. 288)*

As individuals move through adolescence and into adulthood, they continue to engage in all sorts of new social pattern learning including learning the patterns of relationships, work, pastimes, social life, political life, transportation, finance, housing, and even spirituality. While this may at times be taxing to the individual, most of us seek out opportunities out to practice and develop our pattern thinking and we often take much pleasure from doing so:

> We have invented games that encourage us to rehearse our mind-moves, such as chess and Go and poker, as well as prosthetic devices – telescopes, maps, calculators, clocks, and thousands of others – that permit us to apply our mind-moves in ever more artificial and sophisticated environments.
>
> *(Dennett 2017, p. 380)*

Becoming an educated adult is also to become aware, at least to some extent of the limitations of pattern thinking. We learn that our pattern thinking and that of others can be flawed, fuzzy, and ambiguous. We find that our pattern memories can help us to associate, connect, and imagine, but they can also influence, bias, and distract us. The same connections and associations that might support the work of crossword puzzlers, literary scholars, and cryptographers, can distract from the work of engineers, technical writers, and programmers. Developing discipline in thought involves controlling our natural pattern thinking tendencies to be able to function within the field, task, or domain to hand. Indeed, we do not simply learn the patterns of a domain, or learn to use those patterns, we discipline our pattern thinking as a whole to focus on the pattern thinking suitable for our domain(s) of interest. Those who seek to develop their creativity may seek to extend the shimmering and unexpected associative nature of pattern thinking. Rather than disciplining their minds to the patterns and patterning of particular domains, they may seek to develop and hone the broad and unexpected associativity of their individual pattern thinking. It could be argued that this too is a particular kind of disciplinary approach. Either way there would seem to be some domains of pattern thinking that are more expansive creative and others that are more focused and disciplined.

Although those engaged in creative forms of work and expression often seek to develop the associative capacity of their pattern thinking, this cannot be an endless and random explosion of associations. Indeed, if it were, that might constitute a form of insanity. As Shermer observed:

> Some people are ultraconservative in their patternicity, see very few patterns, and are not very creative, while others are indiscriminate in their patternicity and find patterns everywhere they look; this may lead to creative genius or conspiratorial paranoia.
>
> *(Shermer 2011, p. 127)*

Rather than suggesting that there are two distinct solitudes of creative and disciplined pattern thinking, we might be better to consider each different domain of human activity as having its own balance or signature of creative and disciplined pattern thinking. To be an expert pattern thinker, therefore, we need to be able to control and direct the nature of our pattern thinking in ways that can harness its associative nature.

Broadly speaking, as we age, our capacity to learn new things decreases as does our ability to engage in many aspects of pattern thinking that involve translating or transforming pattern knowledge. However, neurodegenerative diseases notwithstanding, while much of our pattern memories seem to survive well into old age, albeit in altered ways, do years of pattern elaboration lead to a blurring or fracturing of some memories? Can our minds become over-patterned? Do patterned minds become more palimpsestic over time? Perhaps we lose our capacity for abstraction as we age but compensate through better associative pattern thinking and affective pattern thinking. (Peters et al. 2008).

Whatever the dynamics of an aging on our patterned minds, even memory declines eventually, particularly the associations and connections between our pattern memories. Indeed, much of the changes in pattern thinking are signatures of aging (such as distraction and dissociation) and many of the diseases of old age such as Alzheimer's disease and other kinds of dementia negatively impact our capacity for pattern thinking. The arc of our lives might, therefore, be understood in terms of our becoming pattern thinkers, acting as pattern thinkers, and finally losing our pattern thinking skills in advance of our decease. Every human has a different pattern thinking journey some develop more or different skills or capacities relative to others, some acquire or lose their pattern thinking rather early or late in their lives, everyone does it and experiences it differently. This arc, whatever form it takes, would seem to be a fundamental shared experience for all humanity.

Pattern and Learning Theory

Learning theories can provide interesting perspectives on pattern learning. For instance, cognitive load theory proposes that, given our capacity for active thinking and short-term memory is limited, learning tasks should be streamlined to manage the load they impose on learners. Pattern thinking manages the load associated with learning by reusing and elaborating pattern memory rather than forming endless new patterns, something that is reflected in Bruner et al.'s (1959) suggestion that our ability to work with regularities (temporal, spatial, etc.) is essential to learning. Bruner drew on the work of Vygotsky in outlining a theory of scaffolding to explain how learning can be facilitated through various means including teaching, learning materials, and learning environments. The use of schemas and mental models were advanced as ways of drawing on past understanding and developing new understanding within many of these facilitating mechanisms. Schemas are concatenations of memory and experience that can be triggered by stimuli that are similar to those that went into creating it. When a schema cannot be found that matches current situation then we switch to mental model making. Schemas and mental models are in effect macropatterns.

Behaviourist learning theory is about learning that changes behaviour and to that end its focus is on action rather than on mind (Skinner 1968). Behaviours can be conditioned such that the individual develops a conditioned response to a particular set of stimuli. This could be understood in terms of conditioning certain kinds of pattern thinking such that when a pattern is invoked it includes the conditioned response associated with it. Behaviourist approaches can, therefore, be understood both as modifying patterns we already have and manipulating individuals in the patterns they develop. Interestingly, from a behaviourist perspective, patterns are simply a means to an end.

Constructivist learning theory is often contrasted with behaviourist learning theory. Originating in the work of Piaget, constructivist learning theories are based on the principle that we learn by linking new experiences and thoughts to past experiences and thoughts, which in turn means that the same educational activity will likely be experienced differently by different participants and lead to different learning outcomes. This clearly aligns well with a pattern theoretical stance both in terms of matching experience to pattern memory and elaborating and developing pattern memories based on new and variant experiences.

Piaget's work was primarily focused on child development, and to that end he outlined four broad stages humans pass through in developing intelligent thought (Piaget 1952). These can serve as a useful outline of how pattern thinking may develop over time – see Table 5.3.

TABLE 5.3 Piaget's stages of cognitive development mapped to the development of pattern thinking

Piaget Stage	Characteristics	Relation to pattern thinking
Sensorimotor (0–2 years)	Learning from senses and movement, pleasure and pain, experimenting, starting to work with abstractions, words.	Creation of rudimentary patterns and pattern thinking skills by linking experiences, embodiment, feelings, and exploring basic cultural patterns.
Preoperational (2–7 years)	Relying more on abstract and connective thought, explores causality and the thoughts of others, works with language.	Patterns and pattern thinking become the central basis of thought and sense-making, acquisition of cultural patterns drives much of this development.
Concrete operational (7–11 years)	Works more substantially with language, categories, numbers, spatial relationships.	Develops abilities to create and use macropatterns, develops ability to alter pattern thinking according to domain of activity.
Formal operational (11+ years)	Develops more sophisticated abstract thinking, employs hypothetical and logical reasoning, imagination.	Develops full macropattern thinking capacity, is able to move effectively between micro-, meso-, and macropattern thinking.

To be attentive to and purposive with one's cognition is to be metacognitive, or, to put it another way: "metacognitive knowledge involves knowledge about cognition in general, as well as awareness of and knowledge about one's own cognition" (Pintrich 2002). Various metacognitive theories have stressed the active aspect not just of thinking about thinking but of doing so in purposive ways (Schraw & Moshman 1995). This can include developing greater understanding of what pattern and pattern thinking are in general, the role they play in one's own mind, and how pattern and pattern thinking can shape social realities. From this perspective, learning involves us in becoming more aware of our patterned minds and how pattern shapes our thinking, both for good and ill.

Although behaviourist and cognitivist/constructivist theories of learning have somewhat dominated the landscape of the educational sciences, there are many other learning theories, including transformative learning theory, (Mezirow 1997) adult learning theory, (Knowles 1984) self-directed learning theory, (Knowles 1975) self-actualisation theory, (Maslow 1943) and experiential learning theory (Kolb 1984) in which pattern learning is at least implied. I am not, therefore, advancing pattern theory as a new learning theory, rather, I would suggest that it is a theory of mind that can be seen reflected in a great many extant learning theories.

Mentioning David Kolb's work brings me to the thorny topic of learning styles. Although there are many different models and theories of learning styles (such as VARK, Myers-Briggs, and the Kolb LSI), they share a common idea that different minds learn in different ways, that these different ways can be identified, and that, having identified an individual's intrinsic learning style, that instruction can be adapted to match an individual's particular style. However, the evidence for learning styles is generally thin and they have been increasingly deprecated in the context of the educational sciences (Furey 2020). That they continue to be used despite these empirical doubts suggests to me that wrapping the complexities and contingencies of learning into neatly macropatterned models is particularly attractive to educators and hard, therefore, to disavow. Rather than claiming we have intrinsic cognitive dispositions qua "learning styles", we might be better arguing, as Entwistle and colleagues did, that we adopt different pattern thinking approaches (that they also called "learning styles") to match our pattern thinking capabilities and needs (Entwistle 1988).

In summary, pattern learning suffuses and underpins all human experience, it is to be found in most if not all learning, it is reflected in learning practices and learning theories, and it can connect many of these paradigmatic and theoretical positions that are otherwise incommensurable. I would note though that, in the interests of brevity, I have taken a relatively superficial view of pattern learning while noting many issues and questions that deserve more attention than I am able to give them here.

Shared Pattern Thinking

I will deal with the practicalities of externalising patterns beyond the minds in which they formed in the following chapter, but I should first explore the implications of the idea of *shared pattern thinking*. There is a paradox here; patterns are a phenomenon of mind and as such they cannot exist in any other medium, and yet I can clearly share my pattern thinking and so can we all, albeit in a multitude of ways and according to our abilities and opportunities. And yet, if our minds are inaccessible to others and the minds of others are inaccessible to us, how is it that we engage in shared pattern thinking?

Humans are (more or less) social creatures. We participate in various communities and societies that each require engagement in shared pattern thinking. This is not simply about communicating the facts of our pattern memories, it can also include sharing our experiences, feelings, meanings, and significances, and it requires the assimilation of shared pattern thinking into our own idiosyncratic pattern memories and pattern thinking habits. Sharing pattern is not at all uniform or even predictable; some minds may influence shared pattern thinking more than others (such as eponymous "influencers") and some may be influenced more than others. Some individuals may accept the shared pattern thinking of their community, society, or culture "as-is", while others may reject it or seek to change it, and dispositions change such that heretics and discontents may become more orthodox over time, while others may be more heterodox or even heretical with respect to the shared pattern thinking of the collective.

While this can explain how social systems work, there is still the question of where does this collective pattern thinking lie? There can be no patterns outside of our minds so is shared pattern thinking the sum or perhaps the product of all the minds that contribute to or participate in it? Perhaps, if it were not that every mind is its own solitude and what can be expressed and shared of that mind is no more than a pale reflection of its richness. Patterns are emergent and fragile things. In minds they may last a lifetime (although rarely without changing over time), or they may be fleeting thoughts that are quickly lost or subsumed. A pattern in an individual mind is based on the substrate of the brain. There is no single substrate for shared patterns other than the community of minds that share them, which, if anything, should make them even more emergent and fragile. However, since shared pattern thinking does not depend on any one individual mind, even if all participating minds are lost the products of their shared pattern thinking may persist. You can visit any museum of antiquities or any archaeological site to experience this first-hand. If individual pattern thinking leaves traces in the things that individuals do and create, the traces of shared pattern thinking are all the more striking and long-lasting.

Shared pattern thinking is clearly temporally situated. The shared pattern thinking of our parents' generation conditioned our own minds as children and shaped our social reality as adults, even as their minds were in turn conditioned by their parents' shared pattern thinking and our own shared pattern thinking conditions the minds of our children. This is not just an endless chain of reproduction and repetition; shared pattern thinking is dynamic, it changes through use and in response to environmental, social, political, and technical challenges. This reflects a memetic sense of shared pattern thinking, something that continues, that exists through time and reproduces itself from one mind to another. However, neither temporality nor memetics can answer the question of where shared pattern thinking is, or what it is; they only describe its propagation and development. This idea of culturally shared patterns is reflected in Churchland's work on folk psychology and in Sellars' concept of "manifest image"; those things that are "obvious to all, and everybody knows that it is obvious to all, and everybody knows that, too. It comes along with your native language; it's the world according to us" (Dennett 2017, p. 60).

Shared pattern thinking can also help us to make sense of what others do and how we connect or relate to each other:

[Humans] not only encode and process patterns representing their own experiences, but also the experiences of their family, friends and workmates. Social interactions require processing of information regarding the histories, behaviors and thoughts of many other individuals ... inter-personal [pattern processing] is critical for success in most aspects of life, including acquiring and retaining friends, a job and a mate.

(Mattson 2014)

Certain shared patterns may be taken to be canonical, axiomatic, and non-negotiable, such as those of religion and law. Other patterns may be shared as memes, ideas and concepts that perfuse through populations. Indeed, memes need not be useful, identical, or even rational, they simply reproduce across minds (Blackmore 2000). Social patterns may also be negotiated or open to interpretation and personalisation. For instance, many behavioural norms (manners, politeness, respect) are expressed in quite individual ways. That patterns can be negotiated, tempered, compromised would seem to offer another fruitful avenue of investigation:

> Our habits of self-justification (self-appreciation, self-exoneration, self-consolation, self-glorification, etc.) are ways of behaving (ways of thinking) that we acquire in the course of filling our heads with culture-borne memes, including, importantly, the habits of self-reproach and self-criticism.
>
> *(Dennett 2017, p. 341)*

Not only are social macropatterns useful to the individual in terms of functioning in and understanding social systems and structures, they can also be shared, they can be developed in collaboration with others, they can help guide the thinking and actions of others, and they can form the basis of a shared field of practice. Shared pattern thinking is implied in the social network analysis of Bruno Latour, in the morphogenetic theories of Margaret Archer, in the realist context-mechanism-outcome configurations of Ray Pawson, and in the various flavours of grounded theory. When Bourdieu (1977) talked about praxis as the nature of a particular field of activity, I would argue that this too reflected pattern thinking. There are also many connections with pattern reasoning, pattern learning, and pattern affect (Mattson 2014). That we can both project and read emotions and have empathy and even compathy with others are not just aspects of a shared humanity and concern for others; they are fundamental parts of our pattern thinking universe.

What about social misuses and abuses of pattern thinking? I earlier noted some of the ways that patterns might be imposed on people or used for other questionable ends. If we can be manipulated such that patterns are forced on us, then is our strength in pattern thinking not also a significant vulnerability? For instance, Feldman Barrett noted the potential for coercive imposition of social reality (shared pattern thinking) on others such that we mistake social reality for physical reality:

> By virtue of our values and practices, we restrict options and narrow possibilities for some people while widening them for others, and then we say that stereotypes are accurate. They are accurate only in relation to a shared social reality that our collective concepts created in the first place.
>
> *(Feldman Barrett 2017, p. 291)*

Might we understand indoctrination, radicalisation, and the phenomena of cults and sects in terms of the manipulation of shared (and thereby imposed) pattern thinking? Do we not have a personal responsibility to engage in critical and rational thought regarding our pattern thinking? Accepting harmful and manipulative patterns at face value without challenging them could be seen as reflecting Arendt's concept of the banality of evil, and yet challenging dominant or imposed pattern thinking is not without risk of harm or ostracisation. Indeed, some people will defend their patterns to the death. If a child is raised to internalise a particular system of pattern thinking then whose responsibility is it to challenge that? After all, the biases and structural violence of any society are typically conditioned patterns that it takes effort and courage to reject or refute. As a result, as societies we collectively reproduce the patterns of social relations that include those that may be considered positive (obeying the law, kindness to others, paying taxes, maintaining the peace, and participating in peaceful political protest) as well as those that are more negative (racism, sexism, ableism, xenophobia, distrust, paranoia, homophobia, hostility, violence, wilful destruction, and disregarding the rule of law). Social patterns typically seem "normal" to us, particularly if nobody we know challenges them, and yet when we do become aware of the imposition of pattern thinking our reactions are often less than positive.

Examples of resistance to orthodox pattern thinking include the Levellers of 17th-century England, the Bohemians of 19th-century Paris, the and the many counterculture groups of 20th-century youth (beatniks, hippies, punks, goths, etc.). It is interesting to note that they were all based on rejecting one orthodoxy by embracing another. Perhaps we can no more escape our shared pattern thinking than we can escape our history; we can only strive to make more responsible use of pattern thinking, both individually and collectively.

Limits to Pattern Thinking

> Our marvellous minds are not immune to fads and fancies that bias our self-redesign efforts in bizarre and even self-defeating ways.
>
> *(Dennett 2017, p. 391)*

I have already noted some limitations associated with pattern thinking: pareidolia (seeing patterns that are not "there"); apophenia (beliefs and actions based on non-real patterns); the blurring of form, order, regularity, and pattern; provisional and satisficing pattern perception; bias in and manipulation of pattern thinking. Even though pattern thinking confers so many advantages, is it worth it given its many failures and limitations? Important as it is, asking this question is moot. We evolved as pattern thinkers and cannot readily abandon something that is so fundamentally a part of us. However, evolutionary advantage is rarely uniformly positive; all evolutionary adaptations and developments have costs and trade-offs, both in terms of the immediate compromises and costs and in terms of the roads not taken. This is, again, reflected in the likely evolutionary basis for human pattern thinking and the trade-off of tending to make Type I errors over Type II errors:

> … is part of the general vigilance that we have inherited from our ancestors. We are automatically on the lookout for the possibility that the environment has changed. Lions may appear on the plain at random times, but it would be safer to notice due to the fluctuations of a random process.
>
> *(Kahneman 2011, p. 115)*

The neurophysiological capacity for pattern thinking is based on the structure and function of our brains. The evolution of folded cortices would seem to have allowed for significant increases in pattern processing capacity. And yet there are limits to the size of human brains including the energy a larger brain would require, the maintenance of connectivity, and the practicality of birthing young with larger skulls (Hofman 2014). Even within existing brain structures and physiologies there are limits to how fast we can think with patterns, how much effort we can or like to invest in pattern thinking, and to the utility and reliability of the patterns we think with:

> There is a limit to how fast you can switch tasks, because changing context requires desaturating the neurons before resaturating them with other thoughts ... The more patterns you have loaded up into multiple levels of your hierarchy, the faster you can make progress, as you recognize patterns everywhere and everything seems to connect to everything else. But with that momentum you sacrifice agility.
>
> *(Forte 2018)*

There are also variances in pattern thinking abilities between individuals (reflecting the variances of humans in general), differences in individual pattern thinking over time (both long term from cradle to grave and short term from one moment to the next), and differences according to context (culture, environment, personal circumstances, etc.). These differences may be a matter of personality or choice, or they may be something more substantial. For instance, there are disorders of the mind that are closely connected to pattern thinking:

> A key feature of schizophrenia is a blurring ... between patterns that are real and those that are mentally fabricated.
>
> *(Mattson 2014)*

Another concern is that we seem to have many pattern memories that we cannot directly access that nevertheless influence our pattern thinking. Exploring these 'hidden' patterns was important to early psychoanalysts such as Freud and Jung and it still plays an important role in modern analytic practice (Barden & Williams 2007). As Jung observed:

> [A symbol] has a wider "unconscious" aspect that is never precisely defined or fully explained. Nor can one hope to define or explain it. As the mind explores the symbol, it is led to ideas that lie beyond the grasp of reason.
>
> *(Jung 1964, p. 4)*

There are limits to pattern thinking related to individual capabilities and physiologies, to aging, to injury and trauma, and to neurological disorders such as stroke and Alzheimer's disease. However, using the clinical term of "disorder" simply to refer to challenges and variances in pattern thinking ability can be problematic.

Pattern thinking evolved to be fast and efficient, but not particularly precise or accurate. We have learned to introduce logic and rigour into pattern thinking (through deliberate macropattern thinking) but that involves discipline and practice, and many of us are happily ill-disciplined as pattern thinkers. Moreover, according to Shermer, irrational pattern thinking increases in the face of uncertainty, danger, fear, and other intense emotional experiences. At these times we are more prone to what he called "magical thinking", both seeing or intuiting patterns (connections, associations, regularities) that are not there:

> We find magic wherever the elements of chance and accident, and the emotional play between hope and fear have a wide and extensive range ... We do not find magic wherever the pursuit is certain, reliable, and well under the control of rational methods and technological processes. Further, we find magic where the element of danger is conspicuous.
>
> *(Shermer 2011, p. 77)*

Patternicity as a tendency to magical thinking is reflected in irrational beliefs, conspiracy theories, and invisible powers, a tendency to believe in and even perceive ever more impossible things, and a refusal to let go of these patterns or what they produce. That said, is patternicity really a cognitive weakness? After all, patternicity, for all its shortcomings, would also seem to be what drives creative thinking. We cannot say that unfounded beliefs are cognitive problematic in and of themselves, only that beliefs that turn into delusions can become problematic.

Shermer also considered bias as a limitation of pattern thinking, which might mean anchoring on a particular pattern to the exclusion of others or seeking for patterns of a particular kind while dismissing others out of hand. Pattern thinking can explain many biases: recent patterns and more often used patterns come to mind more easily, a pattern does not need to conform entirely to the situation it just needs to be a heuristic fit (satisficing) to be useful, and deliberate pattern thinking takes effort (see earlier comments on cognitive load). Moreover, we can slip into pattern cascades when we perceive anticipated patterns as confirmation of their causative characteristics.

A more recent and more substantial challenge has come from neuroscientists Donald Hoffman and Anil Seth. Hoffman (2022) argued that our perceptions are shaped by our cognitive filters and attentions that have in turn be shaped by evolutionary forces over time such that we do not (and cannot) perceive reality "as it is", only as evolutionary advantage has shaped our minds to perceive what is around us. On this basis, our minds select some things and ignore others in creating an awareness and understanding of our surroundings. Hoffman called these representations "perceptual symbols". Seth argued along similar lines to Hoffman regarding the illusory nature of perception:

> Instead of perception depending largely on signals coming into the brain from the outside world, it depends as much, if not more, on perceptual predictions flowing in the opposite direction. We don't just passively perceive the world, we actively generate it.
>
> *(Seth 2017)*

While I agree that our minds are selective and interpretive, I disagree with this binary real/unreal, true/false reading of what our minds perceive. Clearly, our pattern perception is founded in reality. If it were not then our actions and the feedback that we receive from them would in no way correlate. Moreover, pattern perception is not based on single snapshots of perception but on continuous multimodal perception and feedback. Where perception and feedback converge perception solidifies, where they diverge perception shifts to seek alternatives.

Another limitation to pattern thinking is that we may depend on or give too much credit to our pattern perception and the veracity of the patterns we draw on, such that we take them to be the authority to which reality must conform. To an extent this explains pareidolia in that we perceive our patterns as being superimposed on reality such that it does appear to conform to our expectations and understanding:

> Concepts are vital to human survival, but we must also be careful with them because concepts open the door to essentialism. They encourage us to see things that aren't present ... Concepts also encourage us not to see things that are present.
> *(Feldman Barrett 2017, p. 288)*

This again reflects aspects of magical thinking such that anchoring our beliefs in pattern despite the lack of evidence can explain individual and group beliefs in magical (and therefore) invisible forces and agents, in the forces and agents that are held responsible for imagined conspiracies, and in the multitude of other human tendencies to perceive and believe in unreal things.

I described earlier how reasoning and critical thinking can balance out some of our tendencies to magical thinking. However, pattern reasoning can also go awry, particularly when the effort of doing so limits our preparedness to be rigorous in evaluating pattern solutions and implications:

> One problematic corollary of this passion for patterns is that we are the most advanced species in how elaborately and extensively we can get things wrong ... We are so keen to search for patterns, and so satisfied when we find them, that we do not typically perform sufficient checks on our apparent insights.
> *(Bor 2012, pp. 147–148)*

I would also note the placebo effect in this context; by anticipating something our minds can condition our bodies to reflect a predicted change whether or not that change actually happens. If our minds expect or anticipate something and activate patterns that reflect it then our physiological responses can change to reflect this (Meissner et al. 2011).

Sometimes we can pay so much attention to one pattern perception that we fail to notice others. This helps us to be selective in our attention, but it may also mean we are oblivious to patterns that are obvious to others, or we may share a pattern blind spot (through enculturation, memes, etc.). Other examples of pattern error can be found in analogical thinking if the analogy made does not properly connect memory with current experiences or if it transfers too much or inappropriate implications from the pattern to the percept. We know that a pattern does not have to be a perfect match to a current situation, it only needs to be a partial match, but if the wrong or inappropriate aspects of the partial match are included then pattern thinking can be misleading – see Figure 5.2.

Our willingness, intent, and purpose of engaging in pattern thinking can also be directed, sometimes in ways that help us, and sometimes in ways that limit, hinder, or even harm us. If we choose not to be pattern thinkers and do something else instead, then what is those what are those other things that we select in lieu of pattern thinking? When do we do this? Why do we do this? Can we ever truly stop, or are we ignoring the pattern engine that never stops or perhaps squashing it?

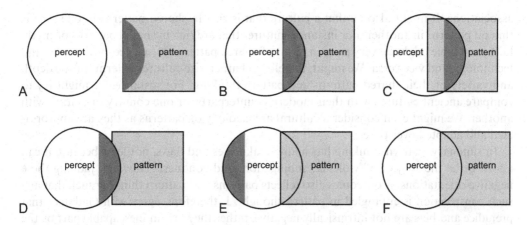

FIGURE 5.2 Examples of pattern-percept overlaps and errors that can occur. In A, a pattern seems to have a partial match (indicated by the darker semi-circle) to a percept. In B, that match is confirmed and appropriate. In C, more of the pattern is applied than is appropriate. In D, less of the pattern is applied than is appropriate. In E, the wrong parts of the pattern are applied. In F, the original assumption (as per A) that there is an overlap is not actually the case.

Diagram prepared by the author.

While deception can include self-deception (seeing what we want to see or resisting what we fear seeing), it can also include deceiving others by manipulating their pattern thinking. We tend to follow the pattern suggested by our minds or that is suggested to our minds. This can be used by magicians (of the entertainment kind) to carry off their illusions and sleights of hand. Pattern deception can also be used by gamers, gamblers, fighters, and others engaged in conflict or competition to deceive or misdirect their opponents. There are whole bodies of thought regarding tells, feints, and bluffs that reflect this use of deceptive and projected kinds of pattern thinking.

Another limitation that I noted in the section on pattern learning, is that patterns can be forced on others through indoctrination, socialisation, and/or conditioning. That this is even possible might be considered a limitation of pattern thinking, except that being able to learn, teach, and share patterns and pattern thinking requires malleability, and this typically confers greater advantage than disadvantages. Nevertheless, we might consider epistemological violence connected to pattern thinking, such that it is manipulated in support of an agenda not held by the pattern thinker.

This brings me back to issues of culture and limits to pattern thinking. We are limited by the patterns we have available to us. Participating in a culture or domain of activity that is rich in patterns and pattern thinking is likely to be able to afford and support greater thinking in all the individuals that participates within it. By contrast, cultures and contexts whose patterns are thin and underdeveloped are more likely to be poor environments in which to develop or practice substantive or rich pattern thinking. The imagination and creativity of the individual can mitigate this though. Indeed, one of the main ways in which cultures and domains become more richly patterned is through the creativity, innovation, and imagination of its members who can go beyond the mundane and the routine pattern thinking of their peers to establish new and richer patterns and pattern associations. It might also be that a culture that is rich in shared patterns in one area is thin on patterns in another. For instance, cultures that are rich in shared patterns of myth, faith, and fable might be very poor in terms of shared patterns of science, philosophy, and technology, and vice versa. We might, therefore, characterise cultures in terms of the depth and variety of their shared patterns and pattern thinking. For instance, we might try to compare ancient cultures with their modern counterparts, or one country or culture with another. We might even consider a cultural archaeology of patterns as they are appropriated and adapted over time.

In summary, pattern thinking has many weaknesses and flaws, not least because it can afford or be the subject of violence, subjugation, and conquest. However, judging these negative orientations or outcomes also reflects patterns and pattern thinking such that any such construction is entangled in pattern thought. I, therefore, agree with Gadamer that prejudice and bias are not intrinsically negative, rather they are an inescapable part of the patterned human mind:

> Prejudices are biases of our openness to the world. They are simply conditions whereby we experience something – whereby what we encounter says something to us.
>
> *(Gadamer 1977, p. 9)*

Artificial Pattern Thinking

Having considered the human mind at some length, I end this chapter with a consideration of ways in which scientists and engineers have sought to reproduce pattern thinking.

As I described in Chapter 2, Grenander et al.'s work on "general pattern theory" (GPT) focused on formal representations of pattern structures, often as a way of getting machines to reproduce human capabilities. Their approach was (broadly) to first develop a pattern theoretic model and then a template that reflected the pattern phenomena of intertest. Instances of the pattern were then created using the template and the latter adjusted to produce more appropriate instances. Their final step was to specify in what ways and to what extent pattern instances might be transformed while still reflecting the underlying pattern.

In terms of artificial intelligence research in pattern recognition and its different theoretical perspectives. Bunke & Sanfeliu distinguished between two broad strategies:

> Decision-theoretical models are based primarily on using numerical-valued features as a means for distinguishing one class of patterns from other classes. By contrast, syntactic and structural methods are based on explicit or implicit representation of ... the characteristic way in which the sub patterns (or elements or components) of a pattern are related or configured together.
>
> *(Bunke & Sanfeliu 1990, p. 4)*

Based on this, GPT is (or at least it seems to be) a syntactic/structural approach while template matching, recognition-by-components theory and their ilk are decision-theoretical approaches. I would note too that decision-theoretical approaches have tended to be more focused on detecting and interpreting order and regularity than on pattern *per se* (Chetverikov 2000). Toussaint made a similar distinction in approaches to that advanced by Bunke & Sanfeliu, albeit with a different terminology. He also suggested that the two kinds of approaches are not necessarily incompatible but may in fact work synergistically:

> Systems that process patterns by analysing the data or input information with ever-increasing levels of sophistication are called data-driven or bottom-up systems. Those that start from overall expectations and work down are called conceptually-driven or top-down systems. It appears that for solving difficult problems efficiently context may have to be used with both bottom-up and top-down processing taking place simultaneously.
>
> *(Toussaint 1978)*

Not surprisingly, much of the research in the computing and artificial intelligence sciences has been experimental, often focusing on specific challenges and approaches, such as developing machines able to identify recurring phenomena in ambiguous images or sounds, or in the ability to tell phenomena apart reliably and meaningfully from various kinds of data. Although much of this work has been highly specialised and specific to scientific subdisciplines and discourses, some broader treatises on artificial pattern thinking have been developed. Of note is Schank's 1986 treatise on "Explanation Patterns", which he described as:

> … a fossilized explanation. It functions in much the same way as a script does. When it is activated, it connects a to-be-explained event with an explanation that has been used at some time in the past to explain an event similar to the current event.
>
> *(Schank 1986, p. 110)*

Schank suggested that the elements of an explanation pattern include a reference to the phenomenon it explains, its parts and connections, the contexts where the pattern explanation applies, the origins or causes of the phenomenon, examples or instances of the pattern being used to explain similar phenomena, and any rules or exemplars the pattern explanation observes. Interestingly, Schank also described phenomena that seem to me to be very similar to concepts of pattern elaboration, pattern efficiency, and pattern development, and the heuristic nature of pattern thinking that I also previously described. That this work was published four decades ago and focused on computers rather than human minds is also notable.

Dennett described the ways in which human minds are both similar to and different from artificial intelligences. A key difference he noted was that machine minds can neither repair themselves nor can they reproduce themselves or do many other things we do as humans:

> Human chess players have to control their hunger pains, and emotions such as humiliation, fear, and boredom… By taking these concerns off their hands, system designers create architectures that are brittle (they can't repair themselves, for instance), vulnerable (locked into whatever set of contingencies their designers have anticipated), and utterly dependent on their handlers.
>
> *(Dennett 2017, p. 158)*

Not only is pattern thinking in humans inescapably embodied, situated, constructed, and experienced (things that are usually not included in computational models), it would seem that it is precisely these things that drive our development of pattern thinking capabilities. Indeed, human pattern thinking can only have evolved to work within the many constraints that humans face including our physiological limitations (tiredness, hunger, pain) and cognitive limitations (bias, inattention, distraction, emotionality). These are not simply boundaries to pattern thinking, they can become existential threats, particularly when we are dependent on pattern thinking to help us avoid or mitigate danger, seek out food and shelter, or otherwise help us to survive. This is different for machines that have no stake in their existence or function. Dennett (2017) illustrated these differences by hypothetical Darwinian, Skinnerian, and Popperian creatures. Darwinian creatures are automatons, they understand nothing, they learn nothing, they simply react. We might say that individual cells and basic single cell creatures are Darwinian in this sense. Skinnerian creatures can adapt but only through trial and error, and reward and punishment. Popperian creatures can learn and adapt, and they can engage in analytical and abstract thought, reasoning, and hypothecation. They are pattern thinkers. Humans are pattern thinkers, but are machines also pattern thinkers?

There are two key recent developments in the development of artificial intelligence that may usefully inform our theory of pattern thinking – reinforcement learning and deep learning. Reinforcement learning reached broad public awareness with the development of Deepmind in the 2010s and more recently with controversies associated with AI services such as ChatGPT. In describing reinforcement learning, Zai & Brown distinguished between "supervised learning" which is about sorting correct and incorrect examples and reinforcement learning (RL) that is based on training through rewards:

> Ordinary image classification-like tasks fall under the category of supervised learning, because the algorithm is trained on how to properly classify images by giving it the right answers … In contrast, in RL we don't know exactly what the right thing to do is at every step. We just need to know what the ultimate goal is and what things to avoid doing.
>
> *(Zai & Brown 2020, p. 8)*

In reinforcement learning, there is an agent (typically an algorithm or system) that interacts with an environment which generates data that the agent can use. The agent processes the inputs it receives in terms of model or hypothesis building and then takes an action based on its provisional model or hypothesis. The agent's actions change the environment which in turn changes the inputs and sends a reward signal (positive or negative) that informs the next cycle of input evaluation and reaction. At a simplistic level a positive reward reinforces the action and the model or hypothesis that informed the action while a negative reward weakens them. A more sophisticated reading of feedback may suggest what aspects of the model/hypothesis were more or less correct allowing them to be adjusted to better fit the inputs being received from the environment. Over multiple (abductive) cycles a robust and reliable model is developed and the results of actions become more predictable – see Figure 5.3. Interestingly, this is essentially conditioning, which reflects behaviourist theories of learning as well as more reflective models such as Kolb's (1984) learning cycle theory. More importantly is that this provides a general theory for how pattern perception and pattern reasoning work, a theory moreover that has allowed for the development of ever-more sophisticated forms of artificial intelligence. Central to this is feedback and reinforcement.

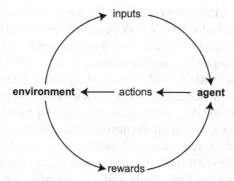

FIGURE 5.3   Reinforcement learning cycle. An environment generates inputs for the agent which then processes the input and acts accordingly. The agent's actions change the environment which changes the inputs and sends a reward signal (positive or negative) that informs the next cycle of input evaluation and reaction.

Diagram prepared by the author.

However, while iterative and abductive reinforcement learning is a critical part of developments in AI, it is not enough to develop the subtlety of human thought or even to deal with the imperfections, partialities and confusions of real-world environments and the inputs they provide.

The other key component that has contributed to the effectiveness of contemporary AI and that aligns with pattern theory is that of deep learning. Having conducted my doctoral studies in Edinburgh, "deep learning" to me means the learning style theories of Entwistle and colleagues where deep learning is where a learner pursues a topic to understand its underlying (or deep) causes and behaviours. However, deep learning from an AI perspective is not about the pursuit of depth but about the layered and recursive nature of reinforcement learning:

> Deep neural networks have many layers … This layered representation is a form of compositionality, meaning that a complex piece of data is represented as the combination of more elementary components, and those components can be further broken down into even simpler components, and so on, until you get to atomic units.
>
> *(Zai & Brown 2020, p. 4)*

This means that rather than there being one agent and single cycles of reinforcement learning, there are a great many interconnected agents and cycles such that the inputs, actions, and rewards can develop complex interrelationships that cumulatively lead to pattern solutions to complex challenges. It would seem that this mirrors our own pattern thinking minds where patterns pull on each other such that these interactions and associations can generate complex and adaptive pattern solutions.

Combining reinforcement learning and deep learning in AI has led to rapid development in thinking and practice and has underpinned the development of some astonishingly effective technological implementations. However, to what extent and in what ways human minds truly reflect deep reinforcement learning structures or principles remains unclear. This is, therefore, another area ripe for research and development.

In summary, artificial pattern thinking can offer useful perspectives on pattern thinking in general. Although Grenander et al.'s work on general pattern theory was used to create simple explicit models of individual patterns, it was inescapably reductionist and seemingly unable to capture the complexity and interconnectivity of a pattern thinking substrate (whether mind or machine) within which a pattern is no more than a fluid nexus. This complex and emergent nature of pattern thinking has been more successfully realised in engineering solutions that employ deep reinforcement learning principles, which in turn show the importance of feedback and reinforcement, of cumulative abductive reasoning and selection, and of multiple layered cycles and their interconnectedness. We might also consider Dennett's observations that the human embodied experience is an essential difference between human minds and machine minds, and how the necessary priorities of resilience, adaptability, and survival are also woven up with our pattern thinking.

Chapter Summary

In this chapter I have necessarily taken a somewhat high-level trajectory in considering a wide range of topics, most of which could benefit from much deeper and more careful examination. I started by considering the ways and extents to which we vary in our *pattern thinking* capacities and approaches, both in terms of our abilities and dispositions and of our socialisation to and participation in our cultures and societies. Although they may be similar (not least when referring to the same phenomena) our patterns are unique to us, woven as they are into the fabric of our memories and life experiences. Indeed, our patterns and our pattern thinking might be unique enough to be used as a biometric if they were at all readable. If they were then they might also be impersonated or simulated.

I described ways in which pattern thinking is aggregative, cumulative, and interconnected. I considered some of the foundations of pattern thinking, in particular the connections between pattern and language, suggesting that language and pattern likely co-evolved, the former a primary use of and benefit of the latter. Reprising the important point that we do not hold discreet patterns in our minds, I described them in terms of a complex and intertwined multidimensional fabric of interconnected pattern elements. Patterns in are not discrete clusters of memory, they are simply denser nexuses. From this, I expanded on the metaphor of *pulling* on a pattern part as the way a broader pattern nexus can be activated or invoked along with all the parts and other nexuses it is connected to.

I considered the concept of pattern reasoning in terms of reasoning towards pattern and reprised the interlinked concepts of pattern recognition and pattern development in exploring the reasoning processes that might be involved in each. I also considered how we reason from pattern in terms of patterns conferring meaning and significance, and how significance can demand and direct our attention, in great part through emotions interwoven with pattern. This led me to a consideration of how pattern reasoning relates to the concept of affordances, and to how patterns relate to risk assessment and prediction.

I considered the role of emotion in pattern thinking and explored some of the ways in which our patterns can please and disgust us and how they can both comfort and alarm us. I explored relationships between pattern thinking and learning at different life stages and in different aspects of our lives, arguing that the development and use of patterns and pattern thinking are synonymous with both learning and its pursuit in systems of education. I also explored relationships between pattern thinking and learning theory, contrasting cognitivist, behaviourist, and constructivist models of learning. This led me to sociocultural pattern thinking, both in our participation in societies and cultures and in the social uses of pattern to direct, constrain and even control others. I followed this with an exposition on some of the limitations of pattern thinking including neurophysiological limitations (limits of brains, pathologies and damage), cognitive limitations (biases, errors, irrationality, magical thinking), and failures in logic and reasoning using patterns. Finally, I took a brief wander through artificial pattern thinking, primarily focusing on deep reinforcement learning principles, involving feedback and reinforcement, cumulative abductive reasoning, and interconnected and layered learning cycles.

Before closing this chapter, I would note the paradoxical side of thinking about pattern thinking – what we might call *pattern metacognition*. Let me illustrate this with the following example. Theory generally has three kinds of goals or purposes: description, explanation, and prediction. Sometimes scientists focus on description, as with much of Darwin's fieldwork, but they are often dissatisfied with pure description as they seem compelled to explain things (reflecting that difficulty we have with perceiving form without also perceiving order and reflections of our patterns that I described earlier). Explanations can lead to predictions, although they do not have to. Sometime scientists have strong predictions with weak explanations (such as planetary motion) and sometimes they have strong explanations with weak predictions (such as the behaviours of markets or societies). We might draw these ideas together from a pattern theoretical stance on the progression of inquiry – see Figure 5.4. Although this is perhaps a little simplistic and linear given what we know of the more and fluid nature of our minds. I include it to illustrate the inescapable circularity of thinking about pattern thinking that is also the product of pattern thinking.

Writing this chapter was challenging in that the linearities of language made it difficult to represent the deep interconnectedness of pattern phenomena. Pattern thinking is ubiquitous and multifaceted. It cannot easily be reduced to discrete topics any more than it consists of discrete patterns. Indeed, it would seem that pattern thinking is not simply one of the things the mind does, it *is* what the mind does.

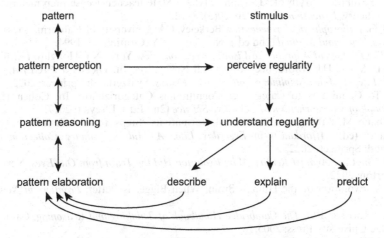

FIGURE 5.4   A possible alignment between pattern thinking and the conscious development of theory and understanding. Diagram prepared by the author.

References

Barden N, Williams T. *Words and Symbols: Language and Communication in Therapy*. Maidenhead, UK: Open University Press: 2007.

Baron-Cohen S. *The Pattern Seekers: How Autism Drives Human Invention*. New York, NY: Basic Books: 2020.

Blackmore S. *The Meme Machine*. Oxford, UK: Oxford University Press: 2000.

Bor D. *The Ravenous Brain: How the New Science of Consciousness Explains Our Insatiable Search for Meaning*. New York, NY: Basic Books: 2012.

Borghi AM. Merging Affordances and (Abstract) Concepts. In: Djebbara Z (ed.) *Affordances in Everyday Life: A Multidisciplinary Collection of Essays*. Switzerland: Springer: 2022.

Bourdieu P. *Outline of a Theory of Practice*. (Trans. Nice R). Cambridge, UK: Cambridge University Press: 1977.

Bower GH. Mood and Memory. *The American Psychologist* 1981; 36(2): 129–148.

Bruner JS, Wallach MA, Galanter EH. The Identification of Recurrent Regularity. *The American Journal of Psychology* 1959; 72(2): 200–209.

Bunke H, Sanfeliu A. *Syntactic and Structural Pattern Recognition: Theory and Applications*. Singapore: World Scientific: 1990.

Caputo JD. *Hermeneutics: Facts and Interpretation in the Age of Information*. London, UK: Pelican: 2018.

Chabris CF, Simons D. *The Invisible Gorilla: And Other Ways Our Intuitions Deceive Us*. New York, NY: Harmony: 2010.

Chetverikov D. Pattern Regularity as a Visual Key. *Image and Vision Computing* 2000: 18(12), 975–985.

Chomsky N. *Rules and Representations*. Oxford, UK: Basil Blackwell: 1980.

Collins A, Burstein M. Afterword: A framework for a theory of comparison and mapping. In: Vosniadou S, Ortony A. (eds.) *Similarity and Analogical Reasoning*. Cambridge: Cambridge University Press: 2009: 550–552

Crick F, Mitchison G. The Function of Dream Sleep. *Nature* 1983: 304 (5922): 111–114.

Dennett DC. *From bacteria to Bach and back: the evolution of minds*. New York, NY: WW Norton: 2017.

Edelman GM. *Remembered Present: A Biological Theory Of Consciousness*. Basic Books: 1990.

Entwistle NJ. *Styles of Learning and Teaching*. London, UK: David Fulton: 1988.

Feldman Barrett L. *How Emotions Are Made: The Secret Life of the Brain*. New York, NY: Houghton Mifflin Harcourt: 2017.

Forte T. *A Pattern Recognition Theory of Mind*. Forte Labs. April 28, 2018. Online at https://fortelabs. co/blog/a-pattern-recognition-theory-of-mind/ accessed 31 July 2022.

Furey W. The Stubborn Myth of "Learning Styles": State teacher-license prep materials peddle a debunked theory. *Education Next* 2020; 20(3): 8–12.

Gadamer H-G. *Philosophical Hermeneutics*. Berkeley, CA: University of California Press: 1977.

Gadamer HG. *Truth and Method* (2nd ed.). New York, NY: Continuum: 1994.

Gibson JJ. *The Ecological Approach to Visual Perception*. New York, NY: Houghton Mifflin: 1979.

Gregorians L, Spiers HJ. Affordances for Spatial Navigation. In: Djebbara Z (ed.) *Affordances in Everyday Life: A Multidisciplinary Collection of Essays*. Switzerland: Springer: 2022.

Harnad S. To Cognize is to Categorize: Cognition is Categorization. In: Cohen H, Lefebvre C. *Handbook of Categorization in Cognitive Science* (2nd Ed.). Elsevier: 2017.

Heras-Escribano M. Affordances and Social Normativity: Steps Toward an Integrative View. In: Djebbara Z (ed.) *Affordances in Everyday Life: A Multidisciplinary Collection of Essays*. Switzerland: Springer: 2022.

Hoffman D. *The Case Against Reality: Why Evolution Hid the Truth from Our Eyes*. New York, NY: W W Norton: 2022.

Hofman MA. Evolution of the Human Brain: When Bigger is Better. *Frontiers in Neuroanatomy* 2014; 8: 15.

Holyoak KJ, Morrison RG. *The Cambridge Handbook of Thinking and Reasoning*. Cambridge UK: Cambridge University Press: 2005.

Inchaustegui YA, Alsina A. Learning Patterns At Three Years Old: Contributions of a Learning Trajectory and Teaching Itinerary. *Australasian Journal of Early Childhood* 2020, 45(1); 14–29.

Jung CG. *Man and His Symbols*. London, UK: Aldus: 1964.

Kahneman D. Thinking, Fast and Slow. Doubleday 2011.

Knowles M. *Self-directed Learning: A Guide for Learners and Teachers*. Cambridge Adult Education: 1975.

Knowles M. *Andragogy in Action*. San Francisco: Jossey-Bass: 1984.

Kolb, David A. *Experiential Learning: Experience as the Source of Learning and Development*. Englewood Cliffs, NJ: Prentice-Hall: 1984.

Kolvoort IR, Rietveld, E. Affordances for Situating the Embodied Mind in Sociocultural Practice. In: Djebbara Z (ed.) *Affordances in Everyday Life: A Multidisciplinary Collection of Essays*. Switzerland: Springer: 2022.

Krueger J. Affordances and Absence in Psychopathology. In: Djebbara Z (ed.) *Affordances in Everyday Life: A Multidisciplinary Collection of Essays*. Switzerland: Springer: 2022.

Maslow AH. A Theory of Human Motivation. *Psychological Review* 1943; 50(4): 430–437.

Mattson MP. Superior Pattern Processing is the Essence of the Evolved Human Brain. *Frontiers in Neuroscience* 2014; 8: 265.

Medina J. *Brain Rules: 12 Principles for Surviving and Thriving at Work, Home, and School*. New York, NY: Pear Press: 2008.

Meissner K, Bingel U, Colloca L, Wager TD, Watson A, Flaten MA. The Placebo Effect: Advances from Different Methodological Approaches. *The Journal of Neuroscience* 2011; 31(45): 16117–16124.

Mezirow J. Transformative Learning Theory to Practice. *New Directions for Adult and Continuing Education* 1997; 74: 5–12.

Nussbaum M. *Creating Capabilities: The Human Development Approach*. Cambridge, MA: Belknap Press: 2011.

Parnafes, O., Disessa, A. Relations between Types of Reasoning and Computational Representations. *International Journal of Computers for Mathematical Learning* 2004; 9: 251–280.

Peters E, Diefenbach MA, Hess TM, Västfjäll D. Age Differences in Dual Information-processing Modes: Implications for Cancer Decision Making. *Cancer* 2008; 113(12 Suppl): 3556–3567.

Piaget J. *The Origins of Intelligence in Children*. New York, NY: International Universities Press: 1952.

Pinker S. *Rationality: What it is, why it seems scarce, why it matters*. New York, NY: Viking: 2021.

Pintrich PR. The Role of Metacognitive Knowledge in Learning, Teaching, and Assessing. *Theory Into Practice* 2002; 41(4): 219–225.

Rips LJ. Similarity, Typicality, and Categorization. In: Vosniadou S, Ortony A. (eds.) *Similarity and Analogical Reasoning*. Cambridge University Press: 2009.

Sampson G. *The "Language Instinct" Debate* (revised edition). New York, NY: Continuum: 2005.

Schank RC. *Explanation Patterns: Understanding Mechanically and Creatively*. Hillsdale, NJ: Lawrence Erlbaum Associates: 1986.

Schraw G, Moshman D. Metacognitive Theories. *Educational Psychology Review* 1995; 7(4): 351–371.

Seth A. Your Brain Hallucinates Your Conscious Reality. TED Talk, Vancouver BC, April 2017 - https://www.youtube.com/watch?v=lyu7v7nWzfo&t=6s accessed 2 Feb 2024

Shah S. The Animals Are Talking. What Does It Mean? New York Times, Sept. 20, 2023 - https://www.nytimes.com/2023/09/20/magazine/animal-communication.html accessed 20 Sep 2023

Shermer M. *The Believing Brain*. New York, NY: St Martin's Press: 2011.

Skinner BF. *The Technology of Teaching*. New York: Meredith: 1968.

Taylor C. *The Language Animal: The Full Shape of the Human Linguistic Capacity*. Cambridge, MA: Belknap: 2016.

Toussaint GT. The Use of Context in Pattern Recognition. *Pattern Recognition* 1978.10: 189–204.

Wilczek F. *A Beautiful Question: Finding Nature's Deep Design*. New York, NY: Penguin: 2015.

Zai A, Brown B. *Deep Reinforcement Learning in Action*. Shelter Island, NY: Manning: 2020.

6

EXTERNALISING PATTERNS

In which I consider different ways in which patterns escape our minds.

In the last two chapters, I described patterns as nexuses of connected memories, impressions, thoughts, and feelings that can help us (and sometimes hinder us) in perceiving and understanding the world around us and within us. I have been emphatic that patterns are not 'out there' no matter how much they seem to be. If patterns are indeed built out memories, and can only be a phenomenon of mind, then how is it that we can share them? We clearly do share our patterns, if we did not then there would be no language, culture, or society. You would not be reading this book, and I would never have written it.

Externalisation of pattern does not mean that our memories can be extracted or duplicated. As much as the Pensieve in the Harry Potter stories is a charming idea, we cannot pull memories out of our minds let alone transfer them to other minds. The best we can do is to use the means of expression available to us to translate or reflect our patterns and pattern thinking. Moreover, as much as spoken and written language, imagery, music, and physical expression allow for a vast range of expressive possibilities, they can never truly capture the rich and idiosyncratic nature of the patterns and pattern thinking that guide them. Nevertheless, we do try to share our patterns in different ways. To that end, I differentiate between pattern expressions, pattern representations, pattern systems, and pattern languages. Before I work through these concepts, I first need to clarify the concept of *pattern instances*, which, although they are not shared, is an important concept in understanding shared pattern thinking.

Pattern Instances

At first glance, an instance of a pattern (a *pattern instance*) would seem to be a simple concept; it is something that we perceive as related to a pattern memory. This thing is perceived as an instance of the pattern of doors, that thing is perceived as an instance of the pattern of cats. Pattern recognition of familiar objects often happens faster than our conscious

DOI: 10.4324/9781003543565-6

minds notice such that we just know that this is a door and that is a cat. Note that recognising something as an instance of a pattern is not the same as establishing a specific identity, we can recognise a cat without having to recognize any one specific cat. We perceive sufficient similarity to assign this thing to the family of things we have brought together as other instances of the pattern.

For something to be perceived as an instance of a pattern it needs to be perceived as having some, but not necessarily all, of the characteristics of that pattern. As an example, in order to perceive something as a cat we might expect it to have fur, four legs, a tail, pointed ears, whiskers, claws, and so on, and, even then, we might perceive this as an instance of a dog or a racoon. On the other hand, this thing may be missing a leg or a tail, be hairless, or have other variant characteristics and yet still be clearly perceived as a cat. Many of the running gags associated with the Warner Brothers' cartoon character Pepé Le Pew are based on whether or not he or other animals are correctly perceived as skunks.

Which characteristics matter in establishing pattern membership can differ between instances and between observers. It may be that some characteristics are more important than others in an individual being willing and able to perceive a phenomenon as an instance of a particular pattern. It may also be that the more a pattern's characteristics are abstracted or the wider or looser the specificity of their characteristics the more they allow for different phenomena to be perceived as pattern instances. An instance may have non-pattern characteristics that may or may not compromise its being perceived as an instance of that pattern. For instance, the winged monkeys in the Wizard of Oz are perfectly recognisable as monkeys despite being blue, wearing waistcoats, and being able to fly. That said, sometimes the presence of certain non-pattern characteristics does prevent something from being recognised as an instance of a particular pattern, particularly when the additional characteristics suggest an alternative pattern. Characteristics of some patterns may intersect in perceived instances without interfering with each other while other pattern characteristics may be exclusive of each other. For instance, the pattern characteristics of costumes and cats do not particularly interfere (a cat costume is a legitimate aggregate), while the characteristics of cats and dogs are more exclusive (we expect something to be one or the other rather than some combination of the two).

Coming up with a set of definitive criteria or a precise pattern recognition algorithm for a pattern can be challenging as there are almost always exceptions to be found to any pattern rules we might devise. Moreover, we might perceive the same thing as an instance of different patterns on successive encounters, what I perceive as an instance of my particular pattern may not align with what others perceive based on their unique patterns, and the pattern instances that I do perceive may shift according to the information available to me. Perceived instances of patterns would seem to reflect rather subjective and fluid thresholds, modified by factors such as context, precedent, and emotion. Something we encounter, therefore, need only have some characteristics of a pattern to be perceived as an instance thereof.

To perceive something as an instance of a pattern we need to be able to sense the thing, we need to have a pattern to which our senses can be matched, and, in the moment, we need to perceive sufficient characteristics, albeit in different combinations, to establish a degree of certainty as to whether something counts as an instance of the pattern or not. These principles might suggest that pattern coherence is more about continua, such that no one element is sufficient in itself to resolve the question of "is this an instance?". Indeed, is there a threshold at which something that was perceived as an instance of a pattern loses this connection or identity? Is the transition from "is not an instance" to "is an instance" symmetrical with the transition from "is an instance" to "is not an instance"? Is there a continuum of being more or less of an instance with no distinct inflection or tipping point?

The complexity of perceptual thresholds notwithstanding, we do seem to make somewhat binary pattern identity judgements when we do not have the time, energy, or commitment to analyse the extent to which something really is an instance of a particular pattern. I would argue that this again reflects the heuristic characteristic of pattern thinking; something that is fast, and lean but also provisional and prone to bias and error. Indeed, our apparent need to resolve "is this or is this not an instance of pattern X?" raises broader questions regarding our responses to ambiguity. It might be that we have evolved to instinctively resolve these perceptual ambiguities given our favouring of Type I errors. That pattern instances do not have to be identical to each other, reflects Wittgenstein's concept of "family resemblance", the seeming paradox that we perceive a cluster of things that share no one common characteristic as members of the same family (or pattern). This reflects a complex faceted and interpretive epistemology in which pattern identity is emergent and contingent on each observer's pattern memories and pattern thinking. Rather than thinking through a comparison item by item, do we (as deep reinforcement learning models might suggest) rapidly assess multiple impressions and gestalts to arrive at a sense or feeling of something being an instance (or not)? That is not to say that characteristics (elements, relationships, etc.) do not matter. We can return to the concepts of form, order, and regularity, as they play a key role in connecting the patterns in our minds with things beyond our minds.

As I argued earlier, form is the specific qualities and essences of a thing; form is the way things are rather than what they appear to be or what they are like. Form, ideally, is perception stripped of meaning, understanding, or any other pattern-generated cognitive phenomena we habitually draw However, it can be hard to perceive form without also thinking about orders and regularities, and this requires patterns. Indeed, it can be our perception of order and regularity that drives our perception of pattern instances. Order is about the perceived presence, kinds, extents, decays, and transformations of ordering principles (including disorder). All order is regular in one or other (and sometimes both) senses of the word, but not all regularity is ordered. Regularity can refer to repetition, recurrence, and to control. However, I would posit that the regularity that triggers pattern thinking is more often related to perceived recurrence across instances rather than to repetition within them.

In terms of regulation as control, again I would propose two further distinctions: causation and conformance. Causation refers to the perceived presence of some guiding force or principle (or the supposition that there is some such presence) that has precipitated and shaped the regularity. Causation can be considered in terms of why there is a regularity at all and why the regularity takes the particular form it does. Conformance on the other hand refers to perceived degrees of regulation, to what extent and in what ways something does and does not follow its causative influences. This can include perceptions of how regulated or unregulated something is (freedom to vary), as well as degrees of compliance (extents and kinds of variance). Finally, conformance may also reflect other characteristics that are not related to the pattern.

Perception of repetition (symmetry, geometry, etc.) can be a stimulus for pattern seeking and it might also be used in finding a suitably matching pattern, but it is not a necessary part of either. Perception of recurrence on the other hand plays a more central role as it is what connects current experiences with past experiences. If there is no recurrence (no pattern recognition) then the phenomenon is perceived as something completely new and pattern development is required. If there is recurrence (at least to some extent) then the phenomenon is perceived as a pattern instance and the pattern's elements, connections, and associations can be drawn on to understand the phenomenon at hand. However, rather than assessing whether something is or is not a pattern instance in any absolute sense, perception of an instance is often a matter of degrees of certainty or likelihood. This is the heuristic gaze again; the pattern match need only be good enough; it will satisfice unless or until new information suggests alternative pattern readings. In part this is down to imperfect or incomplete information, we work with what we can access, and this is usually far from optimal. It is also down to degrees of perceived conformance between the putative instance and the pattern.

Causation is different again. Sometimes we may perceive direct causative interactions (like billiard balls hitting each other and rebounding), but much of the time we infer causative and regulative factors. Moreover, causation is not a necessary part of pattern thinking; much of the time we can get by with perceiving the presence of pattern instances (predators, shelter, food, etc.) without needing to know or understand their origins. However, if patterns do suggest causes then our understanding of the phenomenon as a pattern instance is deepened, and new avenues of thought or action may become possible. Given that the perception of causation can deepen the connections we perceive between an instance and its referent pattern, then it is the pursuit of causation that can help to drive our pattern thinking from micropatterns to mesopatterns and from mesopatterns to macropatterns.

A pattern instance is not an externalisation of pattern. However, it is a relationship we perceive between some outside thing and a pattern memory in our minds. It is not a large step from passively perceiving pattern instances to our actively projecting pattern identities onto the world around us. This is why I have first defined pattern instances before considering externalisations. I now turn to the main focus of this chapter, the ways in which we externalise our patterns.

Pattern Expressions

Pattern expressions are intentional instances of a pattern. As an example, sometimes, if I fancy a cup of tea (which I often do) then I will go and make myself one. At other times I might be offered a cup of tea, or I might go out and buy one. These various cups of tea are all expressions of somebody's pattern for a cup of tea. Someone who has no pattern for a cup of tea cannot create an expression thereof (although they might follow a recipe and in so doing establish a new pattern in their mind). As with pattern instances, there can be much variance between expressions of the same pattern. The tea may be made using different containers (cups, mugs, thermos flasks, etc.), there may be different kinds of tea (black, green, Earl Grey, oolong, rooibos, etc.), the tea may be loose leaf, powdered, or in teabags, a teapot or an urn may be used or the drink prepared in the cup in which it will be served, the tea may be served hot or cold, and it may be served with or without additions (milk, lemon, sugar, sweeteners, etc.). Cups of tea can take many different forms and yet still be legitimate expressions of a common pattern. Although this is also a quality of pattern instances, it is the intentional making of an instance that qualifies it as a pattern expression. Raindrops and kittens are pattern instances if they are perceived as such, but they are not pattern expressions because they occur with no conscious thought (setting aside the will of divine beings and other teleological entities). Copper kettles and warm woollen mittens are pattern expressions because they were manufactured by humans with minds that held a pattern for those things and that used that pattern in creating them.

Some expressions may reflect more of a pattern than others. For instance, somebody who is imitating a cat for a fancy-dress costume may well express rather less of a cat pattern than an artist painting a beloved pet. Moreover, they are likely to be expressing different aspects of a cat pattern, partly through what seems important and partly because their pattern memories cannot be identical. Not only is any given pattern expression not an expression of every part of a pattern, given that a pattern is no more than a nexus of aggregated memory then there can be no such thing as a complete pattern to be expressed. Moreover, because we cannot directly, consciously, and objectively access whole pattern memories, what we can recall and use to shape an expression will always be a subset; there is always selectivity, partiality, and judgment involved in creating pattern expressions.

Externalising a pattern as an expression requires action and intent, and it typically requires one or more patterns that are being expressed and other patterns that guide the expression. For instance, let me take the example of the cover of this book that shows a painting I made of one of the peonies I grew in my garden. I have a pattern for peonies, and I also have a pattern for flower paintings (such as those by Georgia O'Keefe). I also have a pattern for drawing and another related one for painting that I used in creating the painting. That there is more than one pattern involved in an expression is important to note as the *how* patterns may be as important as the *what* patterns in pattern expressions. After all, we can only use the expressive repertoire available to us or that is specified for the particular kind of expression at hand. For instance, a young child's drawings of cats, people, and houses may be simplistic because the child's patterns and pattern thinking are still developing and the media available to them are also rather basic. However, many adults would struggle to do much better than the child, not because their patterns of cats, people, and houses are not more fully developed but because their patterns for drawing have not developed much since they were children. A trained artist should have very well-developed patterns for drawing and be much better able to create sophisticated expressions of whatever patterns they focus on. Simply put, pattern expressions are not simply a matter of expressing a subject pattern, they also draw on patterns of the media in which the expression is made and patterns of craft and creativity in rendering the expression.

Pattern expressions do not have to originate entirely in their creators' minds; some aspects of the expression may be guided by templates, blueprints, or algorithms. For example, there is a difference between somebody improvising a jazz solo and somebody performing a solo by reading a score and playing it note for note. The latter expression is patterned in the sense that there will still be cognitive patterns expressed that refer to reading scores and playing instruments, but the patterning of what notes to play and when and how to play them that is present in the first case is not present in the second. As another example, are manufactured things created from prefabricated parts on production lines with automated processes pattern expressions? Although a robot may build or manufacture a thing, human minds were involved in deciding that they wanted to produce certain things, other minds that specified the design of the thing to be manufactured, and yet other minds that set the manufacturing process up and then operated it. The mind of every person involved will have a pattern for the thing being produced such that they express that pattern in their contributions to the manufacturing process. Their patterns need not and indeed cannot be identical, but they should be aligned. If their patterns are misaligned, their ability to contribute to a collaborative pattern expression would seem limited and may even prove disruptive.

Pattern expressions may involve different levels and forms of interpretation of the pattern(s) being expressed. We may express a pattern with relatively high fidelity to our pattern understanding (such as solving a mathematical problem), we may take it as a general frame of reference around which we may improvise (such as making a spaghetti sauce), or we may dynamically mix and merge different pattern expressions from one moment to the next (such as participating in a conversation). Variation in expressions may also reflect a selectivity in which characteristics of a pattern are included in any given expression thereof. For instance, an individual might tailor pattern expressions to respond to a particular context such as using more or less technical or idiomatic forms of expression. An individual may shape their pattern expressions to meet the needs of their audience. For instance, expressions of the same pattern may vary significantly if they were being shared with young children, an intimate partner, or the local police force.

Someone who has created a pattern expression will likely remember the making of it, which means that whenever someone creates a pattern expression (or contributes to one) this will also result in elaboration of the pattern(s) being expressed to memories of the specific pattern expression, of the context of expression, and memories of the ways in which it was expressed. There is, therefore, an intimacy between patterns and their expression for creators that others cannot share. We recognise that intimacy and often seek to experience it vicariously either through working with a pattern expression creator, through watching them generate pattern expressions, through collecting or appreciating these expressions, or through consuming biographies and critiques of their work. Indeed, pattern expressions may be one of the most compelling ways to share one's pattern thinking with others. If we do not have access to a particular expression creator (particularly those who are long dead), we can only interpret their pattern thinking from the expressions themselves (if they are in the form of artefacts or recordings), the writings of the creator (diaries, letters, autobiographies), the observations of others, or the contexts in which the expressions were created. Much of history, archaeology, literature studies, and art criticism (and many other academic disciplines) engage in what we might call "pattern reengineering" through reading and interpreting artefacts of various kinds to extrapolate the thinking of those who created them, all of them limited by a lack of access to the original sources. There can also be an interpretive and diagnostic aspect to this when an individual's pattern expressions are used to appraise their pattern thinking. For instance, a psychologist might analyse a person's drawings or their writing to identify underlying trauma or psychopathy. Clearly, there are many ways in which we can analyse pattern expressions for the implied pattern thinking of their creators, an issue I will return to later in this volume.

Pattern Representations

When an architect designs a building, when a composer writes an orchestral score, or when a cook develops a successful recipe, they are not constructing the building, performing the piece, or making the dish, they are telling other people how to do these things for themselves, they are creating *pattern representations* for others to use. Simply put, pattern representations are intended to capture the structure and meaning of certain parts or characteristics of a pattern.

However, the patterns in our minds are not stored as convenient lists or diagrams that we can simply reproduce. A pattern representation is necessarily reductive and abstractive in that we select those aspects of the pattern that are (or seem) important while omitting others that are less relevant. Quite how much aggregation and abstraction are necessary in translating a pattern to a representation can vary significantly. Thus, while pattern representations are more likely to be of macropatterns than of micropatterns, there is no absolute or consistent level of abstraction, objectivity, or any other marker of rigour involved in their creation. Indeed, pattern representations may be simplistic (such as a child's drawing) or elaborate (such as an architectural plan), subjective (such as a poem) or objective (such as a mathematical model). Not only are all pattern representations reductions of the patterns they references, one pattern may be represented in more than one way.

A pattern representation might include pattern elements, their topologies and characteristics, and any associated meanings, understandings, emotions, and implications, it might focus on all or only some of them, and it might present them in many different ways. Indeed, a pattern representation could focus on how the pattern is typically used, how it is linked to other patterns, or how it has developed over time. Which aspects are included in a particular pattern representation and how they are presented depends on many factors, including purpose, audience, and medium, and convention, which means that, as with pattern expressions, there could be multiple representations of the same underlying pattern nexus.

There are important differences between expressions and representations. One is that pattern representations will usually require more thinking and more interrogation of the pattern than expressions do. Indeed, a pattern representation requires direct consideration of the nature of the pattern to be expressed whereas pattern expressions are generally less analytical and more performative. This necessary pattern analysis and mindfulness required of pattern representations is likely to result in elaboration of the patterns involved, not just in terms of adding to pattern memories but in developing their macropatterned characteristics. We may also invoke, adapt, or create categories as indexes of patterns to add structure and delineation to our own pattern thinking. The creation of a pattern representation is, therefore, a definitional act. By thinking about a pattern, we define it, we give it shape and meaning, we may even reshape it in response to our seeking to represent it so that it is more coherent, better connected, and more aligned with other patterns.

If our goal is to share our pattern representations with others, we might express them in ways that reflect what we perceive to be how these others think, which in turn draws on other patterns that reflect our understanding of a pattern representation's audience. The purposes of pattern representations need careful consideration. For instance, an individual might create a pattern representation to help them to analyse and clarify their own thinking. This book is an example of this; the words and sentences you are reading were not fully formed in my mind just waiting for me to type them out, my thinking developed through iterative cycles of writing, editing, and reflecting on the ideas as they developed. A pattern representation may be created to help develop or align similar patterns in other minds. Teachers do this all the time through learning activities, schemas, examples, mnemonics, and other techniques to help their students to in acquiring knowledge. Again, this book is an example of this; I hope the ideas it presents find their way into the minds of its readers, albeit in idiosyncratically patterned ways. A personal pattern representation will tend to be more subjective and incomplete than one created for others to use. Indeed, there is an extra layering in those pattern representations created for other people in that they draw on patterns that reflect their creator's understanding of their audience's pattern thinking.

There are many other possible purposes of pattern representations. A pattern representation may be created as a way of recording an individual's thoughts, or of sharing and contrasting the thoughts of multiple individuals. A pattern representation may be used as a way of synthesising ideas or as a way of opening ideas to critique, verification, or validation by others. A pattern representation may be created as a way of lending the pattern agency or authority (such as writing down rules or laws). A pattern representation may be used to appraise or evaluate instances, or even to generate instances by using the representation as a template such as in the dressmaking sense of a pattern: a guide, generator, or parent that gives form to that which is derived from it. A pattern representation may be shared, learned, enforced, adapted, co-opted, sold, or traded, or otherwise exchanged, mediated, or implemented through various social interactions and systems. Indeed, much of our shared pattern thinking would seem to be based on exchanging individuals' pattern thinking in the form of pattern representations and expressions and the conventions (that are themselves pattern expressions) of doing so. In developing pattern representations, we may struggle to distinguish between what is specific to our own patterns (albeit aligned with that of others to some extent) and what is truly shared. Much of human disagreement and conflict may have its origins in the messy and lossy exchanges of pattern representations.

Pattern representations are consumed by others such that they are in some way woven into their own pattern thinking. How much of a pattern is captured in a representation and how much is lost, altered, or augmented in translation is another topic that could fill volumes. For now, I will simply observe that pattern representations are not perfect, they are lossy and transformative – see Figure 6.1. It is also important to note that although a pattern representation can be assimilated (read, listened to, thought about) into the pattern matrices of one or more minds, it still exists externally as a representation. My 'devouring' a book leaves the book intact while my devouring a sandwich does not.

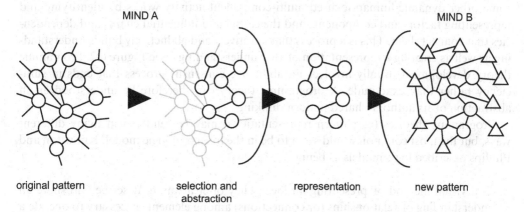

MIND A   MIND B

original pattern   selection and   representation   new pattern
                            abstraction

FIGURE 6.1 An individual decides to create a representation of one of their patterns. To do that certain elements are selected and abstracted and then rendered external to the mind. Subsequently, another mind encounters the representation and assimilates it into their own pattern thinking matrices.

Diagram prepared by the author.

The purpose of pattern representations is (or at least it should be) related both to audience and context. For instance, pattern representations that are intended to help schoolchildren to learn should be aligned with their developmental capacities and the goals of their teachers and of the curriculum. A pattern representation used as a template for creating pattern instances on the other hand will likely include instructions related to the production of an instance but include relatively little on its meanings and implications. A pattern representation used as a foundation for others to use as part of professional activities, such as in mathematics, medicine, engineering, or law, will likely be rigorous and formally structured with as little ambiguity as possible. Scientific pattern expressions, such as frameworks, models, practices, and theories, which have been (one hopes) rigorously tested, evaluated, and critiqued as part of the scientific process, are prime examples of this.

The medium of a pattern representation will also shape what it contains and how it is expressed. For instance, mathematical and musical pattern expressions have a whole symbolic language (or languages) to draw on. Different media have different capabilities in terms of what and how pattern representation ideas can be expressed, and they have different cognitive affordances in terms of what they can get across to those who consume them. For instance, diagrams may be much more useful in guiding others in pattern representations of spatial relationships than prose. Similarly, a symbolic pattern representation might be more useful where logical analysis is required. Indeed, there are many pattern representation conventions involved in rendering or communicating pattern representations, such as those used in engineering diagrams, dressmaking patterns, and architectural blueprints. Some pattern representation conventions might be concerned with how the representation should be read or applied, while others might provide a shorthand for making representations more efficient.

The means of expression are particularly important for pattern representations intended to support collaborative working. As an example, let us consider the development and use of programme theories and their expression in the form of logic models as shared pattern representations (Funnell & Rogers 2011). A programme theory collates and integrates the pattern thinking of its developers regarding how a process, a system, an organisation, or some other dynamic human-focused multi-component activity works by identifying and representing factors and components and the causal and influential chains and dependencies that connect them. This is a process that iteratively and abductively builds understanding as well as creating a representation of that understanding – see Figure 6.2. Programme theory development usually involves an abductive consensus process that ends in sufficiency rather than exactitude. A programme theory can therefore be understood as an abstraction of its authors' shared pattern thinking

Programme theories (as pattern representations) may be rendered in many different ways, but the most common would seem to be in the form of a logic model. Knowlton and Phillips described logic models as being:

… a visual method of presenting an idea. They offer a way to describe and share an understanding of relationships (or connections) among elements necessary to operate a program or change effort … The development of models (or the modeling process) provides an opportunity to review the strength of connection between activities and outcomes. Through the experience of critical review and development, models can display participants' learning about what works under what conditions.

(Knowlton & Phillips 2012, p. 4)

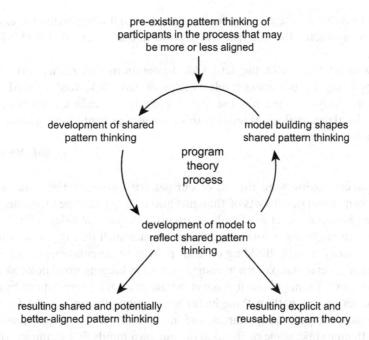

FIGURE 6.2   Steps in building a programme theory from a pattern theoretical perspective where both the representation and the pattern thinking it represents are developed through building and aligning collective pattern thinking.

Diagram prepared by the author.

A logic model is another kind of pattern representation. Despite the qualifier, logic models do not need to be logical in any formal sense. They represent their authors' reasoning with no expectation that the reasoning was of a particular kind or met any particular standard, which reflects broader discourses of "logics" as distinct from formal logic (Thornton & Ocasio 2008). I would also note that logic models are often rendered in the form of directed graphs, which mirrors the use of graphs in General Pattern Theory to model patterns (Grenander 1989). That said, the elements in logic diagrams are highly unlikely to represent a single pattern (unless it is a well-developed macropattern) let alone individual pattern elements.

Mindmaps are similar to logic models in that they are directed graphs that constitute nonlinear networks of ideas represented visually. Although Mindmaps is the name given by Buzan and Buzan (1993) to these network representations of knowledge, they have been used for centuries and they continue to be used in many academic and non-academic contexts. While logic maps are generally analytic and consensus-based, mindmaps tend to be more personal and idiosyncratic (unless developed collaboratively – most sources on mind mapping have focused on individual mapmaking). I would note in passing that the product of mindmapping seems to be less useful than the process, which may also be true for other kinds of pattern representations.

One last form of pattern representation to consider reflects the fluid concept of external cognition, the premise of which is that humans both can and readily do extend their cognition beyond that of their brains, for instance, through using technologies of various kinds. Clark encapsulated this in a theoretical position that he called "EXTENDED":

> According to EXTENDED, the actual local operations that realize certain forms of human cognizing include inextricable tangles of feedback, feed-forward, and feed-around loops: loops that promiscuously criss-cross the boundaries of brain, body, and world. The local mechanisms of mind, if this is correct, are not all in the head. Cognition leaks out into body and world.
>
> *(Clark 2008, p. xxviii)*

Clearly, we can externalise some aspects of our pattern memories. For instance, a photograph album can stimulate all sorts of thoughts and memories; these memories are still in my mind, the photographs act as stimuli, shortcuts, catalysts, or indexes for these memories. Similarly, we might access other people's memories such that they constitute part of our pattern memory matrix. Building on this, pattern representations could perhaps be created and used as external pattern memory, such as in keeping notebooks, sketchbooks, diaries, and journals. We might create more elaborate pattern representations by externalising the processes of our pattern thought (or abstractions thereof) in the form of algorithms, equations, computer programmes, and instructional materials (to name but a few) and then use them to take some of the load off our own minds. For example, when I use a tool to perform a statistical analysis, I do not need to remember the mathematical theory or the equations used, I just need to know which test to select, how to operate the tool, and what to do with what it produces.

It is important to remember that pattern representations are not patterns, they are texts, pictures, diagrams, recordings, performances, or some other reflection of a pattern. No one pattern representation is necessarily definitive, partly because the pattern it represents is dynamic, partly because the creation of a representation is intrinsically selective and interpretive. As our pattern thinking continues to change in response to our ongoing experiences, the representations we make of our pattern thinking will change to reflect this. It may be that the more a pattern is macropattern-like (deliberative, abstract, generalised) the more stable, complete, and logically continuous with itself it might be, although that is by no means a given. Thus, while pattern representations allow us to transfer and elaborate patterns between minds, this is an imperfect and provisional undertaking, subject to entropy and emergence, as well as to our whims, biases, and idiosyncrasies. There are parallels here with memetic theory that differentiates between "copy-the-product" and "copy-the-instructions" (Blackmore 2000). Blackmore argued that "copy-the-product" was more concrete but also open to error in reproduction as there was only the example and no rules or templates to follow, while "copy-the-recipe" was less error-prone given the necessary presence of rules or templates. Yet again, there is much work to be done exploring the questions raised by taking a pattern theoretical perspective.

Pattern Systems and Pattern Languages

In Chapter 2, I described Alexander's work developing a *pattern language* for architecture and his use of *design patterns* as the core constituent parts of that pattern language. According to Alexander et al., design patterns (usually simply referred to as "patterns") are ways of solving a particular kind of problem. Essentially, a design pattern is a particular kind of pattern representation that is specifically intended to be used for guiding the creation of solutions to recurring problems in a particular domain.

Winn and Calder (2002) proposed nine essential characteristics of design patterns in the context of programming languages; they are grounded in a particular domain, they imply that something will be created from them and that the pattern will be manifest in those instances, they work at different levels of abstraction, they have some kind of rationale, they focus on solutions to problems, and they are validated by their use (they have value if and only if they are useful), while Bayle et al. (1998) argued that design patterns afford a way of sharing ideas and understanding, that they deal with problems at different scales, and that they reflect values and practices. The macropatterned nature of most design patterns reflects their authors' focus on providing broad solutions to broad problems that are often encountered. The breadth of their solutions reflects both abstraction and aggregation in the pattern thinking that generated them. Design patterns based on mesopattern representations would be less well worked out, less abstract, and less theorised than macropattern-based representations. They would also tend to be more heuristic, arbitrary, and subjective. Again, there is no absolute binary separation here, different design patterns would likely fall along a continuum between mesopattern and macropattern representations.

Design patterns have a clear utilitarian focus; they are intended to provide generic solutions to recurring kind of problems. The audience for a particular design pattern is made up of the individuals working in the domain who are facing these problems (or anticipate them) and need (or will need) the solutions the design patterns provide. Clearly, not all pattern representations are about problem-solving but invoking them in service of some need will likely increase the focus on problems and solutions.

Design patterns are typically focused on a particular domain of practice (architects, engineers, artists, etc.) such that the language and anticipated user needs reflect the norms for that domain and do not necessarily generalise beyond that domain. That said, the shared pattern thinking in a domain typically cannot be captured in a single representation. Rather, it takes many design patterns to capture the recurring problems encountered by practitioners in any given domain.

A *pattern language* is the collection of the design patterns relevant to a particular domain. Alexander et al.'s pattern language contained 253 design patterns while Gamma et al.'s contained just 23 design patterns – does this mean that architects typically deal with ten times as many problems as software engineers? Perhaps, but this should not in any way be inferred from the number of design patterns in their respective pattern languages. Quite how many design patterns are needed to encompass a domain is a matter of interpretation and judgment not a measure of complexity or need. Collectively, the design patterns in a pattern language are expected to connect and build on each other to solve ever larger problems. Design patterns in a pattern language need to be expressed in ways that allow for this aggregation and interdependency. Pattern languages are also pattern representations (albeit of domains of practice), and they too can be aggregated:

> Different patterns in different languages, have underlying similarities which suggest that they can be reformulated to make them more general, and usable in a greater variety of cases... Gradually it becomes clear that it is possible to construct one much larger language, which contains all the patterns from the individual languages, and unifies them by tying them together in one larger structure.
>
> *(Alexander et al. 1977, p. 330)*

How far this aggregation might go in any formal sense is unclear although it might be a rather creative (albeit exhausting) way of generating a Grand Unified Theory of Everything, or at least of the shared pattern thinking of everything. It seems likely that, like patterns in the mind, while collections of pattern representations can be connected in very large assemblages (that are also pattern collections), many of the connections between them become tenuous and provisional. After all, our minds connect all our memories and thoughts in ways that we typically experience as a synthetic whole and yet still seems to consist of discrete patterns.

Can we say that all collections of design patterns *qua* pattern representations constitute pattern languages? In answering this question, I note the distinction made by those who have described "pattern catalogues" as loose aggregations of design patterns (Dearden & Finlay 2006; Coplien & Schmidt 1995). However, since a catalogue is a list or index referring to things held elsewhere, I will instead use the term *pattern system* to refer to a collection of pattern representations pertaining to a particular domain that can collectively represent the problem-solving pattern thinking of that domain.

From this, we can understand *pattern languages* as pattern systems with additional grammar and syntax that define how the pattern representations are to be expressed in aggregate. A pattern language's *grammar* is made up of the rules that must be followed in combining its pattern representations. For instance, there may be temporal or causal relationships between representations that must be observed such as in building a house where one must start from the foundations and can only add a roof late in the process once the walls have been constructed.

A pattern language's *syntax* reflects the idiomatic ways in which its design patterns can be used and combined. For instance, a pattern language that represents the shared pattern thinking of chefs could be used to reflect many actual and possible dishes that they might be asked to make. Within this range of dishes there may be discrete syntaxes according to preparing the ingredients (meats, vegetables, grains) or that reflect different cuisines (Italian, Creole, Thai). Note that, typically, there can only be one grammar per pattern language, but there may be many syntaxes.

The coherence of a pattern system or language is in part reflected in expressing each component design pattern (representation) in a standardised way. For instance, Alexander et al. (1977) defined a design pattern as "a relation between a certain context, a problem, and a solution". The template they used provided an exemplar image, descriptions of the contexts in which the problem occurred, the problem faced, the solution to the problem, and notes on related patterns. Gamma et al.'s (1995) template included a name and classification followed by descriptions of its intents, synonyms, motivations, applicability, structures, participants, collaborations, consequences, implementations, sample code, known uses, and related patterns. Different pattern systems and pattern languages might use different templates depending on the conventions of the domain and the pattern thinking of its authors.

Pattern systems and pattern languages should be complete in that the collection of design patterns is sufficient to describe or represent all possible pattern instances in its domain. For example, a pattern system for automobiles would have design patterns for their construction, their functions, their aesthetics, their uses, their dependencies, and so on, such that a) there is a design pattern for any aspect of importance of automobiles, and

b) collectively the design pattern can represent any automobile phenomenon (vehicles, design, infrastructure, economics, fashions, etc.). As Alexander observed, this:

> ... allows its users to create an infinite variety of those three-dimensional combinations of patterns which we call buildings, gardens, towns ... both ordinary languages and pattern languages are finite combinatory systems which allow us to create an infinite variety of unique combinations, appropriate to different circumstances at will.
>
> *(Alexander et al. 1977, pp. 186–187)*

Alexander et al. also outlined principles that should apply to all pattern languages, and, by implication, to pattern systems too – see Table 6.1.

Although design patterns are pattern representations, this does not mean that we can extrapolate what we know about design patterns to all pattern representations. For instance, pattern representations may be created for reasons other than generalising and codifying solutions to problems, they do not all have to follow a common format, and they might be collected in ways that do not reflect a complete outline of a domain of human activity. An individual might create pattern representations that reflect their idiosyncratic understanding of certain things (such as fabulous beasts or kinds of clouds), they might express their representations in whatever ways come to mind, and they might collect or associate representations based on their state of maturity or their state of mind when they created them. However, there should be some kind of aggregating (ordering) principle involved in collecting these pattern representations together.

The most common aggregating principle in the extant pattern language literature is that of problem-solving in a particular domain of human activity, but there are others. For instance, a different (and broader) aggregating principle might group representations that contrast the shared pattern thinking of a particular network of practice, a profession, or of a particular social or political class. There is no intrinsic reason why we might not also consider these aggregations as pattern systems or pattern languages. Building on this last point, although pattern systems and pattern languages have been explicitly identified as such and purposefully developed in domains such as architecture and computer science, there are

TABLE 6.1 Principles of pattern languages (after Alexander et al. 1977, pp. x–xiii)

Principle	Description
Structure	Each design pattern description is based on the elements that are sufficient to describe it.
Application	Design patterns describe a recurring problem and a core solution to the problem that can be adapted to fit different contexts
Relationships	Design patterns are never isolated, they are always supported by other patterns and in turn they support other patterns.
Authority	No pattern language is definitive as it reflects a body of practice that is variant across contexts and over time. Other pattern languages are possible, likely even.
Invariance	Some design patterns are invariant (the only way to solve the problem) while others may admit to alternative ways of solving the problem.
Density	A pattern language may express in ways that range from simple and syntactic (prosaic) to complex and semantic (poetic).

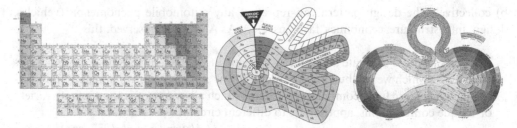

FIGURE 6.3 Different syntaxes of a periodic table of elements. Left to right: the Mendeleev periodic table, the Benfey periodic table, and the Hyde periodic table.

Sources: left: https://commons.wikimedia.org/wiki/File:Periodic_table_large.png (public domain), centre: https://en.wikipedia.org/wiki/Periodic_table#/media/File:Elementspiral_(polyatomic).svg (CC BY-SA 3.0 by DePiep), right: https://commons.wikimedia.org/wiki/File:The_chemical_elements_and_their_periodic_relationships.svg (CC-BY-SA 4.0 by Jeremy Sachs).

other expressions of pattern thinking that reflect many of the qualities of a pattern system or pattern language that were not identified or developed as such. Consider Mendeleev's periodic table of elements as an example of this. The periodic table represents the regularities of individual elements and the regularities of similarities and differences between them. To that end, there are pattern representations for an element, for behaviours, and for families and groups, and a grammar and syntax for how elements are portrayed and related to each other. Moreover, although its grammar is fixed (reflecting basic elemental chemistry and physics), there is more than one syntax that can be applied – see Figure 6.3. That this is a patterned view of the physical world is reflected in its exclusion of some details (it does not account for isotopes), its simplification (exact atomic weights are rounded to the nearest median integer), and its layering on of structure (such as families of elements).

Pattern systems and pattern languages are also likely to differ according to their intended audience. Those that are intended for their creator's own use will probably not make the idiosyncratic pattern thinking of the author explicit, while pattern systems or pattern languages that are intended to be used by others will also draw on patterns regarding the understanding, knowledge, and culture of their putative audience(s). This emphasises the fact that, like the pattern representations they contain, pattern systems and pattern languages are synthetic. Indeed, we can consider layers of synthesis; the synthesis of individual patterns in the form of pattern representations, the deliberate and purposive synthesis of multiple pattern representations to create pattern systems, and the layering on of additional synthetic structures such as grammars, syntaxes, idioms, and contexts to create pattern languages. The different renderings of periodic tables noted earlier reflect the synthetic nature of a pattern language or system; some of their structures are direct reflections of physical phenomena while others reflect attempts to render the regularities of these phenomena comprehensible and useful. Indeed, we can consider many putative pattern systems and pattern languages that are more or less implied or explicit in the cultures and activities of groups and collectives in human society (Dearden & Finlay 2006).

In exploring pattern systems and pattern languages I will differentiate between five different approaches: speculative, normative, naturalistic, historical, and coherence.

Speculative Pattern Languages and Pattern Systems

A pattern system or pattern language that is developed before any instances are (or even can be) created from it or guided by it is clearly speculative. This is not a common approach, not least because it is hard (but not impossible) to model phenomena, practices, and/or thinking that do not (yet) exist. Speculative pattern systems and pattern languages are to be found in the world-building of fantasy and science fiction writers, particularly those that involve multiple works and extensive worldbuilding. For instance, the Star Trek and Dr Who universes and the genres of zombies and vampires all have well-developed and nuanced pattern languages regarding what things are, why they are the way they are, what they can and cannot do, and how they interact. Even when pattern systems and pattern languages are developed to model phenomena that we are unsure whether they actually exist (such as dark matter or superstrings) they can still draw on knowledge and understanding of extant phenomena. Similarly, in worldbuilding, the new may well be based on variations on existing phenomena. Speculative pattern systems and pattern languages can be used to create something entirely new, but they can also draw on or interact with pattern systems and pattern languages of everyday life (such as in the Harry Potter and James Bond franchises). While noting the characteristics of speculative pattern systems and pattern languages and the possibilities of developing and exploring them as such, they are not my main area of interest in this volume.

Normative Pattern Languages and Pattern Systems

A second approach involves making pattern thinking explicit in an existing domain of thinking or practice with the expectation that the resulting pattern system or pattern language will be used to define standards and guide practice. Rather than seeking to capture "how things could be," the focus is on "how things should be" (Manns & Rising 2005) and these pattern systems and pattern languages are developed to share and guide design practices, to serve as lexicons or reference works, to capture organisational or community memory, and/or to serve as the basis for coherent design conversations (Dearden & Finlay 2006). Given that their authors typically aspire to defining ideals and "best practices", these are *normative* pattern systems or languages. Because the vast majority of pattern systems and pattern languages to date have been normative in nature, it is not surprising that normative thinking has dominated pattern language thinking and practice.

Iba (who developed a number of pattern languages) outlined three approaches to pattern language development and application: the first pattern languages focused on physical forms (i.e., architecture); the second focused on non-physical forms (i.e. software design), and the third focused on human actions (such as education, innovation, and communication) (Iba 2017). Finidori et al. (2015) categorised first generation pattern languages as being focused on "why" questions, the second generation on "how" questions, the third on "what" questions, and the most recent as being focused on dynamic human systems (social change) that combined the why, how, and what foci of earlier pattern languages. In another context we might build on this to consider an analytical history of pattern languages, but I will leave that for others to pursue.

The two things that unite these different approaches is their acknowledgment of the pioneering work of Alexander and a focus on improving and optimising practice.

This reflects a strong ideological, ethical, aesthetic, and in some cases even a moral and political focus of their authors on the anticipated uses and impacts of their pattern language work. However, although many authors of normative pattern languages have approached their work with the expectation that their language will somehow transform mainstream thinking and practice, outside of software engineering, this has not been substantially realised. Late in his career, Alexander observed:

> When I started out twenty-five, thirty, years ago, I really thought that I would be able to influence the world very fast. Especially when I got to the pattern language. I thought, boy, I've really done it. This is going to work. No problem. The patterns are self-evident and true. They will spread. And, as a result, the world of buildings will get better. Hey presto. But it hasn't yet worked out like that. In practical terms, so far, I've done almost nothing ... I believe that the cultural process of influence is simply too slow to be able to take care of this problem.
>
> *(Alexander 1996)*

If a normative pattern language represents the putative best practices of a domain then experienced practitioners will already be aware of them and, therefore, find little new in them. Similarly, those learning a domain have many ways of acquiring knowledge open to them (such as training programmes and textbooks) that would also reflect (more or less) the implied pattern language of that domain. A formalised pattern language would only add value if it afforded better instruction or learning opportunities (as it seems to have done in software engineering). Indeed, it may be that, although our thinking is intrinsically patterned, we resist constraints to our pattern thinking, preferring our own idiosyncratic approaches over formalised and structured approaches such as pattern systems and pattern languages. There might be more value in pattern language development as a way of analysing a domain than in supporting practice in that domain. Pattern systems and pattern languages might be useful as thinking tools for the individuals developing them, but rather less useful or interesting to others. Yet again, more research is needed.

As much as normative pattern systems and pattern languages tend to be couched in terms of ideal and affirmative pattern representations, there has been some acknowledgement of the practical realities of their use in practice, for instance, in the development of the concept of "anti-patterns" where:

> An anti-pattern has an overall negative effect, by producing more problems than it solves ... there is also a type of anti-pattern that is a fake solution; one which seems to solve a problem while actually making the situation worse.
>
> *(Leitner 2015, p. 73)*

This would seem to be an important addition to the design (normative) pattern repertoire (albeit one that has not been widely adopted), particularly for those considering the pattern thinking of real world activity rather than some ideal version of it. After all, a pattern representation may be used for other reasons than as a solution to a particular problem and it may yield far from optimal results. A bigger issue from a pattern theoretical perspective is exactly what it is that pattern representations in normative pattern systems and pattern languages represent. For instance, the implied pattern systems and languages of that collective might be analysed as a way of understanding convergence in the pattern thinking of individuals engaged in collective action. Indeed, without pattern thinking convergence or at least complementarity then collective action would be hard to achieve or sustain.

Consider a musical group with members who play different instruments (drums, bass, guitar, keyboards, saxophone, etc.). Each musician necessarily has different patterns and pattern thinking appropriate to their instrument that includes (or should include) how they can play with other instruments (pattern complementarity). When they come together as a band to play the same songs, there should be convergence around the patterns all of the musicians draw on in performing the song (melody, harmony, rhythm, idiom, etc.). There should be convergence (although not necessarily identity) in the shared pattern thinking of people who perform particular roles and complementarity in the thinking of those who perform different roles alongside others within the same activity, such as in sports teams, healthcare, and higher education. Indeed, I would argue that pattern complementarity is to be found in any collaborative activity that involves specialism and/or division of labour because without it there could be no meaningful collaboration. I raise this point in the context of normative pattern systems and pattern languages because they typically represent convergent and complementary shared pattern thinking (or at least they attempt to do so).

One could develop a normative pattern system or pattern language that represents the pattern thinking of a single individual. However, normative pattern system and pattern language development has focused on collective knowledge and practice and as such the patterns that these systems and languages represent are an implied aggregation of pattern thinking across many minds rather than being extant in any one mind. This presents a challenge for pattern system and pattern language developers as they try to represent something that does not exist beyond the implied patterns and pattern thinking of those individuals engaged in a collective enterprise. Indeed, it is perhaps doubly challenging because the authors of normative pattern systems and pattern languages seek to capture the optimal or best aspects of this implied collective pattern thinking, a double extrapolation that could severely tax methodological rigour. A naturalistic representation of implied collective pattern thinking on the other hand needs only to encompass the implied nature of collective pattern thinking.

Naturalistic Pattern Languages and Pattern Systems

A *naturalistic* approach to pattern systems and pattern languages involves making explicit the implied pattern thinking in an existing domain as a way of understanding or modelling it as-it-is rather than as-it-should-be. Focusing on what actually happens (or what has happened) rather than what should happen (or should have happened), these pattern systems and pattern languages may reflect negatives, conflicts, waste, banality, irrationality, misunderstandings, and other suboptimal expressions that would not be a part of normative pattern languages and pattern systems (except perhaps as "anti-patterns"). A naturalistic approach should, therefore, reflect a more balanced appraisal of existing thinking and practices than a normative approach, and as such it is well placed to explore possible weaknesses, harms, and deficiencies in collective pattern thinking. That said, I have been unable to identify any naturalistic pattern languages outside of my own work and that of my colleagues, and as such the following exposition is based on my own sense of the potentials and possibilities of this kind of work rather than on established practices.

Given that all pattern languages and systems are synthetic and artefactual, the authors of naturalistic pattern languages and pattern systems need to be attentive to the implied nature of the patterns and pattern thinking they are seeking to represent. While this is true to some extent for all pattern system and pattern language developers, there are particular opportunities and challenges in pursuing normative pattern analysis. On one hand, the developers of normative pattern systems and languages must extrapolate the implied pattern thinking of a collective and then identify its best or most optimal expressions, while naturalistic pattern system and language developers need only extrapolate the implied pattern thinking of a collective. On the other hand, normative system and language developers can fall back on the problem-solution metanarrative of design patterns while naturalistic system and language developers have no such recourse. Indeed, normative problems are not implied, they are concrete practical problems that are faced by those working in the domain. As Alexander (1979) argued "the problem is real". A normative approach, therefore, reflects a realist perspective; it leans toward real and concrete things rather than ideas or abstractions. Naturalistic pattern languages and pattern systems represent how people think about things and how they organise and act in response to their thinking, and therefore lean more towards constructivist and interpretivist perspectives. Such thinking and action may include problem-solution dyads, but they are not constitutive of pattern languages and pattern systems. While normative pattern representations are based on problem-solution dyads, naturalistic pattern representations may take whatever form is appropriate both to the domain and to the purposes of those analysing that domain.

Alexander argued (and I agree) that pattern languages are inevitably provisional and subjective in that they can change as understanding changes and as different developers or users prioritise or organise pattern representations differently. I would argue, therefore, that naturalistic pattern languages and pattern systems are theories rather than guides, audits rather than templates. They represent their authors' understanding of the domain they reflect. While they describe, explain, and/or predict, they are also inescapably tangential and provisional. It I also important to note that naturalistic pattern systems and languages need not be useful in an instrumental sense, only in the sense that they contribute to knowledge and understanding.

Not only do all pattern languages and pattern systems have theory-like characteristics, normative approaches are also necessarily grounded in the social sciences, not least because they deal with the pattern thinking of collectives. I would argue, therefore, that naturalistic pattern languages and pattern systems should be developed with a theoretically grounded sense of the kinds of representations that they employ, something that is not well developed or even needed in many normative approaches. This in turn suggests the use of methods in developing naturalistic pattern languages and pattern systems that are more attentive to hermeneutic, social constructivist, and other theoretical orientations to knowledge and understanding. This includes attending to the implications of rendering an implied social system, the roles and reflexivity of the analysts/authors, and the knowledge claims that they make.

For all the differences between normative and naturalistic approaches I have described, they do share a common procedural and conceptual landscape. For instance, pattern languages and pattern systems only make sense in the context of human action or human thought, primarily but not exclusively collective action and thought. You could not develop a pattern language of the behaviour of bismuth or tidal currents, but you could develop a pattern language of how and what scientists think about the behaviour of bismuth or tidal currents. A key principle then is that pattern languages and pattern systems cannot meaningfully reflect something that does not involve pattern thinking. This is a notable departure from understanding theory as applying directly to a phenomenon rather than to the ways in which we think about a phenomenon. Even the various strands of theoretical philosophy focus on how we think in general rather than on how we think about a specific phenomenon or domain. We might, therefore, provisionally consider pattern representations and their collection in pattern languages and pattern systems as metatheoretical in nature. Each pattern representation can be understood as a metatheory of one or more theories within a particular domain. Each pattern language and pattern system can be understood as a collection of pattern representation theories that collectively represent broader metatheories of the theories of practice or action in a domain.

Historical Pattern Systems and Languages

Another common characteristic of all pattern languages and pattern systems and their constituent pattern representations is that they should be understood as being partial and approximate representations. Not only may better theories be found, even the ones we have are generally representative of a collective body of pattern thinking rather than an individual one. As I noted earlier, there is no reason why pattern languages or pattern systems that reflect an individual's pattern thinking relative to a particular domain could not be developed, or that, depending whose views are included or excluded on behalf of a collective (or when or how), that the resulting pattern systems or pattern languages may diverge. A consequence of this approximation is that the translation from a pattern system or language to the pattern thinking of individuals that interact with the pattern language or system will (as with every other interaction or experience) be woven into their individual patterned minds. Alexander reflected on these individual aspects of pattern languages, arguing that without them a pattern language does not and cannot "live":

> A language is a living language only when each person in society, or in the town, has his own version of this language ... to reach this deeper state, in which each person has a pattern language in his mind as an expression of his attitude to life, we cannot expect people just to copy patterns from a book ... a living language must constantly be recreated in each person's mind.
>
> (*Alexander et al. 1977, pp. 337–338*)

This reflects ideas of heuristic sufficiency; our patterns and the things we make from them do not have to be exact or precise, they need only be sufficient to the task at hand. This sense of imperfection, approximation, and transitiveness of pattern representations, languages, and systems again reflects hermeneutic and heuristic principles.

We should also, therefore, consider the temporal framing of practice in a pattern system or pattern language. Most pattern languages reflect a snapshot of thinking and practice bounded by the time during which the language was being developed. Longer-living languages might go through multiple revisions over time, or their authors may attempt to organise a pattern language as a living representation of practice through folksonomy techniques such as using wikis. However, the vast majority of extant pattern languages (almost all of which are normative) seem to have been one-off developments. Alexander continued to develop his ideas (for instance, in his multi-volume "The Nature of Order") but there was no second edition of Alexander et al.'s "A Pattern Language", while Gamma et al.'s "Design Patterns" is still in its first edition (although with several reprints) three decades after its original publication. It might be argued that these (and all the other) one-off pattern languages simply got it right the first time, and it may be so, but this hardly reflects Alexander's principles of collaboration and learning, nor does it reflect the shifting nature of human understanding and interest.

Other than the effort involved, there is no good reason why pattern systems and pattern languages might not be developed to reflect pattern thinking in the past. However, such a pattern system or pattern language would not be normative other than as a reflection of its historicity, nor strictly would it be naturalistic as the patterned minds are only accessible through the artifacts and records that they left behind. We might, therefore, consider historical pattern systems and pattern languages as a fourth approach. To an extent, systems and languages that are synthesised from the research literature over decades (or any other transtemporal source) are historical in nature even if this is not acknowledged or explicitly factored into their development (see Chapter 8 for more on this issue).

A historical approach (perhaps drawing on concepts and approaches from the field of historical linguistics) could also include pattern systems and languages that reflect how pattern thinking has changed over time. After all, we might expect that some patterns would appear while others disappear, some might split, some might merge, and some may shift their meaning. This would only make sense for a body of practice that has existed in a recognisably coherent form over an extended period of time, such as medicine, mathematics, or the law. Newer domains, such as computer programming, genetics, or social media are likely to have less historical depth to plumb or represent, although one might prospectively capture as much about the development and evolution as possible as material for a future pattern system or pattern language.

Completist Pattern Systems and Pattern Languages

A fifth kind of pattern system and pattern language approach is a *completist* approach. A naturalistic approach may identify many shared patterns and render them as pattern representations based on what is discussed in a particular field. However, it could not capture patterns that are shared but are not discussed, that are tacet or assumed, or that are socialised on entry to a field but not mentioned thereafter. There are parallels here to Hall's concept of high-context cultures where patterns are shared through mutual understanding and cultural participation but rarely communicated explicitly (Hall 1990). A completist approach, therefore, focuses on making the implicit shared pattern thinking of a domain explicit. This work is likely to be challenging as it involves working with hints and absences. Nevertheless, a completist approach can draw on logical completeness (patterns that must be present to complete the problem-solving raised in other more explicit patterns), insider knowledge (participants in a domain drawing on their experiences), and patterns that are hinted at or implied.

A completist approach might be used on its own or used with other approaches. If it is used on its own, then the result is likely to be a collection of representations of tacet shared patterns. If a completist approach is used to complement other approaches, then it can help to produce a complete pattern language or pattern system which would otherwise be incomplete or less complete. When used with other approaches a completist approach should follow their focus, i.e., it should be normative when combined with a normative approach and it should be naturalist when used with a naturalist approach. I would also note that working with tacet shared patterns may require rather more interpretation than working with more explicit patterns and pattern thinking. Given that investigator pattern thinking must play a role in this kind of structured inquiry then we can productively understand such pattern analyses as reflecting Gadamer's concept of interpretive horizons, such that meaning emerges from the interaction of two (or more) conceptual systems.

A Repertoire of Pattern System and Pattern Language Approaches

Rather than treating different pattern system and pattern language approaches as distinct and exclusive solitudes, we might consider combining different approaches in useful and productive ways. For instance, while normative approaches focus on regularities in the practicalities, quandaries, and problems in a domain, naturalistic approaches might be used to focus on the pattern thinking that both generates and subsequently responds to those practicalities, quandaries, and problems. Different pattern system and language approaches might also be combined such that a normative model affords insights on the development of a naturalistic model or vice versa. A naturalistic approach might become more speculative- or normative-like if used prospectively rather than retrospectively or reflectively. Rather than relying on my rather blunt typology, we might instead consider specific pattern languages and systems as being situated in shared conceptual space. In Figure 6.4 I have set out one (but not by any means the only) way of modelling this based on three continua: emic/etic, normative/naturalistic, practical/theoretical such that a "pure" normative approach could be understood as being emic-normative-practical while a "pure" naturalistic approach could be understood as etic-naturalistic-theoretical. This kind of mapping could help to conceive of many more approaches that constitute some hybrid of different pattern system and pattern language approaches. Similar hybrids might be explored combining speculative and/or historical approaches with normative and naturalistic approaches.

This model also raises questions regarding the situatedness of the pattern analyst(s). For instance, how much knowledge and experience does the analyst bring to the process in developing pattern representations, pattern systems, or pattern languages? A pattern analyst may already be a part of a community of practice or professional group and familiar with the pattern thinking that they seek to analyse. An analyst of this kind may "hit the ground running" but they may also be biased as a result of their participation, both in terms of what they pay attention to and what they do not, and in terms of their assumptions and ideological positioning in relationship to the collective they participate in. This reflects an emic perspective, one where the perspectives of the group or community delimit the study and where no generalisation is sought beyond the group or community. On the other hand, a pattern analyst who comes a professional group or community as an outsider with little or no experience of their shared practice has more work to do to access and understand the shared pattern thinking but may be more objective and critical in their analyses, although they too can have their own paradigmatic or professional biases. This reflects an etic approach, one where the perspectives of the group or community are taken

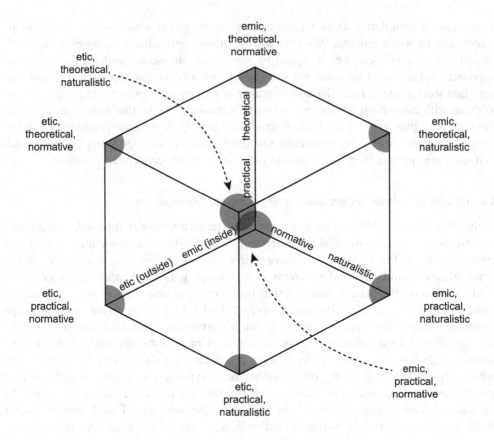

FIGURE 6.4 A conceptual space for modelling and comparing different pattern language and pattern system approaches.

Diagram prepared by the author.

as examples of a broader social reality and where generalisation is sought beyond the particular groups or communities studied. It would seem that most normative pattern language development has been emic in nature, such as architects building architectural pattern languages, computer scientists building pattern languages for use in their community and should be appraised in that light. Although etic pattern analysis does not seem to have featured in the provisional pattern sciences I have outlined, it might be that a more generalised pattern theoretical approach provides the conceptual context for developing such approaches.

We might also ask whether we might apply the concepts of pattern systems and pattern languages (or something like them) to our patterned minds. Certainly, patterns relative to the same contexts or phenomena are likely to be connected because they coincide through experience. This means that we are very likely to have groups or clusters of patterns relative to certain contexts (such as social behaviour in the workplace) such that associated patterns are more likely to be invoked in their particular kinds of context than in others. Perhaps "system" is not the right term here though as there need not be any systematicity in their association other than that of a shared context. Given that external pattern languages add grammar and syntax to a pattern system, the cognitive equivalent is perhaps more systematic in that there are rules and idioms for how patterns in the language work together. We might also expect to see more of a blurring between cognitive pattern systems and languages given that individual process elements may be shared and invoked more fluidly than in externalised forms.

What makes something a "collection" is also a more challenging concept for cognitive pattern systems and languages. After all, our patterns associate to reflect our experiences and thoughts regarding coincidence and connection between things without our having to make them do so. This does not mean that associations between patterns in a system or language are never deliberate; patterns may be acquired or built together relative to a particular context, for instance, in learning subjects at school. The analogy between externalised and cognitive pattern systems and languages is also stretched when there are no hard or definite boundaries to our pattern associations. At best we might say that patterns in a cognitive system are more likely to be invoked together.

If individual pattern expressions reflect individual patterns that are then associated by including them in an external system or language, then the basis for collecting the pattern representations together is based on criteria such as logic, empiricism, or arbitrary decisions of the authors. If an external pattern system or pattern language is built to reflect a cognitive pattern system or language, then both the component patterns and the pattern system as a whole are representations of pattern thinking. There are ontological and methodological consequences of this that I explore in the next chapter.

In summary, there are many possible approaches to developing pattern systems and pattern languages, each with its strengths and limitations. Rather than seeing the five kinds of pattern systems and pattern languages I have described as distinct and mutually exclusive, we might be better placed to consider this as a faceted conceptual space such that any given pattern system or pattern language may employ and combine different approaches to suit their needs and goals. The provenance of any resulting pattern systems or pattern languages should be clearly described, and their knowledge claims appraised in this light.

Pattern Creativity

I had originally intended to discuss *pattern creativity* as one of the many modalities of pattern thinking in the preceding chapter. However, so much of what I needed to say on this topic depended on having defined and explored pattern expressions and representations. Having now provided this outline, I can now more substantially consider the concept of pattern creativity. Indeed, it has been argued that pattern is intrinsically wound up with creativity (Mattson 2014).

Creativity would generally seem to be considered a good thing to have, and it would seem that there are many of us who seek to be more creative as reflected in the large quantity of books, videos, and other resources on the subject of creativity. In many contexts a lack of creativity is considered a serious flaw, not just in the work of artists, musicians, and writers, but also in the work of scientists, engineers, doctors, and leaders (Amabile & Khaire 2008). The breadth of applications of creativity is reflected in the many debates as to what creativity is and how to judge whether someone or something is or is not creative. While expressions of creativity may look very different in different contexts, we can consider common characteristics of creativity whatever the context or mode of expression. For instance, Simonton argued that:

> A creative idea must be original and useful and surprising. Each separate criterion is thus rendered necessary but not sufficient.
>
> *(Simonton 2018)*

Let us consider these criteria; originality, utility, and the ability to surprise. First, I would differentiate between originality as novelty (I have not seen anything quite like this before) from originality as the deliberate creation of novelty (nobody has done anything quite like this before). The former is simply the sense we get when we cannot match a pattern to current perception. The latter is a form of pattern recognition in that we recognise that the thing has been deliberately created and the novelty of the specific creation is modified thereby. Indeed, I would argue that a perception of creativity can emerge from the tension between recognising that something has been created (and perhaps recognising the signature of the creator based on their past work) and the discovery of what it is that differs from previous works.

Utility seems questionable as a criterion for creativity as it seems to be an altogether separate issue. While creativity and utility can be combined (particularly in the many aspects of industrial, product, and graphic design) there is much creativity that is not particularly useful beyond the pleasure it might give to its makers and to others. Instead, I would suggest that creativity depends on a connection between the creative act or what it produces and the perceptions of those who witnessing or interact with the act or product. This reflects noticing creative acts and the significance thereof, which in turn draws in Simonton's third criterion of creativity needing to be surprising. Creativity is perceived in acts and products that our patterned minds notice, that we find significant because of the patterned way we experience and interpret them. This in turn raises the issue of who it is that decides that something or someone is creative. Presumably different people may perceive the same action or product differently in this regard, some perceiving substantive creativity while others perceive a lack thereof. Anyone who has read a review of a book, play, film, or musical performance will likely recognise this variability. Interestingly, Simonton addressed this point in contrasting personal creativity (small "c" creativity) with consensual creativity (big "C" Creativity):

> Personal creativity ... [is where] individual creators are basing the creativity assessment ... on their own subjective experiences during a given episode ... consensual creativity does require the assessments of others besides the idea's creator.
>
> *(Simonton 2018)*

The creative individual makes interpretations and abstractions of certain aspects of their own pattern thinking through their actions and productions. In doing so they attract attention to their pattern thinking such that it is scrutinised and even commodified by their audiences. This would seem to be an asymmetric relationship both in terms of numbers (one creative individual interacts with many audience members) and in terms of depth (the creative individual bares their pattern thinking but the patterned response of their audience is limited to applause and/or attendance or reward). There may be some interplay of pattern thinking between creative individuals and their critics but the creative pattern thinking of the audience remains largely undisclosed. It might be a mistake, therefore, to only focus on those relatively few individuals whose creative pattern thinking is sought out and rewarded, but if we accept the criteria of novelty, utility, and being surprised then creativity can only be realised in the relationship with an audience and only those individuals who have an audience can be considered creative.

Rather than agreeing that creativity is only realised in the relationship between creator and audience, I think we also need to consider what it is in an individual's or collective's pattern thinking that is creative independent of any attributing or validating audience. To that end, I would suggest three areas in which a pattern theoretical perspective can be used to describe or anticipate different kinds of creativity: creativity in pattern thinking, creativity in expressing patterns and pattern thinking, and creativity in reacting to creativity.

Creativity in Pattern Thinking

Creativity in pattern thinking is about how an individual forms, associates, and otherwise draws on their patterns and pattern thinking. Our patterned minds are constantly involved in low level and involuntary creativity, for instance, as the basis of our moment-to-moment understanding. More actively pattern thinking can include creative association (connecting or associating patterns in new and unexpected ways, such as in free association), creative dissonance (connecting or associating patterns in dissonant or paradoxical ways, such as in the use of metaphor), and creative stacking (using one pattern or group of patterns to interpret or reinterpret other patterns, such as in translating ideas between genres or idioms). Creativity in pattern thinking may reflect possible future things (such as social developments or innovations) and imaginary things (such as stories and magical thinking). Indeed:

> While the ability to reproduce perceived patterns in drawings and maps is robust in humans, so too is the ability to create new patterns of entities that may or may not exist in the real world.
>
> *(Mattson 2014)*

Creativity in pattern thinking is not just a matter of fanciful imagination, it can also play an important role in scientific, critical, and analytical thinking. For instance, Geoffrey-Smith suggested that:

> Great scientists are opportunistic and creative, willing to make use of any available techniques for discovery and persuasion. Any attempt to establish rules of method in science will result only in a straitjacketing of this creativity.
>
> *(Geoffrey-Smith 2003, p. 111)*

There are echoes of this in Dewey's (1929) exhortation that "every great advance in science has issued from a new audacity of imagination". Schank outlined a more comprehensive sense of pattern creativity in the context of reasoning and problem-solving, ideas that align with pattern reasoning as much as they do with pattern creativity:

> We find ourselves wondering why something has occurred. We look for a set of beliefs or rules that would explain this event. But, it often happens that we don't have such rules. (If we did, there is a good chance that we would have been able to predict this event in the first place.) So, the next step is to attempt to find rules from some other

domain that might fit the case at hand ... [these] generalizations from only a few instances, far from being a bad thing, are actually quite significant with respect to the issue of creativity. Creativity requires the ability to make an explanation, especially a pattern-based explanation, where we are seeking new knowledge rather than correcting misinformation.

(Schank 1986, pp. 47–48)

Does this kind of creative thinking reflect Simonton's criteria for creativity such that it is original, useful and surprising? Originality is a difficult concept in the context of pattern thinking given that all of our patterns and pattern thinking is built from individual experience. Can our minds ever create an entirely new pattern nexus that is not based on or connected to any other nexuses or parts thereof? Originality in this context is, therefore, about weaving new memories and associations into the fabric of a patterned mind. What about "utility"? I have argued several times that patterns are developed because they are useful in perception, understanding, implication, and so on. A pattern we hold does not have to be useful in an instrumental sense only in the sense that it can contribute or has contributed to our lives in some way at some point in time. This reflects Damasio's argument that pattern thinking can be banked against future use as well as being useful in the moment:

When we relate and combine images in our minds and transform them within our creative imaginations, we produce new images that signify ideas, concrete as well as abstract; we produce symbols; and we commit to memory a good part of all the imagetic produce. As we do so, we enlarge the archive from which we will draw plenty of future mental contents.

(Damasio 2021, p. 47)

As to "surprising", if we take the term to include aspects of noticing, significance, and tractability, and so on, then we might say that creative pattern thinking has the potential to be surprising. However, it may be creative and yet not surprising in that such thinking elicits different emotional responses, such as relief at a sense of danger or uncertainty relieved when what seemed to be a threat is recognised as benign. I would, therefore, argue that deliberate pattern creativity within a patterned mind is novel in a constructivist and expansivist sense, that it has utility albeit possibly in a more or less abstract sense, and that it can have significance but not necessarily in ways that are surprising. While all such thinking may be an end in its own right, the question of utility begs the question as to what it is we do with all this creative pattern thinking.

Our minds are not cameras, they do not register let alone store everything they are exposed to. Rather we filter what we perceive and what we subsequently cast into memory. I would argue that there can be creativity in this selection and weaving of memory every bit as much as there can be in recalling and associating memories. Creativity in pattern thinking may be spontaneous or it may be pursued more deliberately, or it may involve some combination of the two. Of course, spontaneity in creativity may simply reflect pattern thinking that is faster than the conscious mind is aware of or can follow, and this may reflect a particularly active or energetic capacity or disposition for pattern thinking and association. Was Coleridge simply reflecting his capacity for creative pattern thinking when he is reported to have observed:

... against my better interests I was carried away with an ebullient fancy, a light and dancing heart, and a disposition to catch fire by the very rapidity of my own motion, and to speak vehemently from mere verbal associations.

(Nicolson 2020)

Deliberate creativity can be pursued as a practice, but, as with other expressions of systems 1 and 2 thinking, deliberate creativity (system 2) can be conditioned (for instance, through training or mindfulness) such that it becomes both reflexive and expert (system 1.5). The ability to 'think things through' is an example of this; pattern creativity involves engaging in speculative pattern association to explore its implications. In other words, advanced pattern thinkers can to an extent simulate and model their own pattern thinking and to an extent that of others.

Creativity in Expressing Patterns and Pattern Thinking

We can also consider the role of creativity in the ways we express our patterns and pattern thinking in guiding our actions (performances, makings, crafts) and productions (inventions, artefacts, creations). After all, this is where much of the lay sense of pattern seems to lie; pattern as something that is used to create or to guide the creation of something else, and pattern as a characteristic of something that has been created. This sense of thought translated into action is reflected in Shermer's description of pattern thinking (or rather his concept of "patternicity") that involves both thought and production:

Creativity involves a process of patternicity, or finding novel patterns and generating original products or ideas from them.

(Shermer 2011, p. 123)

As with any other pattern expression, the performances, actions, and products that are guided by patterns and pattern thinking of necessity reflect the patterned minds of their creators. However, there can be creativity in the ways in which these expressions are brought about. French Impressionists painted mundane subjects but in remarkably creative ways; jazz musicians play the same songs over and over but breathe creativity into every unique performance. Many theories of creativity have focused on how creative individuals express themselves, the audiences for their creativity, and the contexts in which their creativity is expressed and experienced (Rhodes 1961; Glăveanu 2013). Wallas' (1926) model of creative problem-solving is an interesting exception in that it described creativity as starting with preparation (understanding the problem), then incubation (indirect thought, rumination), illumination (insight), and finally verification (testing the new idea). The four stages cycle until a verifiable solution is "found".

We might instead describe creativity of pattern expression in terms of phase changes: disposition into thought, thought into action, and action into production. The first, disposition into thought, is about developing or articulating a creative mindset, one that is open to and even trained to creative thought. You might argue that this should be a part of the previous section, and I would not disagree in principle. However, I include it here both as a connection to the previous comments on creative pattern thinking and as a necessary precursor for of thought into action and production. By disposition I mean pattern attention, pattern mindfulness, the development and exercising of seeking out pattern experiences and processing them in the interests of future creative action. This is a necessary part of being a creative person, whether a designer, artist, scientist, or some other role, as they all need deliberate engagement with and drawing upon their respective pattern repertoires. For instance, Alison (2019), in presenting a series of design patterns for crafting written narratives, argued that authors need to be "alert to patterns" not just in their own writing but in the world around them so they can assimilate new pattern ideas in their work.

The second phase change, thought into action, is reflected in the concept of *design*, the planning and conceptual shaping of creative acts. Design is about turning ideas into plans and pre-emptive problem-solving within those plans. And yet, as with so many concepts in this book, design has many meanings in many different contexts. We might draw on Owen's (2007) model of "design thinking" that he contrasted with "science thinking" using two orthogonal axes. The first axis reflected the process of creativity, ranging from analytic (creative discovery and explanation of the unknown) to synthetic (creatively synthesising what is known in order to create and invent things), The second axis reflected the contexts of creativity, ranging from symbolic (abstract, theoretical) to real (concrete, applied). Placing these axes orthogonally, within this model for thinking about different kinds of creativity Owen argued that scientific thinking was symbolic-analytic, medical thinking was analytic-real, artistic thinking was synthetic-symbolic, and design thinking was synthetic-real – see Figure 6.5. This typology notwithstanding, it is important to note first that design can also play a role in science (such as designing experiments), in medicine (such as coming up with a therapeutic plan), and in the expressive and plastic arts (such as planning the practical aspects of a piece of work or performance), and second that there are many ways of moving from thinking about a creative act to operationalising those thoughts, all of which can involve pattern thinking. Design pattern thinking includes repertoires of "how to patterns" as many pattern systems and pattern languages illustrate, (Lidwell et al. 2003; Martin &

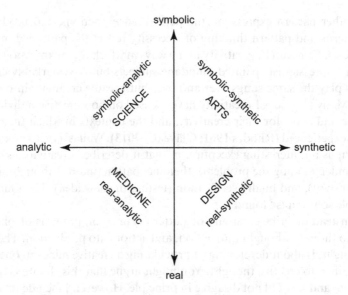

FIGURE 6.5   A biaxial model of approaches to creativity (after Owen 2007).

Diagram prepared by the author.

Hanington 2012; Alexander et al. 1977; Gamma et al. 1995) as well as associative and implicative aspects of pattern reasoning and thinking more generally.

Design may involve (and, depending on the domain in which it is conducted, it may require) the creation of plans, blueprints, engineering diagrams, and maps. As such, design can be a creative undertaking in its own right as well as a means to guide subsequent creative activity. Different design media reflect different patterns that in turn reflect different conceptions and conventions of representation and meaning. A blueprint or template is not itself a pattern (although I observe that many people would argue that it is exactly what a pattern is) but it is a pattern representation (or possibly a pattern system if it is more complex and variable) in that it captures using a design process the thinking and intention of the designers in what it will be used to produce. I should, therefore, also include pattern representations, systems, and languages in this category, as they too involve creative acts that generate artefacts.

The third phase change, from action to production is where practical problem-solving 'craft' skills of enacting and building things comes in. Again, this reflects pattern thinking in both its regular forms and its more creative forms (the difference being an arbitrary one based on originality, utility, and surprise as described earlier) but in different ways according the medium, domain, creators, and other factors. Something might be created with little design or planning based on the creator's behalf, for instance, in improvised dance or music, or it might be closely or directly based on a pattern representation, language, or system, such as in engineering or architecture. Either way, pattern thinking is involved, not least in the training and continuing professional practice of creative people. For instance, although improvised music or dance is not designed as such, the performers may have had years of pattern-based training and experience that they draw on as they improvise. If there is a deliberate or substantive connection between a person's pattern thinking and that which they create, then what they produce is by definition a pattern expression. I would argue, at least based on my own experiences as a musician, artist, and writer that pattern thinking is always present in acts of creation, indeed that creation could not occur or be meaningful as such without pattern thinking. I should also note that any creative act may be an end in itself, a means to an end, or some combination of the two.

Creativity in Reacting to Others' Creativity

A third aspect of pattern creativity can be found in the pattern thinking of those interacting with and reacting to those things that creative individuals or groups produce. There may be creative pattern recognition in terms of what the thing is, and what styles, idioms, or creators were involved. There may be pattern development in terms of expanding, adapting, and elaborating existing pattern understanding to accommodate the new. And there may be pattern thinking more broadly in terms of assessing similarities and differences between the current thing and experiences of past similar things, and in terms of connections and associations, for instance, between schools or groups of creatives or experiencing unexpected or unfamiliar (hence surprising) in terms of connections and associations.

Shimamura (2013) differentiated between the art of seeing where "a coherent spatial world is created from a jerky, ever-moving jumble of images", the art of knowing which involves understanding the current situation based on prior experience, and the art of feeling that involves weaving, stimulating, exploring, and responding to emotional triggers and state changes triggered by aesthetic experiences. This parallels MacCallum's typology of how art is understood: its physical matter, its conceptual matter, and its emotional matter (MacCallum 1930). All three reflect aspects of pattern thinking; seeing or otherwise interacting with form is about making sense of available stimuli, knowing or conceptualising is about understanding what the stimuli mean and how they relate to previous experience, and emotion is tied to meaning and significance. That the experience of others' creativity is patterned is also reflected in Bateson's observations that we perceive pattern in the specifics of creative expressions things and in its associations and connections such that we need:

> ... to have a conceptual system which will force us to see the "message" (e.g. the art object) as both itself internally patterned and itself a part of a larger patterned universe.
>
> *(Bateson 1972 p. 132)*

Our responses to the creativity of others are both patterned and creative in their own way. This is reflected in critiques of shows and performances, and in the analyses of influences or meanings detectable in the work of others. Indeed, scholars like Barthes and Baudrillard built their careers around such activity. Although some individuals may do this critical work with greater skill (and reward) than others (such as Barthes, Baudrillard, and Hitchens) creative criticism is something we all engage in from time to time. Not only do we read pattern expressions in terms of what they mean and how they make us feel, we often seek to understand the intentions, beliefs, and character of their creators.

As much as there is a particular focus on art, literature, and performance arts (music, dance, theatre, cinema) in much of this work, it applies every bit as much to buildings and communities, to organisations and institutions, and to cultures and societies. As Shimamura argued, the aesthetic experience is universal:

> Aesthetics is not tied to art. I have defined aesthetics as a hedonic evaluation based on pleasure or interest. As such, we apply aesthetics on a daily basis as we are always determining what we like or don't like or what is interesting or not. In an evolutionary sense, one could argue that aesthetic responses are essential as survival depends on our ability to determine what is good or bad for us.
>
> *(Shimamura 2013, p. 260)*

I note my earlier comments about developing pattern representations, systems, and languages by observing actions and/or products in a particular domain and reverse engineering or interpreting the pattern thinking that created them. These too reflect hermeneutic fusion of the pattern thinking of the creators and participants in the domain of interest and the creators of the pattern representations, systems, and languages. Pattern languages and systems are effectively third-party representations of others' pattern thinking as much as they are representations of their authors' pattern thinking.

In summary, pattern creativity is something we all have, that we all use, and that we all depend on, both individually and collectively. As Damasio observed:

> We, owners of the patterns, can mentally chop them in parts and rearrange the parts in myriad ways to yield novel patterns. When we attempt to solve a problem, reasoning is the name we gave to the cutting and moving about that we engage in as we pursue a solution.
>
> *(Damasio 2021, p. 36)*

Chapter Summary

As much as patterns and pattern thinking are intrinsically cognitive, we can externalise our patterns and pattern thinking, and we can do it in different ways. In this chapter I have distinguished between *pattern expressions*, *pattern representations*, *pattern systems*, and *pattern languages* as forms of externalisation, and I contrasted these with *pattern instances* as the perception of pattern in external things. A pattern instance is a cognitive phenomenon where sensory stimuli are matched to memory and perception of something is the result. An instance is, therefore, the product of mind and not a property of the thing being perceived. Although there may not logically be any one point where perception flips between being and not-being an instance, we do seem to make judgements in our perceptions that are more binary in nature than continuous. The ability to resolve pattern instances and what they resolve to varies between individuals and within the same individual over time. Perception of pattern instances tends not to be rules-based but rather seems to be gestalt- and utility-driven, and heuristic rather than analytical.

Pattern instances are perceptual moments that involve close couplings between external phenomena and a mind's perception of them, while pattern expressions, pattern representations, pattern systems, and pattern languages are somewhat decoupled from the pattern thinking that spawned them such they can exist independently of a mind. Another key difference is that perceiving something as an instance of a pattern does not materially change the thing being perceived but it can change (and very likely will change) the perception a mind has of that thing. That is not to say that the subsequent actions of an embodied mind informed by pattern perception cannot change the thing, only that perception of something as a pattern instance does not.

Pattern expressions are the result of pattern thinking guiding someone's actions such that something is created or performed in some way that reflects one or more patterns. Creators of pattern expressions are selective in terms of what pattern characteristics they include and the ways in which they are included. The pattern thinking expressed in a pattern expression may be perceived differently by others encountering that pattern expression even though a pattern expression may be used to experience or analyse the creator's pattern thinking.

A pattern representation is an expression of the thinking that constitutes a pattern in our minds. Because of the fluid and unbounded nature of pattern nexuses, a pattern representation need not reflect the whole pattern, only parts thereof. While pattern representations will tend to be more macropattern-like, there is no consistent level of abstraction or objectivity, or any other marker of rigour involved in their creation. Pattern representations are not perfect, they are lossy and transformative, and they can have many purposes.

Pattern systems and pattern languages are collections of pattern expressions. Pattern systems are simple aggregations while pattern languages include grammar and syntax as to how their pattern representations are to be used. I described speculative approaches that model anticipated pattern thinking in novel circumstances, normative approaches that are intended to capture best practices, naturalistic approaches that capture the actual pattern thinking in an existing domain as is, historic approaches that capture past pattern thinking in a given domain, and completist approaches that fill in the gaps in apparent shared pattern thinking. Although there are many differences between these approaches, they share a common procedural and conceptual landscape such that they can be understood as collections of pattern theories that collectively represent broader theories of the theories of practice or action in a particular domain.

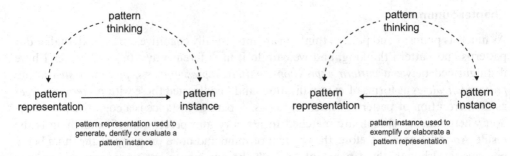

FIGURE 6.6   Examples of relationships between pattern representations and pattern instances.
Diagram prepared by the author.

Putting the definitions and ideas of externalised pattern phenomena from this chapter together, there would seem to be fundamental connections between external stimuli and pattern thought that I have called pattern instances. Then there are four kinds of deliberately externalised pattern phenomena: pattern representations (renderings of a pattern in terms of its components, associations, and variations); pattern expressions (pattern instances created or contrived based on a pattern or a pattern representation); pattern systems (collections of pattern representations relevant to a particular domain); and pattern languages (pattern systems with grammar and syntax). These are all linked through pattern thinking and we can move backwards and forwards between them with some skill albeit often with little awareness of doing so – see Figure 6.6.

Pattern creativity is a particular kind of pattern thinking, drawing on various creativity theories and the characteristics of creativity they outlined. While concepts of design and production were recurring themes and my examples were drawn from more traditionally creative domains, I noted the ubiquity of pattern creativity in all aspects of our lives. If we consider pattern thinking to be an essential part of thinking as a whole, then all of our actions that are guided by or respond to that pattern thinking might also be considered externalisations of that thinking. The question, therefore, shifts from asking how do we externalise our pattern thinking to asking when and how are we not doing so?

References

Alexander C, Ishikawa S, Silverstein M. *A Pattern Language: Towns, Buildings, Construction.* New York, NY: Oxford University Press: 1977.

Alexander C. *A Timeless Way of Building.* New York, NY: Oxford University Press: 1979.

Alexander C. The Origins of Pattern Theory, the Future of the Theory, And the Generation of a Living World. 1996 *ACM Conference on Object-Oriented Programs, Systems, Languages and Applications (OOPSLA)*, San Jose, California. Transcript: https://www.patternlanguage.com/archive/ieee.html

Alison J. *Meander, Spiral, Explode: Design and Pattern in Narrative.* New York, NY: Catapult: 2019.

Amabile TM, Khaire M. *Creativity and the Role of the Leader.* Harvard Business Review, October 2008. https://hbr.org/2008/10/creativity-and-the-role-of-the-leader

Nicolson A. *The Making of Poetry: Coleridge, the Wordsworths, and Their Year of Marvels.* London, UK: MacMillan: 2020.

Bateson G. *Steps to an Ecology of Mind: Collected Essays in Anthropology, Psychiatry, Evolution, and Epistemology.* Chicago, IL: University of Chicago Press: 1972: pp 132.

Bayle E, Bellamy R, Casaday G, Erickson T, Fincher S, Grinter B, et al. Putting it All Together: Towards a Pattern Language for Interaction. *SIGCHI Bulletin* 1998; 30(1): 17–33.

Blackmore S. *The Meme Machine*. Oxford, UK: Oxford University Press: 2000.

Buzan T, Buzan B. *The Mind Map Book: How to Use Radiant Thinking to Maximize Your Brain's Untapped Potential*. New York, NY: Plume: 1993.

Clark A. *Supersizing the Mind: Embodiment, Action, and Cognitive Extension*. Oxford, UK: Oxford University Press: 2008: pp xxviii.

Coplien J, Schmidt D. *Pattern Languages of Program Design*. Reading, MA: Addison-Wesley: 1995.

Damasio A. *Feeling and Knowing: Making Minds Conscious*. New York, NY: Pantheon: 2021.

Dearden A, Finlay J. Pattern Languages in HCI: A Critical Review. Human Computer Interaction 2006; 21(1); 49–102.

Dewey J. *The Quest for Certainty: A Study of the Relation of Knowledge and Action. Gifford Lectures*, 1929. New York, NY: Minton, Balch: 1929.

Finidori H, Borghini SG, Henfrey, T. Towards a Fourth Generation Pattern Language: Patterns as Epistemic Threads for Systemic Orientation. *PURPLSOC Conference* 2015, Krems, Austria. https://www.researchgate.net/publication/308100015_Towards_a_Fourth_Generation_Pattern_Language_Patterns_as_Epistemic_Threads_for_Systemic_Orientation

Funnell SC, Rogers PJ. *Purposeful Program Theory: Effective Use of Theories of Change and Logic Models*. Jossey-Bass: 2011.

Gamma E, Helm R, Johnson R, Vlissides J. *Design Patterns: Elements of Reusable Object-oriented Software*. Reading, MA: Addison-Wesley: 1995.

Geoffrey-Smith P. *Theory and Reality: An Introduction to the Philosophy of Science*. Chicago, IL: University of Chicago Press: 2003.

Glăveanu VP. Rewriting the Language of Creativity: The Five A's Framework. *Review of General Psychology* 2013; 17: 69–81.

Grenander U. Advances in Pattern Theory. *Ann. Statist.* 1989; 17(1): 1–30.

Hall ET. *The Silent Language*. New York, NY: Anchor: 1990.

Iba T. Generations of Pattern Language: Architecture, Software, and Human Actions. *MiniPLoP 2017, Programming 2017 conference*, Brussels, Belgium, 2017. http://web.sfc.keio.ac.jp/~iba/papers/PURPLSOC15_IbaKeynote.pdf

Knowlton LW, Phillips CC. *The Logic Model Guidebook: Better Strategies for Great Results* (2nd ed.). Thousand Oaks, CA: SAGE: 2012.

Leitner H. *Pattern Theory: Introduction and Perspectives on the Tracks of Christopher Alexander*. Spartanburg, SC: CreateSpace: 2015.

Lidwell W, Holden K, Butler J. *Universal Principles of Design*. Gloucester, MA: Rockport: 2003.

MacCallum HR. Emotion and Pattern in Aesthetic Experience. *The Monist* 1930; 40(1): 53–73.

Manns ML, Rising L. *Fearless Change: Patterns for Introducing New Ideas*. Boston MA: Addison-Wesley: 2005.

Martin B, Hanington B. *Universal Methods of Design*. Beverley, MA: Rockport: 2012.

Mattson MP. Superior Pattern Processing is the Essence of the Evolved Human Brain. *Frontiers in Neuroscience* 2014; 8: 265.

Owen C. Design Thinking: Notes on its Nature and Use. *Design Research Quarterly* 2007; 2(1): 16–27.

Rhodes M. An Analysis of Creativity. *Phi Delta Kappan* 1961; 42: 305–311.

Schank RC. *Explanation Patterns: Understanding Mechanically and Creatively*. Hillsdale, NJ: Lawrence Erlbaum Associates: 1986.

Shermer M. *The Believing Brain*. New York, NY: St Martin's Press: 2011.

Shimamura A. *Experiencing Art: In the Brain of the Beholder*. Oxford, UK: Oxford University Press: 2013.

Simonton DK. Creative Ideas and the Creative Process: Good News and Bad News for the Neuroscience of Creativity. In: Jung RE, Vartanian O. *The Cambridge Handbook of the Neuroscience of Creativity*. Cambridge, UK: Cambridge University Press: 2018

Thornton PH, Ocasio W. Institutional Logics. In: Greenwood R, Oliver C, Sahlin K, Suddaby R. (eds.) *Handbook of Organizational Institutionalism*. Thousand Oaks, CA: Sage: 2008.

Wallas G. *The Art of Thought*. London, UK: Jonathan Cape: 1926. https://archive.org/details/theartofthought/mode/2up

Winn T, Calder P. Is This a Pattern? *IEEE Software* 2002; 19(1): 59–66.

7

PATTERN PHILOSOPHY

In which I explore some of the philosophical bases of pattern phenomena and pattern theory

I struggled with the title of this chapter. The original title was "Pattern Epistemology" as my plan was to focus on the nature of pattern knowledge. However, I found that, as onto-logical and axiological issues proved so deeply entangled with pattern epistemology, I could not maintain a focus solely on epistemic issues. I changed the title to "Pattern Philosophy" to reflect this broader landscape. I should be clear though that a whole volume could be dedicated to pattern philosophy, and that "Pattern Philosophy" could just as well have been the title for this book. I hope you will forgive me in my being selective in those philosophi-cal aspects of pattern I cover here.

I have long been intrigued by the idea of similar things being grouped and thereafter treated as being essentially the same. It is certainly useful as a social scientist to say that these or those social activities or structures can be grouped despite knowing that they differ in many ways. One advantage of doing so is that it allows us to generalise across low (but not zero) variance contexts (reflecting Merton's concept of "middle-range the-ory") so long as the variances do not compromise the inferences we draw. These caveats notwithstanding, I have found it useful to use pattern systems to provide an epistemic framework for grouping, modelling, and comparing intrinsically variant things as instances of a common pattern by noting they have common kinds of characteristics even though how those characteristics are expressed may differ in every case (Ellaway & Bates 2015). This kind of pattern aggregation requires transparency as to the bases for both aggregation and distinction, and epistemic caution in that such groupings are contrived (they are not natural, we construct them for our own ends), arbitrary (what counts as a pattern instance is subjective and open to debate), dynamic (they can change over time and between contexts), and personal (every individual's patterns are unique to their minds). However, in practice this attention to similar differences and different similarities

DOI: 10.4324/9781003543565-7

can be lacking when groupings and the characteristics on which those groups are based are treated as essential and natural (independent of pattern thinking). Indeed, the nature and quality of knowledge, in particular what counts as reliable knowledge, have long been debated both within and across scientific paradigms, in various philosophies of science, as well as in lay contexts.

I should first outline what I mean by ontology, epistemology, and axiology in the context of pattern theory:

- Ontology is concerned with what things exist, how these things exist, and what kinds of things they are. An ontology of pattern phenomena might be concerned with questions such as: do patterns exist, what kinds of things are patterns, what are patterns similar to and different from, how do patterns work, and how do patterns relate to other phenomena?
- Epistemology is concerned with knowledge and knowing. An epistemology of pattern phenomena might consider different kinds of pattern knowledge, how patterns support or provide knowledge, relationships between pattern thinking and knowledge, and the reliability of pattern knowledge (how and to what extent it is believed, justified, and true). We might also consider how and in what ways our knowledge *of* patterns is distinct from the knowledge that patterns give us.
- Axiology is concerned with values and their expression in ethics, aesthetics, and evaluation. An axiology of pattern phenomena might consider how values are represented in patterns and pattern thinking, the ideological, ethical, and aesthetic dimensions of pattern thinking, and how our values have their origins in, are applied to, or are shaped by our pattern thinking capabilities.

As much as they are presented as distinct concepts, ontology, epistemology, and axiology are intimately connected and largely inseparable. For example, we might use categories to describe the ontological characteristics of different patterns, but the categories are themselves epistemological constructs, that likely reflect (or are used to reflect) various values – see Figure 7.1. I also acknowledge that the uses and meanings of ontology, epistemology, and axiology are individually and collectively contested and as such I use them as concepts to articulate an argument rather than as a commitment to a particular philosophical stance.

Pattern Ontologies

In this section I consider whether and how patterns exist, the ways in which we organise our pattern memories, the existence of patterns outside of minds, the ways in which patterns seem to work, the nature of shared patterns, and the ontologies afforded by patterns and pattern thinking.

Do Patterns Exist?

I have described patterns as nexuses of interconnected memories in our minds. This is a theory, one that explains and connects many phenomena and ideas. To that end I cannot

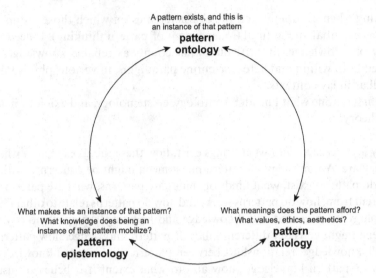

FIGURE 7.1   A relational model of pattern ontology, pattern epistemology, and pattern axiology.

Diagram prepared by the author.

say in any absolute sense that patterns exist, only that we all subjectively perceive and experience things in pattern-like ways. I have grounded this theoretical position in tangential empirical evidence, in others' theoretical and critical thought, and in a shared human experience of pattern thinking. Delving a little deeper we can consider the existence of patterns using the lens of the "mind-body problem" that Westphal described in terms of a tetrad of connected and yet paradoxical propositions:

> (1) The mind is a nonphysical thing. (2) The body is a physical thing. (3) The mind and the body interact. (4) Physical and nonphysical things cannot interact.
>
> *(Westphal 2016)*

Different theories challenge, ignore, or collapse one or more of these propositions to resolve the paradox. For instance, dualists work on the basis that mind and body are intrinsically different and then seek to resolve the paradoxes of interactions (or the absence thereof) between them. A dualist perspective would suggest that pattern is a quality of mind and that it is not continuous with neurological structures or processes. However, we can connect apparent phenomena of mind to observable structures and behaviours of brains, for instance, in suggesting that pattern memory is held in the form of discrete but intimately connected neurons and that patterns are nexuses of memories that are used in support of perception and understanding. As such, pattern theory is not dualist, in purpose at least.

Some theories have sought to eliminate the idea of mind altogether by focusing only on physical phenomena. For instance, behaviourist theories consider mind only in terms of behaviour, while functionalist theories argue that mind is the result of mental computation. There have been other models that have similarly sought to eliminate the concept mind as something other than brain or body:

> Because the hard psychological sciences are exclusively third-person disciplines ... a narrowly construed hard scientific attitude is unwilling to consider as scientific any information about the mind that cannot be eliminatively or reductively explained from an outsider's perspective.
>
> *(Jacquette 2009)*

While I have sought to connect erstwhile physical pattern phenomena with phenomenological cognitive pattern phenomena, I have not felt any need to collapse one into the other. This brings me to emergentist theories of mind that have posited that mind emerges from the physical world (brain) but is not itself physical, although there is an assumed but undefined (and for some an undefinable) connection between them (Westphal 2016). Pattern theory is, to a degree, emergentist in that it suggests that phenomena, such as perception and understanding emerge from a history, density, and accumulation of pattern thinking. That said, I am cautious in equating pattern thinking with consciousness as we seem to engage in pattern thinking whether or not we do so deliberately or consciously. For instance, how does pattern and pattern thinking contribute to our subjective continuities? We have a sense of some (but not all) memories as being temporally situated, and we also have a sense of the recurrence and continuity of things from our memories. Pattern certainly seems implicated in this but what role it plays is unclear.

A computational theoretical perspective is also implicated in pattern theory (although not in its eliminativist sense) in that there must be some kind of processing involved in finding, matching, and sorting pattern candidates in response to current stimuli. For instance, Tononi's (2008) "workspace theory" reflected pattern-like characteristics in that memories are interconnected, they integrate different kinds of mental phenomena, they persist over time, and they work temporally in relating current and past events and anticipating future events. Although there is a potential continuity of pattern thinking, I do not suggest that it only functions as a single integrated whole. Rather, pattern thinking would seem to involve multiple nexuses and processes that seem to run in parallel:

> The sum of the mind's parts can be regarded as the mind, but one does have to recognize that when the mind acts, or the body, it is not the whole mind or the whole body acting, as it were, concentrated at a point, but only one part of it.
>
> *(Westphal 2016)*

The integration of emotion into patterns and the resulting emotional triggering when a pattern is triggered is an important aspect of pattern thinking. Emotional memory would seem to be the basis of noticing and significance that guide us toward or away from things, that help us to prioritise things, and that guide our interactions with things. I would expand on this to include the idea of the embodied mind. Pattern thinking, at least as we understand it, can be a profoundly embodied experience, particularly when pattern memory includes physiological states, actions, reactions, and other sensations.

Neutral theories of mind take neither body nor mind as something to which the other must be subsumed. Westphal considered "neutral monism" to be the best currently available solution to the mind-body problem:

> The view of the neutral monist is that neither mind nor matter is basic, and that both are composed of more basic neutral elements, elements that are in some ways very similar to qualia ... The most important tenet of neutral monism – what makes it genuinely neutral – is that a neutral element, considered in a physical sequence, is physical, but the very same element, considered in a mental sequence, is on that account to be regarded as mental.
>
> *(Westphal 2016)*

Neutral monism aligns with pattern theory in that it suggests that we encode elements of experience and thought in memory (Wagner et al. 1999), and that how we do this varies based on the kind of experience it is (Otten et al. 2001), while simultaneously describing this phenomenologically in terms of the subjective experience of our own minds.

Although the mind-body problem can serve as a vehicle for reflection and reasoning, pattern theory does not align entirely to any one theoretical perspective. Rather, pattern theory aligns with facets of different theories. For example, there are alignments with some aspects of emergentist and computationalist theories in that, while mind is complex and cannot be either reduced to or predicted from its physical basis, there is nevertheless a connection between them. There is also alignment with some aspects of neutral monism in that different theories are needed to explain different phenomenon and that these theories are not necessarily commensurable. Clearly, there are many different aspects of pattern and pattern thinking to be considered including brain as substrate, mind as the medium for pattern, pattern phenomena external to brain and mind, and experiences of those things in the world that patterns refer to.

Using the mind-body problem as a lens also leaves many questions unresolved. For instance, what are the mind-generating qualities and characteristics of the many apparent levels of pattern association and connection? Could our complex morass of pattern associated memories, thoughts, and experiences come together in any combination and still produce mind? Again, there is more here than I can cover in this volume. However, I would argue that there is nothing random about pattern thinking. It might be arbitrary and subject to bias and error, but there is an apparent organic logic to the formation and elaboration of patterns in that they respond to actual things in the world, they are useful in many different ways, and they allow us to think about possible and even impossible things. The pattern associations and connections that form the fabric of our patterns are established out of experience and reflection, out of thought and feeling, out of order and regularity, and out of necessity and imagination.

Returning to the question of whether patterns exist, the most fundamental kind of pattern element is a memory engram, a fragmentary reflection of a past cognitive event or experience. As I noted, there is much to suggest that these memories do indeed exist (in the cortex) and that there are pattern processing areas of the brain (such as the hippocampus) that are independent of but functionally connected to these memory elements (Yadav et al. 2002). A single altered neuron is not a coherent memory although it is a quantum of memory. Indeed, research has shown that some neurons can store the building blocks of memory while others can interrogate it (Rutishauser et al. 2021). Although it would seem that a great many neurons interacting in layered and entangled ways are required for coherent patterns to emerge, there are direct correlates between brain structures and behaviours and the formation and interrogation of memory. On this basis I would argue that patterns do exist in the sense they are the product of neural processes that weave together layers and nexuses of memories.

Patterns are made up of memories. While thoughts, feelings, and emotions play a role in pattern thinking, it is only memories of thoughts, feelings, and emotions that can be integrated into patterns. Moreover, patterns do not suddenly spring into existence; it takes time for patterns to be established or elaborated because thoughts, feelings, and experiences need to become memories before they can be woven together as patterns. Put another way, it is only those aspects of experience and thought that are encoded as memories that can be incorporated into patterns. Pattern is made of memory, but it is not memory. Rather, pattern is the organisational modality of memory and at the organisation of memory is, therefore, pattern.

Do the connections and associations between pattern elements I and others have proposed really exist? Connectivity, as Bateson (1979) (and many others) noted, is an essential characteristic of pattern, and it might be argued that it would need to exist even if no sign of it could be found. However, connectivity does seem to have a tangible if elusive neurological basis (Ison et al. 2015), in particular, in the structure and behaviour of the cortex and the connecting nature of the cortical claustrum (Nikolenko et al. 2021). From a phenomenological perspective, we do experience memories as connected, we do experience continuities of thinking and association, and we do have some experience of working with these associations and connections. However, as before, while a matrix of pattern thought is strongly suggested, while layers and levels of interacting pattern knowledge cycles are implicated, and while a direct complex and emergent relationship between brain and mind can be discerned in pattern theory (albeit in a mirror darkly), it is still as much theoretical as it is empirical.

I close by considering whether and how *shared pattern thinking* exists. I earlier argued that shared practice is established and maintained through the exchange of abstract pattern representations, and the conventions of doing so. Shared practice exists in that it based on activity that multiple individuals participate in and more or less agree what things are and what they are trying to achieve. There is, therefore, a necessary convergence in the shared pattern thinking of people who perform particular roles, and in the shared pattern thinking of those who perform complementary roles within the same activity. So, is shared pattern thinking no more than the aggregation or product of aligned thought? If it were then it would of necessity be interpreted and would, therefore, exist in representation but not in actuality. Although shared pattern thinking is reflected in the interactions, structures, and material expressions that constitute collaborative practice, the patterns that an observer perceives in the shared pattern thinking of others are no more than implied and should be treated as such.

Do Pattern Instances Exist?

Do pattern instances exist? While many of the things that we perceive as pattern instances exist in a real and objective sense (such as dogs, doors, mountains), perceiving something as an instance of a pattern depends on the pattern thinking subjectivities of an individual mind. To perceive that this thing before me is a cat requires that stimuli (such as what I see and hear) trigger a search of my pattern memory for something that matches these stimuli. The pattern this search returns informs my perceptions and my reactions to those perceptions.

None of this means that the thing in front of me is a cat in any absolute sense, only that I perceive it as such and, therefore, treat it as such. Others may perceive it as something else or, not having a pattern for "cat" may struggle to perceive what it is with any specificity and fall back on broader patterns such as "animal", "pet", or "predator" (depending perhaps on what the thing is doing). I could test my initial pattern perception (for instance, by touching it or trying to pick it up) and the additional data this provides may revise my pattern perception accordingly. Pattern instances are ephemeral (what I perceive may change from one moment to the next), subjective (what I perceive is not necessarily what others perceive), and subject to bias and error (pattern perception is heuristic, it is fast but approximate).

Typically, some aspects of a pattern instance can be more objectively assessed, such as its regularities or orders, and sometimes, as in mathematics, these can also be equated to or used as a proxy for the pattern. A computer tasked with "pattern recognition" will typically employ algorithms to select a pattern that fits its inputs but in doing so it is still drawing on objective rather than subjective information. We can say that, while some of the characteristics of a pattern are mind-independent, patterns typically involve both subjective and objective elements. Moreover, pattern conformance is more a matter of judgement than it is of rules-based evaluation. We use patterns all the time for identifying and classifying what things are. Indeed, I would argue that all our classifying and typing activities are pattern based. Every time we say that this thing belongs to this group of things, or that this thing is similar to these things and different from those things, we are using pattern thinking, but the specifics of such thought are unique to the mind in which they occur.

So, is a cat a "cat"? Clearly the thing I perceive is not an instance of a particular pattern in and of itself, the instantiation is constructed in my mind out of the tensions between the stimuli I receive and my pattern memories. Moreover, most of the things I perceive exist (whatever they are) independent of my perceptions of them. I might resolve this inductively by gathering and analysing evidence or deductively by reasoning from what we know, but they still depend on deciding whether the thing I am looking at conforms to a particular pattern in my mind. Even if I agree with others that this is a cat, this only means that we have similar patterns not that this thing is intrinsically a cat. Even if we had to do DNA tests to determine whether this is a cat, this is still a categorical relationship rather than a fundamental reality. This subjectivity is further amplified if the thing being perceived is itself intangible and subjective, such as a sense of happiness, a feeling of pain, or a sense of confusion.

Subjectivity cannot easily be dismissed from pattern ontology. Patterns reflect our experiences and our understandings of these experiences. The ontology of pattern can, therefore, be approached in terms of the relationships between patterns and the things that they refer to. We might draw on concepts of types and tokens as a way of exploring this relationship, such that a type is roughly equivalent to a pattern and a token is roughly equivalent to a pattern instance. The identities of and relationships between types and tokens has long been a topic in metaphysics, linguistics, and in other philosophical domains, and yet with little agreement on what a type or token is, or what the relationship between them is:

> The relation between types and their tokens is obviously dependent on what a type is. If it is a set, as Quine … maintains, the relation is class membership. If it is a kind, as Wolterstorff maintains, the relation is kind-membership. If it is a law, as Peirce maintains, it is something else again.
>
> *(Wetzel 2018)*

Drawing on this and reasoning deductively (i.e., what are the principles of types understood in terms of tokens?), we might argue that a type is defined simply by all the tokens it refers to (memberships), or that it is defined by sorting and grouping similarities and differences (kinds), or that it is defined by essences and regularities (laws). I have argued that pattern in general must be open to new instances and new variations of those instances, which would suggest that a membership definition of pattern would be inadequate as it would always be retrospective and incomplete. I have also argued (and will return to this point in considering pattern epistemologies) that specific patterns cannot be defined in terms of essences and laws due to their dynamic, fluid, and unbounded natures. This would leave similarities and differences as the primary ontological basis of the relationship between pattern and pattern instances.

If we reason inductively (i.e., what are the principles of tokens understood in terms of types), the tentative nature of the relationship between pattern and instance (type and token) is perhaps clearer, and that in turn depends on how one- or two-handed that relationship is. A one-handed perspective would be that a token must reflect enough of its type's characteristics to be identified as such but that its type may be suggested but cannot be directly and fully inferred from any individual token. A two-handed perspective would be that all tokens are exemplars such that their type can be fully inferred from the exemplar. I have argued that pattern is neither fully one- or two-handed but somewhere in between in that we may only be able to understand patterns in terms of their instances even as instances are expected to differ and vary in their differences. Moreover, tokens are not "out there", as instances they are relational between an observer's percepts of a thing and the observer's pattern perceptions of that thing. That is not to say that things do not exist, only that our perception of a thing as an instance is not the same as the thing itself nor is it the same as the pattern.

Do Externalised Patterns Exist?

Pattern expressions, pattern representations, pattern systems, and pattern languages are by definition all external to our minds and as such their ontologies might be more straightforward. Pattern expressions are things we make based on one or more patterns; they form outside of our minds even as that form is guided by the pattern thinking of our minds. Expressions in the form of artefacts may exist longer than performances (unless the latter are recorded) but they still exist outside of our pattern thinking minds. However, if we more loosely define pattern expressions as anything that is derived from our patterns then we should also include the products of our imaginations that draw on patterns but are distinct from them. Or are they distinct? What if we were to consider imagination as a reweaving of pattern thought and memory where connections are made playfully rather than purely in response to external stimuli or practical need? For instance, does asking someone to think of a cat rather than to draw a cat or to do an impression of a cat count as a pattern expression or a pattern elaboration?

Whether someone perceives something as a pattern expression depends on whether they know that it was manufactured or otherwise guided by someone else's pattern thinking. Where expressions are manufactured in the industrial sense of the word, there may be many individual pattern thinking contributions such that no one person's pattern thinking dominates. Whether a third-party observer perceives an expression as an instance of a pattern in ways that are similar to those of the expression's creator depends on whether they hold similar patterns (noting that no two individuals can have the exact same pattern). The question of whether pattern expressions exist is harder to answer than it might at first have seemed. Although expressions can be separable from the mind that created them, there are degrees of separation here as something of the mind of a creator is inevitably woven into the substance of the things they create, whether or not it is perceived by others.

Pattern representations share a similar ontological basis to pattern expressions, albeit with some differences in what is externalised and how it is externalised. One difference to note is that pattern representations might be held to a higher standard of fidelity than pattern expressions. After all, pattern expressions are guided by a pattern but do not have to exemplify that pattern. Pattern representations on the other hand do explicitly replicate the pattern (albeit selectively and approximately) and as such the fidelity of reproduction may be more of a concern.

Pattern systems and pattern languages share much of the ontological character of pattern representations, not least because they are themselves both collections of pattern representations and macropattern representations. However, as I earlier observed, there are potentially many system collections of patterns that were not developed explicitly as pattern systems or pattern languages. On one hand we might say that they do not exist as pattern systems if they were not declared or designed as such, on the other if we accept that pattern systems have certain characteristics then something need not have intentionally been developed as a pattern system to still count as one.

How Do Patterns Work?

My focus shifted from essentialism (what patterns are) to functionalism (what patterns do) at certain points in the previous section, particularly where I described the existence of pattern phenomena in terms of what they do rather than what they are. A functional perspective is important (and difficult to avoid) because pattern is not some random concatenation of memories. Patterns are formed as a resource for our minds to draw on. To that end, patterns and pattern thinking serve many purposes. They organise memories and allow them to be searched and recalled in a coherent way, they are a fundamental part of perception, and they allow us to understand things, to work with meaning and significance, and they can even allow us to make predictions. Patterns and pattern thinking also allow us to fold our thoughts, feelings, and experiences back into pattern memory for future use, and by so doing they form the basis of our learning and of our social lives. Pattern is the bedrock of language, mathematics, music, and many other abstracted modes of human thought and communication. Clearly, it is hard to avoid talking about function when talking about pattern. So, how is it that pattern can do all of these things?

First, we should not consider pattern to be something 'other' with which our minds interact. Pattern is woven into our minds, particularly as educated and socialised beings, such that nobody is separable from their pattern thinking. When the brain is disrupted in its abilities to support pattern thinking, for instance, in Alzheimer's disease or through traumatic brain injury, the result is dysfunction and confusion. This is not an area in which I have expertise so I will leave this for others to consider. My point is that pattern thinking is not something we have or that we go to or access; it is a part of who we are.

In terms of function, I have described perception in terms of searching and matching specific pattern memories to stimuli, which suggested a rather computational perspective on the function of pattern thinking. I also described the difference between pattern recognition and pattern development in terms of speed (conscious vs unconscious pattern recognition) and whether a suitable (satisficing) match can be found. This provides an effective simile for the apparent behaviour of pattern thinking, and one that is both (provisionally) logical and consistent with the evidence we can bring to bear. For instance, in terms of sorting and recognising patterns, Rutishauser et al. found different kinds of neurons that can distinguish between novel and familiar stimuli in the medial temporal lobe of the human brain:

> …neurons in the hippocampus and amygdala that exhibit single-trial learning: novelty and familiarity detectors, which show a selective increase in firing for new and old stimuli, respectively.
>
> *(Rutishauser et al. 2006)*

This, plausibly, may reflect the way (or at least one way) in which we parse pattern memory, but it may simply be a coincidence. I also noted research findings that suggest memory is stored in the and across the cortex but that it accessed and processed via or at least involving the hippocampus (Voss et al. 2017). We cannot say specifically how our brains do this, only that this is what they seem to do. This analogic reasoning suggests that patterns coalesce or accrete based on reinforcement linked to factors such as frequency of triggering and the sense of significance that is woven into them. That groups of neurons connect and change as we learn may be the basis for this coalescence, but researchers have yet to show whether or not this is the case.

What about the ontologies of micropatterns, mesopatterns, and macropatterns? Again, there is little evidence to support their existence other than the experiential, phenomenological, and behavioural evidence that this is what our minds seem to do. Certainly, we experience some patterns as being more elaborate than others, some as being more abstract or generalised than others, and some as being more formalised or rigorous than others. We also experience pattern elaboration (adding detail, nuance, and depth) and aggregation (some patterns being collected as parts of other patterns), again suggesting a continuum from micropattern to macropattern development. However, since one pattern can form the matrix of another pattern, quite which pattern is macro, or whether we have macro systems or macro languages of patterns rather than unitary macropatterns is not something that can be resolved easily or conveniently. As ever, more research is needed.

What about the functionality of the other pattern constructs I have proposed? Pattern expressions function by mirroring or capturing some aspects (presumably the most important, germane, or useful aspects) of particular patterns such that they can be used in place of the pattern, or they can be analysed in ways not possible with the original pattern. For instance, if I make a representation of my pattern of a cat, psychologists or cultural theorists might use that to infer things about my mental state or my cultural conditioning. Looking back at the previous chapter, I seem to have gone into a lot more depth about the functions of pattern systems and languages, for instance, in contrasting normative and naturalistic approaches. Rather than going over that ground again, I will simply say that pattern representations, pattern systems, and pattern languages have many possible functions.

Ontologies from Patterns

What about the ontologies that patterns and pattern thinking can provide? Although I consider pattern knowledge in more depth in the following section, it is useful to consider this question here in the context of ontologies from pattern. Patterns allow us to make sense of the world around us, they allow us to rapidly appraise our changing circumstances, and they can help us to respond to the situations and challenges we face. To use a vernacular concept, we are always looking through 'pattern goggles'. What then are the ontologies they afford? I look around me and I see a room, a chair, a bookcase full of books, a window, a cup of tea. What I actually see is a continuum of shape, colour, light and shade, perspective, and perhaps movement, which might be supplemented by what I can hear, feel, smell, and taste. I can distinguish between things, and I can identify those things I can distinguish between because I have developed patterns and pattern thinking to help me do so. When I see a chair, my mind matches sensory data to my pattern memory and (hopefully) pulls up my pattern for chair, which in turn pulls on its associated memories so that I can put a name to it, I can compare it to other chairs I have encountered, and I have a pretty good idea what it might do and what I might do with it. My ontology of this particular chair is in part grounded in my immediate experiences of it, in part in my pattern memory suggesting meanings, understandings, and possibilities in response to my experience, and in part in my cognitive reactions to both streams of information (effectively what cognitive scientists call "executive functions") (Miller & Wallis 2009).

Where the three begin and end is hard to say as they are all reflected in mind, and they would seem to function together or not at all. After all, what is a mind entirely without sensory input (that includes embodiment, self-awareness, and other conscious thoughts)? What is a mind if it has no way to drive recognition, understanding or meaning? What is a mind that has no way to respond to or act on its perceptions? The interdependence of inputs, patterns, and executive functions is easily illustrated. If, like a wise monkey, I reduce my sensory inputs or if the lights are dim or my senses are otherwise occluded then I have less input to work with. If I have few or ill-formed patterns, or if I have lost some of my patterns through injury or disease then I also have less to work with. If I am an inexperienced pattern thinker or am otherwise inhibited or limited in my executive functions, then I am again less able to work with what my senses and my patterns are telling me. Conversely, I can improve my pattern thinking if I improve my sensory inputs, if I expand and elaborate my pattern repertoire, and if I improve the capacity of my executive functions to make use of my inputs and patterns.

We can understand pattern knowledge as the product of these three broad cognitive functions – see Figure 7.2. However, I should stress that this is a functional rather than a structural representation. Pattern is not an end in itself; there must be something that pattern is informing. Moreover, the processing of patterns needs to be located somewhere. By grouping these corresponding functions as "executive functions", I also acknowledge that there are ongoing debates regarding what executive function is.

Another ontological challenge is whether what we perceive as a result of pattern thinking is real. I have described many ways in which pattern perception can mislead us. We can perceive pattern instances that are illusory, we can misrecognise instances, and we can fail to perceive things at all if we have no pattern to guide us. Moreover, there are many aspects of perception, such as aesthetic experiences and emotional reactions, that are so subjective

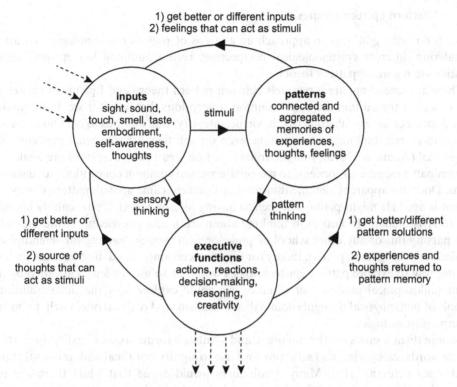

FIGURE 7.2 Process diagram illustrating the suggested relationships between senses, patterns, and executive functions in the mind.

Diagram prepared by the author.

that their existence cannot be resolved by appealing to verifiable realities; we can trace their physiological impacts, for instance, in terms of heart rate, hormone levels, or pupil dilation, but we cannot verify whether they reflect something real beyond these phenomena. That said, if we can think of something, then it is real in our minds whether or not it is real beyond our minds. This reflects what Quine described as ontological commitment – that we can treat things as objects whether or not they exist outside our minds (Wetzel 2009). Without this capability there would be no imagination, no fiction, no fantasy, no deceit, and no theory. Without ontological commitment we would be much less than we are.

Given that we can indirectly perceive things that our senses cannot but that our reasoning, collaborative, and imaginative minds can, rather than asking whether our pattern perception reflects real things, we might, therefore, be better considering the range of subjective and objective realities that patterns allow us to perceive. After all, the macropatterns of science, philosophy, aesthetics, social relations, and cultural expressions can all inform our perception of intangible and to all intents and purposes imaginary things, even though their existence is broadly accepted. Patterns allow us to transcend physical and immanent realities, and they allow us to challenge or break free of them.

General Pattern Epistemologies

There is no one 'right' way to approach an analysis of pattern epistemologies. I start by considering different epistemological perspectives before outlining key epistemological questions in regard of pattern theory.

There are several apparent parallels between pattern theory and Epicurean epistemology. Epicurus argued that there are common conceptions that we all share, both tangible (cats, doors, books) and abstract (truth, virtue, integrity) that allow us to communicate and engage in shared thinking and reasoning, even though they are somewhat imperfect and provisional (Adamson 2015). I might argue that Epicurean preconceptions are analogous to micropatterns, general concepts to mesopatterns, and common conceptions to macropatterns. Does this apparent alignment mean that Epicurus came up with pattern theory two millennia ago? He might perhaps be seen as having foreshadowed it, particularly his solution to Aristotelian absolutism in thinking about truth and knowledge. However, rather than parsing this or any other school of philosophical thought looking for serendipitous parallels or analogues to pattern theory (or for counterexamples and dissonances), perhaps a better way to explore pattern epistemologies would be to use the lenses afforded by different philosophical models. For instance, we could explore how the three traditional schools of metaphysical thought (nominalism, idealism, and realism) might help to outline pattern epistemologies.

Nominalism focuses on the linguistic and symbolic formations of reality, in particular the words, categories, and structures we use to signify both real and universal things (Rodriguez-Pereyra 2019). Many nominalists would argue that while there are real things that exist independently of our words for them, there are no universals, only words and concepts that signify these imaginary things. A nominalist reading of pattern and pattern theory might focus on whether my assertions of what patterns are or how they work are meaningful. I have defined pattern as a phenomenon of mind that cannot be directly experienced except for the phenomenological sense that we have of our own pattern thinking or by the behaviours of others who we perceive to have patterned minds. An aggressive nominalist perspective on this might argue that pattern is a universal construct and, therefore, illusory. A less aggressive approach might accept that the various observable phenomena I have described are real but deny that a connecting principle of pattern is any more than a category we impose to treat similar things as if they were all the same. There may also be nominalist concerns about the polysemy of the word pattern given it is used to refer variously to regularities, instances, expressions, and representations.

Idealist thought has been grouped into two broad themes: ontological idealism (all reality is generated by the mind) and epistemological idealism (there is an external reality but our perception of it is inescapably filtered by our minds) (Guyer & Horstmann 2023). An ontological idealist reading of pattern theory might agree that patterns and pattern thinking play a large part in the creation of the illusion of reality, while an epistemological idealist reading of pattern theory would focus on the ways that patterns and pattern thinking can shape our perceptions and understanding. I have argued that patterns play a critical role in the functioning of our minds but not that patterns are synonymous with mind. After all, we can be playful with patterns, and we can (with a degree of self-discipline) ignore them or manipulate them.

Pattern thinking is invaluable, and it colours much of our perceptions and actions, but are we (or do we allow ourselves to become) slaves to our own patterns? Sometimes it would seem that we do indeed allow ourselves to be held hostage by pattern thinking, for instance, in succumbing to blunt and biased stereotypes or to social conventions that disadvantage or harm us. However, by and large, as adult humans we do not generally see the world solely in terms of archetypes and abstractions. It may be that young children and otherwise less capable minds might tend to fall back on archetypes and stereotypes. It may also be that our pattern perception will tend to become more archetypal or stereotypical at times of uncertainty and stress, such as finding ourselves in a strange or dangerous environment. However, when we are relaxed and things are relatively familiar, then we can be less dependent on pattern archetypes and stereotypes. Or perhaps mental relaxation is in part a reflection of little or no dissonance in the patterns we pull on. Maybe we should take as a mark of a disciplined mind the ability to both to focus pattern thinking and to know when to interrogate and critique it. Of course, I should also question what I mean by "we" in these past few sentences as it might suggest a dualism between self and pattern as if it were some kind of library or guide a separate self refers to. This last point brings me to realism.

Realism is about things that exist independent of our minds. A realist epistemological perspective on pattern theory might, therefore, ask whether our knowledge of pattern exists (or can exist) independent of mind. Although patterns cannot exist outside of our minds, the fact that we can create pattern expressions and pattern representations illustrate that patterns (albeit in translated or transcribed forms) do have some kind of existence outside of our minds. Certainly, our pattern thinking does not require that we understand it or even be aware of it, although we do (or should) develop some deliberate pattern thinking habits as part of our education, both as children and as adults.

There is a potential paradox in a realist position that seeks to explore phenomena of mind that are also independent of mind. One way of resolving this might be to relate phenomenological experiences of pattern thinking to observable physiological phenomena in the brain. While I noted that there do seem to be strong parallels between patterns in the mind and cortical structures and behaviours in the brain, these are provisional at best and no robust connections between the two phenomena have (yet) been established empirically. Another realist question we might ask is whether our pattern knowledge of things is the same as the things it refers to, and of course it is not. We are selective in what we attend to, we can only pursue certain things, and the associations and past experiences are unique to our situated minds.

While this comparison of a philosophical model to pattern theory affords some interesting questions and perspectives, it still feels somewhat random as an approach. Moreover, I have not captured here how many frameworks I considered (there were many) before I found one that offered some interesting perspectives. It is a fascinating approach, but not an efficient one. So, let me take a different approach based on outlining and examining a series of theoretical positions based on pattern theory as I have advanced it up to this point.

Pattern Knowledge

Pattern thinking is so intimately woven into our cognitive makeup as to be inseparable from it. Although human minds are particularly adept at pattern thinking, other species also seem to have pattern thinking capabilities. Pattern thinking, with its constant assessments of similar differences and different similarities, allows us to organise things into categories which can in turn allow us to build conceptual and theoretical understanding into and around our experiences. Patterns are how we identify things, relate things, organise things, and manipulate things in our minds. Patterns can include memories of meanings and feelings, and they can give us a sense of certainty and order. Epistemically, patterns afford a semantically rich and associative system of knowledge. That said, there is a paradox of precedence here; patterns require pattern thinking and pattern thinking requires patterns. Clearly, we need patterns to be able think with them, and yet how do we build pattern but through pattern thinking? What comes first, pattern or thinking with pattern? I have suggested that our infant minds are evolutionary primed to develop pattern thinking. Rather than patterns or pattern thinking developing first or in parallel, they would seem to coalesce as a single cognitive capability such that their integrated use is what develops the capacities of both. However, whatever the modality or pressures on or from pattern thinking we experience, patterns allow us to coalesce ourselves in relation to the world around us, they allow us to bring ourselves into the world, and they allow us to bring the world we perceive into us.

Let us consider four dimensions of pattern knowledge. First, there is the pattern knowledge that tells us what something is. Identity is the primary product of pattern perception; our minds seek a match between sensory input and pattern memory and from this we perceive that something is an instance of a particular pattern we hold in our memories. Our confidence in these perceived identities may vary according to the ambiguity of the situation and our confidence in our patterns and pattern thinking. Second, there is the knowledge of something that we pull from our patterns and apply to our perceptions and thinking about pattern instances. This knowledge can include memories of past encounters with instances of a pattern, as well as associated meanings, implications, and feelings. Third, there is the working pattern knowledge that we build out of our pattern memories and our interactions with and impressions of a thing. This knowledge may also reflect the ways that current experiences can contradict our existing pattern knowledge. Finally, there is the knowledge created when our pattern thinking in working memory is integrated back into our longer-term pattern memories, something I have described as *pattern elaboration*. While we do not remember everything from every encounter (those with eidetic memories notwithstanding), it would seem that we do elaborate our pattern memories to some extent every time that pattern is triggered or pulled upon. Are these four dimensions distinct kinds of pattern knowledge, or would it be better to understand them as part of a cycle of pattern knowledge formation and application? If so, this redefines knowledge not as something we have but as something we do; knowledge from this perspective is a process rather than the product or consequence of a process – see Figure 7.3.

This model of pattern knowledge as a cyclical process would seem to go against traditional philosophical models of knowledge as being true justified beliefs (de Grefte 2023). From that latter perspective, anything that is not true, justified, and believed means that, whatever it is that we do have, it falls short of being knowledge. If my pattern perception tells me that the thing in front of me is a cat, then what does that mean? My current experience of this four-legged furry thing need only be similar to past experiences to be seen as an instance of a particular pattern. While it may be true that I perceive this thing as an instance of a cat, my pattern is an abstract and somewhat fuzzy categorical construct and so (assuming I am not in error or deceived, which would also invalidate conditions for truth) the best I could

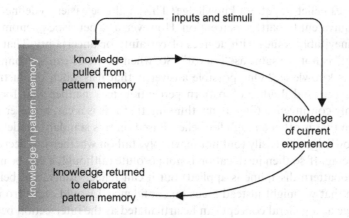

FIGURE 7.3   Different aspects of pattern knowledge, forming an iterative process of pattern knowledge formation and application.

Diagram prepared by the author.

say is that this seems to be a cat. There are degrees of certainty but no absolute binary truths or falsehoods. Given the heuristic provisionality of pattern (with more information my perception may change) and layering of pattern perception (I might simultaneously perceive this to be a domestic cat, a pet, a carnivore, and a source of allergies) then pattern knowledge is not intrinsically exclusive. Of course, some patterns are more exclusive than others (if this is a cat then it cannot be a dog or an octopus) but exclusivity would seem to be a quality of category, not a quality of pattern. On this basis, patterns are not necessarily true in any absolute sense (with a nod to Gettier), but they can be sufficiently true to satisfice. They can be heuristically true. Some pattern knowledge may be more rigorously true, in particular pattern knowledge that is more macropatterned. In this way, we might make stronger truth claims regarding analytical, theoretical, and synthetic pattern knowledge.

What about justification in pattern knowledge? I might empirically justify one sensory perspective by confirming it with other sensory perspectives such that everything I have (sight, touch, smell, hearing) aligns. However, I cannot justify experience by comparing it to my previous pattern memories as that would still be pattern perception. I might justify my perceptions by comparing them to some kind of objective criteria but where would those criteria come from? If I recalled the characteristics of a cat from memory, I would in effect be querying the same pattern that informed my perception. If I had developed it and tested the pattern to be more macropatterned it might be more reliable than one that had not been developed and tested but that is still only a relative degree of reliability. Alternatively, I might justify my experience through discussion with others, but if we had common pattern deficiencies or biases then this might not take us very far either. It would seem that justification is also problematic. Rather than getting caught up in circular arguments, we might be better using utility as a criterion of justifiability, i.e., pattern knowledge is justified if it is useful. This would satisfy the "might be a predator" test (our tendency to err on the side of caution in potentially life-threatening situations). Similarly, for more innocuous pattern perception, justification might be better understood in terms of the utility of my perceiving something as a cat rather than any indisputable confirmation of the truth of a cat.

What constitutes belief in pattern knowledge? This might be easier to define if we were to take it to be equivalent to pattern perception. However, as a heuristic phenomenon, pattern perception inevitably comes with degrees of certainty or uncertainty. That this thing "seems to be a cat" is not the same as "is a cat". So, when do we or can we commit to pattern perception as knowledge? One possible answer is that it is when the justification is strong, but the speed and fluidity of pattern perception may neither wait for nor need robust proofs, only contingent utility. If my thinking that this is a cat, however provisionally, is useful then that is often enough for belief. Based on this standard model of knowledge, pattern knowledge can easily (but not inevitably) fail on whether it reflects truth but it could be satisficing. If so then justification is not absolute (although it can be made more rigorous if macropattern discipline is applied) but it can reflect utility, and belief is also provisional, such that we might instead focus on confidence. I would argue from this that pattern knowledge as a general concept can be articulated as the intersection between satisficing, utility, and confidence – see Figure 7.4.

One might conclude from this that pattern knowledge is pragmatic rather than pure, provisional rather than definite, and contingent rather than universal. While this might be generally so, macropatterns can be developed to be more in accord with the standard model of knowledge as justified true belief. Indeed, the standard model of knowledge is, I would argue, a representation of the pattern thinking of those that generated the model in the first place. From this I might argue that the standard model is a special case (a transform) of the broader pattern knowledge concept. If this is so, then we might anticipate other transforms to represent other specialised kinds of pattern knowledge.

A different definition of knowledge might focus on a sense of necessary depth, for instance, that for something to count as knowledge it requires understanding of the underlying principles of the phenomenon to which it refers. This would suggest that knowledge requires some kind of theory or inferential understanding in addition to any information it

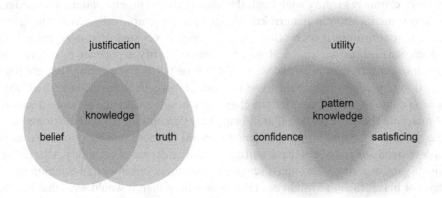

FIGURE 7.4 Left: the standard philosophical model of knowledge in that it is justified true belief, anything not in the intersection of the three characteristics is not knowledge. Right: a proposed model for pattern knowledge based on intersecting degrees of satisficing, utility, and confidence.

Diagram prepared by the author.

advances. From this perspective, pattern knowledge, at least that part of the knowledge cycle that is pulled from pattern memory, would count as knowledge in that it includes connections, associations, significances, trends, and possibilities. Indeed, returning to one of my opening points, one could argue both that pattern knowledge is the basis of theory and that theory is pattern knowledge that has been developed to be more macropatterned (reflective, rigorous, abstract, shared, etc.). Another definition might compare explicit knowledge (that is easily available and shareable), implicit knowledge (that needs to be derived from experience), and tacit knowledge (that is difficult to articulate or share). Applying this model to pattern knowledge, comparing knowledge that is easily available with knowledge that needs working out arguably reflects the difference between pattern recognition and pattern development I earlier described as the two main modes of pattern perception. As to knowledge that is difficult to express or represent, this equates more to degrees of macropatterning, in that the more macropatterned a pattern becomes the more it is made explicit and rendered in abstractions.

While a single memory element relates to a particular experience of a phenomenon, patterns collate memories across multiple encounters and thoughts and, potentially, across multiple layers of abstraction. This allows our minds to group memory elements as instances of the same pattern, both synchronously and asynchronously. That connection and association are essential aspects of pattern was something Bateson (1979) argued for, it is implicit in the graph theory pattern modelling of Grenander and colleagues (2007), and it is found in the explanatory nature of scientific inquiry in the social sciences (Diesing 1971).

A pattern is formed when our minds connect and associate similar experiences from memory, which means that patterns are not independent and exclusive units of thought, they are nexuses of associations. Moreover, pattern knowledge is, theoretically speaking, constructivist in that it arises out of a constant process of weaving together present and past experiences.

Not only is the fabric of pattern aggregative, its application is also aggregative. In arguing this point, I draw on the Stoic position that knowledge is mediated by impressions of things that may or may not be true, and that it is our assent to the suggestions these impressions give us that establishes knowledge:

> ... the cognitive impression is not yet knowledge. It is, rather, the basis for knowledge. We achieve knowledge only when we are systematic, when we combine together many rational impressions into a global understanding of the world.
>
> *(Adamson 2015, p. 62)*

Knowledge from this perspective requires rational thought, analysis, and time. This aligns with my earlier observations that it takes more than a few fleeting impressions to confirm the pattern identity of something. It also aligns with my model of the circularity of pattern knowledge activation and elaboration. This is a departure from my arguments in Chapter 4 though where I described pattern recognition in terms of a simple searches, matches, and returns. If pattern recognition returns are multiple, cyclic, and aggregative, then recognition could be more a matter of accumulation of evidence to inform some kind of organic probability assessment or threshold. Alternatively, if such reasoning is more about learned macropattern thinking then, for most of the time, pattern recognition might be better described in terms of the sorites paradox (how many grains of sand does it take to make a heap) such that we accept there is a fuzzy zone of "maybe an instance" that we seek to resolve to either "is an instance" or "is not an instance" through additional cycles of pattern knowledge. Interestingly, this would also seem to align with the layering and panarchic connecting of a multitude of inference cycles in large language models (Zai & Brown 2020). If there are many simple but interacting pattern matching cycles that overlap and interweave, any of which may suggest different interpretations, then pattern knowledge would seem to arise from aggregation and cumulative interpretation. If this is so, then there is no one point where pattern thinking generates pattern knowledge, rather pattern knowledge is being constantly formed and reformed in our minds. There are implications here regarding interactions between pattern knowledge (and the ways it is generated) and consciousness, but I will again leave that topic for another time.

It is important to again note the heuristic nature of these processes. Pattern knowledge tends to be satisficing rather than confirming. Our confidence in our pattern knowledge broadly increases as provisional and potentially contradictory pattern knowledge accumulates and we abductively work towards "best fit" understanding. Of course, we may act before we are confident in our pattern knowledge, for instance, in the face of possible danger. It also seems likely that stress and other physiological factors shift our relative thresholds of sufficient confirmation to precipitate action, and that building confidence in a pattern perception may require a pattern to be triggered multiple times in response to a perceptual event. Confidence may also be supported through triangulation across different elements of a pattern that is being activated. This is also a factor in pattern knowledge reliability in that the more consistency there is over time (the thing I perceived as a cat yesterday I still perceive as a cat today) the more reliable (broadly speaking) the pattern knowledge. Reliability would likely build confidence in a person's abilities as a pattern thinker and in their pattern memories.

How much connectivity between memories and how many layers of such connectivity are needed to establish a coherent pattern nexus that can provide meaningful pattern knowledge? Even if this could be calculated for one particular pattern in one particular mind, it seems unlikely that this would be constant (echoing the continuum I have suggested between micropatterns, mesopatterns, and macropatterns). Rather than asking how many connections or how many of what kinds of connections between what kinds of memory elements are needed to constitute a pattern, we might instead argue that a nexus of connected memories only becomes a pattern when it behaves as a pattern, i.e., when it is called on and responds in support of perception. After all, many connections between memory elements may be provisional, orphaned, or otherwise lack sufficient coherence to be able to function as parts of a pattern. Rather than patterns simply being nexuses of connected memories, we might redefine pattern as a nexus of connected memories that functions in pattern-like ways, i.e., it supports perception and understanding, it can convey significance and meaning, and it can be elaborated, abstracted, and developed to be or be a part of macropattern thinking. Moreover, given the apparent layering of patterns and the panarchic interactions between these layers, the question of "how many connections makes a pattern?" seems even more simplistic. A better question might be: how many layers and what kinds of interactions between layers of connected pattern memories are sufficient to support pattern thinking? On this basis, we might redefine pattern again as a confluence of layers of connected and interacting nexuses of memories that functions in pattern-like ways. Pattern from this perspective is more about behaviours and affordances than it is about structures and associations. Of course, this means that we would have to consider what these functions are and to that end we might focus on similar differences and different similarities, the relationship between patterns and instances, elaborations from micropatterns to macropatterns, providing the bases for perception (both recognition and development), and providing organisation and structure to memory and its applications.

Based on this model of cycling pattern knowledge in and out of memory and elaborating it as part of this cycling that I described earlier, it seems likely that only a subset of a pattern is drawn on in pattern thinking, i.e., those aspects that match the stimuli and those that are flagged as (somehow) most significant by the pattern's emotional elements. From this we are likely to build a provisionally elaborated pattern subset in combining pattern memory with current experience, and it is this subset that is returned to pattern memory to elaborate the base pattern(s) involved. Based on this, we can revisit the cycle of pattern knowledge – see Figure 7.5. This would challenge my earlier suggestion that patterns are activated when some of their elements are 'pulled on', such that pulling on one part pulls on the pattern as a whole. It would be inefficient for a whole pattern memory nexus to be pulled into working memory every time it is triggered, the idea that a subset of proximal and (putatively) germane pattern parts is what we activate or pull out of memory seems more likely. This in turn suggests that different kinds of knowledge in a pattern knowledge

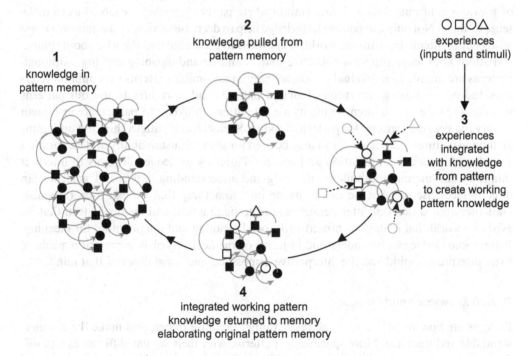

FIGURE 7.5 Reprising the pattern knowledge cycle (in Figure 7.3) knowledge is stored in a dynamic pattern nexus of memory, a subset of the pattern is pulled from memory, new experiences are integrated with the knowledge from pattern to create new working pattern knowledge, and this new pattern knowledge is woven back into pattern memory.

Diagram prepared by the author.

cycle have different epistemologies and different relationships with each other. Not only is what is pulled from memory only germane parts of a pattern, patterns and pattern thinking reuse memories, they can include other patterns, and similar patterns connect and merge. Add to this the layering of symbols and categories and other symbolic and abstracting characteristics, pattern knowledge would seem to afford many important efficiencies.

What are the implications of affective elements in pattern knowledge? Knowledge pulled from a pattern that convey meaning to current experiences by linking those experiences with memories (and abstractions of memories) of similar past experiences. This aspect of pattern knowledge tells us what this thing is, what it might do, how we might interact with it, and so on. While understanding from mesopatterns will likely be relatively impressionistic in nature, understanding from macropatterns will tend to be more explicit and abstract, often taking the form of theories, models, and frameworks. Pattern knowledge can also convey a sense of significance by triggering emotional responses to recognising something about which we feel deeply, which in turn can help us to prioritise alternative patterns, focus our attention, and note implications, risks, and opportunities. We weave our current feelings and significances into our active pattern knowledge and then weave this back into our pattern memories following these experiences. Pattern knowledge can convey a sense of order and confidence by helping to make the unknown knowable, the unfamiliar familiar, and more generally by making things seem less chaotic and more ordered. This is in large part based on memories of previous confrontations with uncertainty where pattern knowledge enabled us to make sense of them. Not only can pattern knowledge help to decrease anxiety and establish a sense of agency in interacting with the world, it allows us to generalise and theorise about things.

Pattern knowledge supports collective understanding and meaning making. Although patterns are unique to individual minds, we may share similar patterns for common referents. Indeed, we learn a repertoire of shared patterns and learn how to use our patterns socially when we are children. Humans are all pattern thinkers; to participate in human society requires the capacity for pattern thinking. Sociocultural pattern thinking underpins all human cultures and societies and may be used for societal sustainability and elaboration as well as for social manipulation and control. Pattern knowledge allows us to interpret things by drawing out and building meaning and understanding in terms of what we can directly recognise or what we can translate into something that we can then recognise. Gadamer argued that our interpretations tend to be practical and nuanced given that we exist in a world that is already infused with social meaning and the potential for meaning. Pattern knowledge can be understood in terms of the horizons of interpretation available to an interpreting mind and the interpretive disposition and capabilities of that mind.

Pattern Knowledge and Category

Patterns are how we identify things, relate things, organise things, and make them understandable and tractable. More specifically, patterns, with their similar differences and different similarities, allow us to organise things into categories which allow us to build conceptual and theoretical understanding around them. As Chi et al. argued:

> Knowledge does not grow as a set of isolated units but in some organized fashion. To capture the organization of ... declarative knowledge, cognitive scientists operate with three distinct representational constructs: semantic networks, theories, and schemas ...

Semantic networks are memory structures, the ways in which memory is organized and connected ... Theories describe relationships between core concepts ... Schemas are bounded units of knowledge.

(Chi & Ohlsson 2005, pp. 374–375)

I have used the term *pattern* to describe various cognitive phenomena that at times involved the use of category. For instance, is perceiving something as an instance of a cat simply a matter of assigning that thing to the category of "cat"? Or, if we were to suppose that something we perceive is indeed a "cat", then what does this categorisation imply? I might suggest that there are two broad kinds of category thinking related to pattern knowledge. The first is based on our already having a categorical system for a particular kind of phenomena. Reasoning involves inductively working out where things fit in it, first by assessing similarity and difference, and then (based on this) allocating something to one or more categories. For instance, for things A, B, and C, I might categorise A as a cat and B and C as dogs, I might also categorise A and B as non-threatening but C (who is snarling and giving me the evil eye) as a threat. The categories of cats, dogs, and threats were well-established in my mind such that all I had to do was to sort those things I encountered into the correct categories. If so then pattern and category are effectively two sides of the same coin. A second kind of category thinking involves the creation and manipulation of categories. After all, we are not born with categories imprinted in our minds, so all of the categories we hold in our minds have been acquired, whether through experience, schooling, reflection, or social immersion, or developed from these things. As an example, learning something new typically involves learning its categories and how they fit into our broader category systems. Even if we already have a category system for a particular set of phenomena it is unlikely to be fixed. We can also be playful with category by speculatively moving things between categories and finding amusement in the inconsistencies and possibilities this suggests, and we can use contradictions in category in support of metaphorical thinking (Lakoff & Johnson 1981).

If categories are abstract ways of organising and relating things in our minds, then are categories and patterns essentially the same thing? They are primarily cognitive, they are abstracting and connecting, and they are both concerned with and dependent on assessments of similarity and difference. Moreover, patterns and categories can be expressed in various ways, and they can be used as part of other categories. However, despite these similarities, there are important differences. Patterns are nexuses of memory and thought while categories are abstract and symbolic, patterns convey meaning, significance, and possibility while categories convey structure and relation. The connection between category and pattern I would argue, albeit provisionally, is that categories are our indices for our patterns.

In Chapter 5, I described how patterns can include symbols and how these symbols can act as part of other symbolic pattern systems in our minds (such as language). It would seem that these symbols and symbol systems form the basis of category. For instance, not only do most of us have a pattern for "cat" but it is very likely that a part of that pattern includes one or more word(s) for "cat" in whatever language(s) one uses. Given that pattern elements can be a part of more than one pattern, then these words are likely to be both a part of the memory pattern of cat-related experiences and thoughts and of the abstract pattern system of language. The connection would seem to work in either direction; if I read the word "cat" (as you just did) then it might invoke memories and thoughts of cats, if I see something my pattern recognising mind tells me is an instance of the pattern of a cat then it can trigger the word "cat", for instance, in thinking "that's a cat!". Arguing counterfactually, patterns without categories would be no more than a chaotic continuum of memory. Categories add structure and definition to our patterns. However, category would not solely seem to be an indexing system; what we call formal category is perhaps better understood in terms of macropatterned connectivity. Indeed, category itself can be categorised, as Medin & Rips suggested:

> A concept is a mental representation that picks out a set of entities, or a category. That is, concepts refer, and what they refer to are categories. It is also commonly assumed that category membership is not arbitrary, but rather a principled matter. What goes into a category belongs there by virtue of some lawlike regularities.
>
> *(Medin & Rips 2005, p. 37)*

If concepts are categories of categories, then maybe, like pattern, our minds form layers of categorical structures that are consonant with the layers of patterns we build. If so, then concepts might be considered to be pointing to macropatterns and categories to mesopatterns. However, I do not think we need such absolute (categorical) distinctions for phenomena that seem fractal, complex, and layered, that differ only in the levels and degrees of abstraction at which they work. Occam would, I think, have argued for a simple system of pattern thinking that (as I suggested in the earlier section on "artificial pattern thinking") involves multiple layered and interacting (even panarchic) cycles of pattern recall, creation, and elaboration.

Without categories how would our minds be able to access their pattern memories? Clearly, we can search and make use of our patterns quickly and efficiently even though our patterns are fluid nexuses of thought that are perhaps called into being by nothing more than density and frequency of connection. Without some kind of symbolic indexing our patterns would be hard to distinguish and we would be unable to make much use of them. It would seem, therefore, that the more easily, consistently, and efficiently we can resolve "which pattern?" questions the more robust our categorical thinking is likely to be, and the more subtlety and nuance in our categorical thinking the more subtle and nuanced our pattern thinking is likely to be.

Although category affords a degree of ease, consistency, and efficiency in pattern thinking, our categories need not be perfect. Indeed, what would constitute a perfect category? One that in every way aligns with the reality it reflects perhaps? Except that reality can be and often is simultaneously interpreted in terms of different and even contradictory categories. A categorical system that allows plurality, partiality, and ambiguity in pattern thinking would seem more useful, and (useful or not) it would appear to reflect our phenomenological experiences of being pattern thinkers. We have little access to each other's minds and even less to those of other species, indeed we are still unsure what non-human mind might be like or even if it exists (depending on how "mind" is defined) but, assuming that many more species than humans are pattern thinkers, the difference between them and humans would seem to be the abstraction and symbolic connection and layering of pattern memory in the human neocortex.

Category would also seem to be the basis of pattern distinctiveness. Why do we not experience our pattern memory as one continuous chaotic ocean of thought? I would suggest that it is category, in pointing to a particular pattern nexus or parts thereof, that affords our phenomenological experience of patterns as being distinct. This is not a matter of individual categories though but of systems of categories. Distinctions between particular categories may still be somewhat arbitrary and reflect relative tensions between their underlying logics, but boundaries nevertheless become apparent when we use categories. Indeed, we might consider categories to be functionally similar to Voronoi diagrams such that they allow us to organise continuities of patterns into distinctions (Aurenhammer & Klein 1999).

If it is our learning of or development of new categories and category systems that adds nuance and distinction to pattern thinking and perception, then how does this work? It cannot simply be the acquisition of a particular symbol or label for a pattern, such as the letters c-a-t in that order, as this has no more intrinsic meaning than n-g-e-r-u, p-a-k-a, or n-e-k-o (the Maori, Swahili, and Japanese words for cat, respectively and phonetically). Clearly, there also needs to be some rationale for what is and is not legitimately to be a part of any given category. If we have a pattern for "cat" and have associated the linguistic symbol c-a-t (or neko, katze, etc.) with that pattern, then pulling on an experiential part of the pattern for a cat will trigger the symbol every bit as much as pulling on the symbol will trigger the pattern. Although the pattern may include macropattern logics it need only be a bundle of associated subjective memories of cat-related experiences to solve the "which category does this belong in?" question.

If category is a kind of taxonomic layer of pattern thinking in our patterned minds, is this the only way that we can work with category? I think not; category can follow a taxonomic logic, but it can also follow other logics. For example, consider the categorical structures of taxonomies and facets. Although as a term taxonomy can be used to refer to any form of classification, its use in the biological sciences has made it synonymous with Linnaean classifications such that everything fits logically into one branch of a common tree. As I described earlier in the context of reasoning, faceted classification is about non-exclusive intersections of characteristics. For instance, a taxonomic category for a garden pea could focus on its being a member of the legume family, its edibility, its size, its colour, and so on. While a taxonomic category seeks a precise location, a faceted approach is open and expansive. Sometimes we use faceted categorical thinking that queries similarities and differences across pattern elements and characteristics and sometimes we use a more taxonomic categorical approach that comes from language and our other symbol systems.

Fuzzy Pattern Knowledge

I have noted many ways in which pattern knowledge might be considered to be *fuzzy*. Patterns have no distinct edges (setting aside the overlaid distinctions of categories), they change over time (memories are added and lost), and we adapt and elaborate on the as we commit our experience to memory. Pattern thinking is not static; it is developed and honed to respond to experience and to allow us to be the best pattern thinkers we can be according to our capacity and opportunity. Pattern thinking can adeptly reflect subjectivities, variances, and individual perceptions. Patterns can become more objective, particularly when shared and elaborated and when they engage and reflect evidence, consensus, and proof. A pattern may be activated in different ways such that starting from different elements can make different pattern characteristics more prominent or dominant. Pattern perception is primarily (but not exclusively) abductive and heuristic in that it works rapidly, often with imperfect or partial information, to suggest provisional pattern solutions. Apparent pattern solutions may change as the information changes and alternative better matches are found. Solutions can, therefore, satisfice without having to be exact or definitive. Moreover, pattern thinking has many limitations; it is subject to illusions, pathologies, bias, and irrationality. Patterns are not precise, they use heuristics, although to different degrees. Patterns are unique to a particular mind; we can share our patterns, but only in lossy and imprecise ways. All of this would again suggest that pattern knowledge has an intrinsically fuzzy epistemology.

In noting both that patterns group similar but non-identical things, and that pattern perception often works with ambiguous and incomplete information, I might describe these characteristics in terms of structural fuzziness, epistemic fuzziness, and procedural fuzziness. I discussed structural fuzziness earlier in this chapter, and I deal with procedural fuzziness in the upcoming section on pattern heuristics. It is epistemic fuzziness, therefore, that I consider here, and this, to be clear, is about the clustering of non-identical things around our patterns. In doing so I return to Wittgenstein's (2009) concept of "family resemblances" (Familienähnlichkeit). Let me explore what happens when I substitute "pattern" for "family" and see how this plays out.

Pompa (1967) argued that rules need to be defined before any family resemblance can be established or evaluated, which would suggest that both entity X and entity Y must be perceived as instances of pattern Z for their family resemblances (i.e., pattern resemblances) to be apparent. This is a somewhat constrained logic however and one moreover that does not apply well to pattern. If, as I have argued, patterns are developed abductively then perceiving instances of a pattern reflects the partial and emergent state of that pattern at the point of perception, not a stable and fully developed set of rules or logics. Indeed, perceiving an instance of a pattern changes the pattern as memories of the instance are woven into it. What if we perceive something as being an instance of multiple patterns simultaneously, some of which may afford family resemblances while others may not? Pattern knowledge is the product of abductive (iterative approximations), probability based (more or less likely), plural (more than one pattern may apply or be applied), symbolic (categories), and affective thinking. Even in cases where pattern knowledge is more formal (macropatterned), I would argue that family resemblance is still intrinsically middle range in that, while not all things in a family share the same characteristics, as a bounded set of things there is a dynamic sufficiency of common characteristics to establish a common pattern identity. Pattern knowledge has both middle range and heuristic characteristics that satisfice without needing to be exhaustive of all possible conditions.

Rosch & Mervis also explored family resemblance, focusing on competing categories rather than categories in isolation from each other. We might say, therefore, that there is an assumption that X must belong somewhere and that our primary challenge is to select the appropriate (and available) category for it:

> The principle of family resemblance relationships can be restated in terms of cue validity since the attributes most distributed among members of a category and least distributed among members of contrasting categories are, by definition, the most valid cues to membership in the category in question.
>
> *(Rosch & Mervis 1975)*

The apparent speed of much of our pattern recognition suggests that pattern matching is not a deliberate and exact process but no more than a rapid scan or trawl (like passing a magnet over dirt to catch any iron fragments) followed by a cascading, aggregating, and satisficing evaluation of the "hits" from this trawl according to the distribution of apparent and significant characteristics across candidate patterns. Of particular interest here is Rosch & Mervis' observation that not all categories resolve based on choice between contrasting categories or by using family resemblances. The higher the level of superordinacy of a category (i.e., the pattern that it refers to is more macropatterned) the greater it depends on family resemblances to resolve what can be considered a part of that category and the less it depends on contrasting categories:

> Superordinate semantic categories … are sufficiently abstract that they have few, if any, attributes common to all members. Thus, such categories may consist almost entirely of items related to each other by means of family resemblances of overlapping attributes … their membership consists of a finite number of names of basic level categories …. [that] … are the level of abstraction at which the basic category cuts in the world may be made. However, basic level categories present a sampling problem since their membership consists of an infinite number of objects [although they] do form contrast sets.
>
> *(Rosch & Mervis 1975)*

This would suggest that the more mesopatterned our knowledge is, the more "which pattern?" questions resolve by comparing alternative patterns and the more macropatterned our knowledge is the more it depends on resolving family resemblances.

That not all pattern thinking is based on resolving family resemblance seems to be an important distinction, both in drawing out the layered nature of patterns and in noting different kinds of processes would seem to apply at different levels. As before, in describing a continuum from micropatterns to macropatterns, there are no clear distinctions between levels other than by relative degrees of abstraction, comparison, and collation, to which we might also add the distinction based on the degree to which comparators or family resemblances are used. Again, there is likely to be no abrupt switch from one modality to the other but rather a situational and heuristic accommodation of the two together. This would support my earlier suggestion that there is no one overall pattern epistemology, rather there are different epistemological positions that may apply heuristically (at need and only as long as they work) or abductively (iteratively adjusted to fit). Pinker's observation of parallel systems within our minds would also seem to support this argument:

> Human rationality is a hybrid system. The brain contains pattern associators that soak up family resemblances and aggregate large numbers of statistical clues. But it also contains a logical symbol manipulator that can assemble concepts into propositions and draw out their implications.
>
> *(Pinker 2021, p. 108)*

However, their differences notwithstanding, both approaches must be based on the same pattern thinking matrix. While macropatterns may tend to be more abstract, they are not intrinsically logical or symbolic. Rather, we might consider different pattern logics that might apply to different kinds and levels of aggregation and abstraction.

This plurality of approaches and considerations extends, I think, to all aspects of pattern knowledge. For instance, there are different bases for assessing similar difference and different similarity that we might use, each of which might return a different result. As an example, homologous similarity considers similar structures but different functions while analogous similarity compares similar functions but different structures (Root-Bernstein 1984). Thus, my earlier comments on pattern matching to stimuli that suggested a simple better/worse matching relationship may be too simplistic. Rather, pattern thinking may involve many different assessments of similarity and difference that run in parallel, in sequence, or in some other combination. Central to this then is a sense of pattern thinking and the pattern knowledge it generates as being based on probabilities (likelihoods, risks, certainties, and doubts). This suggests that fuzziness in pattern knowledge can also be understood, at least in part, in terms of probability. Bayesian epistemology may be useful here as a way of modelling and theorising how we think with truths that are not absolutely or unambiguously true or false:

> The probability calculus is especially suited to represent degrees of belief (credences) and to deal with questions of belief change, confirmation, evidence, justification, and coherence … Bayesian epistemology allows for a more precise and fine-grained analysis which takes the gradual aspects of these central epistemological notions into account.
>
> *(Hartmann & Sprenger 2011, p. 609)*

A key concern in Bayesian reasoning is that it responds to uncertainty rather than chance. This is reflected in Bayes' rule is that an individual's degrees of uncertainty can change based on new evidence (Titelbaum 2022). Given the adaptive nature of pattern perception and the abductive nature of pattern formation and elaboration, a pattern theoretical stance would seem to align with this. A full exposition of pattern knowledge and pattern thinking through a Bayesian lens is out of scope here, but, as an example, we might consider problems associated with poor understanding of prior and posterior problems as our use of emotional signalling and comparison in resolving uncertainty rather than the rigorous techniques of Bayesian analysis. This in turn reflects Pinker's comments:

> [Kahneman and Tversky] suggested we don't engage in Bayesian reasoning at all. Instead we judge the probability that an instance belongs to a category by how representative it is: how similar it is to the prototype or stereotype of that category, which we mentally represent as a fuzzy family with its criss-crossing resemblances.
>
> *(Pinker 2021, p. 155)*

If true, this suggests that pattern thinking is different from and can interfere with more rigorous thinking. However, a focus on priors and base rates assumes that this information is available, that it is accurate and representative, and that the trade-off between a large data set and a specific sample type is appropriate to the phenomenon in question. After all, pattern thinking is heuristic in that it works with incomplete and ambiguous information, it works with different possible interpretations, it often deals with mistakes and misunderstandings, and it often deals with facets rather than categories. We typically reason from what we have and know, not from what we do not have and do not know.

Sound reasoning should involve caution in the face of uncertainty. Uncertainty may be due to inadequate or ambiguous information such that pattern recognition falters or struggles to find a definite match. It may be due to unfamiliarity if the necessary patterns to make sense of something are absent, poorly formed, or underdeveloped. The more uncertain we are the more cautious about committing to one pattern knowledge-informed interpretation and the more open to alternatives we are likely to be.

Of course, it might also be argued that pattern knowledge fuzziness reflects referent fuzziness. Certainly, patterns can refer to things that are changing, uncertain, variant, and fluid. However, I would argue that pattern knowledge is adept at accommodating variances in referent instances without this making that knowledge particularly fuzzy in any of the other ways I have described. For instance, superordinate patterns (or their indexing categories) that are inclusive of wider ranges of variance are fuzzier that those that are more specific ("cats" as opposed to "tortoiseshell kittens") but the perception of particular instances may be fuzzy independent of what the pattern knowledge refers to. The epistemologies pattern knowledge affords are not the same as the epistemologies of pattern knowledge itself, although they are not entirely distinct. For instance, an individual's perceptions and the patterns that underly them may be fuzzy if their patterns are still forming (for instance, in young children), or if they include illusions and misunderstandings.

Although there is variance in form and order in the world, I would suggest that it is our knowledge of things that more usually lacks clarity, precision, and certainty. Although knowledge is always to some extent fuzzy (imperfect, partial, fluid), subjects such physics, chemistry, biology, and medicine, are typically taught as if their knowledge claims are definite and unambiguous. Even the accommodation of errors, miscalibrations, noise, false assumptions, and uncertainty reflect a striving for precision. Heisenberg and others described fundamental limits to knowledge in the world of particle physics, Gödel demonstrated the flawed nature of mathematics, and Shannon described entropy in the context of information theory. If sciences that depend on precise measurements are intrinsically fuzzy (acknowledged in much post-positivist thought), then how much more so are the social sciences with their dependence on hermeneutics, abductive theorising, and middle range theory? How well do any of the sciences accommodate the inescapability of pattern thinking and pattern knowledge? What if we take formal models of knowledge, such as justified true belief, to be special cases, which, despite their heightened rigour, still rest on foundations of pattern knowledge? Rather than saying that pattern knowledge is intrinsically fuzzy it would be better to say that it has the capacity for fuzziness both in different degrees and in different ways.

Macropattern Epistemologies

It would seem that we invest much time and effort in developing macropattern knowledge. Macropatterns constitute our theories of ourselves and the worlds we inhabit, they allow us to communicate and collaborate with each other, and they form the bedrock on which we build so much of human activity. Moreover, macropattern knowledge tends to be abstract rather than fuzzy and is often (but by no means always) subject to rigorous and critical development. In this section, I therefore consider the epistemological characteristics of macropatterns.

A first broad characteristic involves macropattern *aggregation*. Macropatterns are often created from other patterns, albeit in different ways. The most straightforward of these

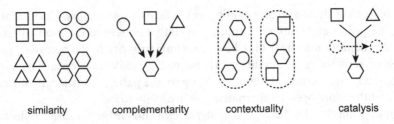

similarity   complementarity   contextuality   catalysis

FIGURE 7.6 Four kinds of macropattern aggregation.

Diagram prepared by the author.

involves superordinate groupings based on connecting similar things. For instance, at some point (although I do not remember quite when it was) I connected my patterns for lions, tigers, cheetahs, and panthers with my pattern for domestic cats to create a macropattern of felines, and then later from felines to mammals, from mammals to vertebrates, and so on. A second form of aggregation involves connecting different patterns based on their complementary nature. For instance, I can understand cats in terms of their similar differences and different similarities to other non-catlike things. Pattern aggregation may also be based on connecting things that share the same or similar contexts such that a context is understood in terms of its various interacting patterns. A fourth form of aggregation involves one or more macropatterns catalysing connections between other patterns to form a new macropattern. For instance, the macropatterns of language, mathematics, and scientific methods can all be used to catalyse the creation of new macropattern knowledge from other patterns. Macropattern aggregation can be deliberate and rigorous, but it need not be. For instance, we can build macropatterns around superstitions, suppositions, idle thinking, or our imaginations. These four kinds of macropattern aggregation are illustrated in Figure 7.6.

There are various possible benefits of pattern aggregation. One is that we tend to build these groupings because the things they refer to rarely if ever exist or are experienced independently. Patterns can be aggregated into macropatterns simply because they reflect phenomena that are themselves aggregated. Another benefit is reflected in the efficiencies that macropattern aggregation can afford. Aggregation allows us to deal better with wholes than parts, it allows us to perceive deeper or more subtle similarities and differences, it allows us to deal with archetypes (and as such it is a major enabler of category), and it allows us build pattern knowledge from knowledge we already have rather than starting afresh with each new encounter. This would seem to afford important efficiencies; we need fewer patterns overall, and we establish putative associations between different macropatterns that have integrated the same component patterns.

A second broad characteristic of macropatterns is *abstraction*:

Abstraction is a distinctive mental process in which new ideas or conceptions are formed by considering the common features of several objects or ideas and ignoring the irrelevant features that distinguish those objects.

(Falguera et al. 2022)

While this may be generally the case, as I earlier noted, comparing common features would seem to become less and less useful the more abstraction a pattern has such that family resemblance replaces comparison as the basis for establishing identity. From this, I would argue that macropattern knowledge is not abstract in any absolute sense, but it is, broadly, the product of abstraction processes in that it groups similar things (macropattern aggregation), and it is selective in which characteristics are used in these groupings. The principles of abstraction are simplification and unification (Cheng 2022). Abstraction is not an end point but a process that can continue to refine or alter its products, connecting and merging with other abstractions. Pattern knowledge cycles produce knowledge that can be returned to pattern memory that is more (or differently) abstract than the knowledge it pulled from pattern memory. Such cycles might typically make small abstraction changes, such as adding memories of a new instance to an existing pattern and accordingly they likely go unnoticed. However, if the difference in abstraction is larger, for instance, in learning a new concept or in experiencing a significant revelation then we are more likely to notice as the effort required to accommodate the changes in pattern memory are more substantial and more prone to fatigue and error.

A third broad characteristic involves macropattern *theorisation*. Much of our pattern thinking is focused on associations, origins, causes, and regularities. When they are found, we can elaborate our existing macropatterns based on abductive translation of similarities and differences between instances into principles, tendencies, and rules. The resulting pattern knowledge can help to describe, explain, and/or predict existing and future encounters with the things it refers to. Essentially, different apparent or possible relationships are considered, and, if they fit the evidence (or are otherwise satisficing), then they are retained or promoted while those that do not fit the evidence are dropped or demoted. This kind of abductive thinking can also help us to identify and assess probabilities, confounders, and limitations to the relationships or connections proposed. This is, in essence, theory generation (depending, of course, on how you define theory). From this perspective, macropattern knowledge helps us to describe or explain what we have already experienced, it can help us to describe or explain our current experiences, and it can help us to predict (to some extent) what is likely to happen next. Each time it is triggered it can be developed and refined further, proceeding through multiple and iterative abductive cycles. The theory or theories that best fit our perceptions, past and present, will be what is reinforced as pattern knowledge and what is woven back into pattern memory. I note parallels in this to the assertion associated with Kant that "theory without experience is empty, but experience without theory is blind", which I translate to suggest that our patterned thinking shapes our actions and perceptions even as those actions and perceptions shape our patterned minds. Does this mean that all macropatterns are theoretical or infused with theory? If there is any sense of meaning, significance, explanation, causality, origin, import, or probability then I would argue that indeed all pattern knowledge derived from this is indeed theory rich. However, while true for macropatterns, the more micropatterned things are the less this would seem to be true. We are, therefore, left with another partiality, another continuum.

A fourth broad macropattern characteristic is *generalisation*. Generalising involves extrapolating knowledge from a small number of examples to apply it to other examples, to other contexts, or to whole populations. We can contrast nomothetic generalisability of knowledge which should apply to all instances everywhere (as in physics and chemistry) with middle-range generalisability that applies in certain similar contexts but not necessarily beyond (Merton 1949). Generalisability is an assessment, sometimes undertaken deductively (doing all the things that should increase or ensure generalisability) sometimes inductively (testing whether something does work or apply elsewhere) and sometimes abductively (something generalises to some extent, although it still needs to be adapted to fit a particular context or population). Every pattern perception we have is a test of generalisability in that it is predicated on finding a pattern memory to match current experience even if the fit is not ideal. Pattern matching is not a matter of finding the 'right' pattern but of abductively finding a satisficing pattern. If the fit is still not good then the pattern might be updated to accommodate the new experience, a variation on the pattern might be created, or a new pattern nexus might be developed. Generalising macropattern knowledge can involve anticipation in that, to be of any use, my pattern knowledge should be able to be applied to instances and contexts not yet encountered. Rather than saying that macropattern knowledge is intrinsically general, I think it would be better to say that generalisability in the context of pattern knowledge is the result of a pragmatic process, a dynamic trend towards better generalisability based on experience and apparent need.

Before moving on, I would briefly note differences between concepts of generalisation, translation, and transfer as the implications for pattern knowledge are different for each. Generalisation refers to the breadth or range of examples and variations to which a pattern can meaningfully apply; the pattern does not change, rather, generalisation is a product of the necessary degree of fit to count as an instance and the range of things that have that degree of fit or better such that they count as instances. Translation is required when the fit is close but insufficient to count as an instance such that changes to the pattern are required to achieve a satisfactory fit. In translation the pattern changes to accommodate the new variance. Transfer is when the fit is poor enough that a variant pattern is required but near enough such that aspects of existing patterns can form a part of the new pattern nexus.

A fifth broad characteristic of macropattern is *signification*. Explanations and predictions can spark revelations (both positive and negative), they can have aesthetic or ethical implications, and they may help us to deal with other issues of value, all of which can evoke emotional responses. In the context of macropatterns, this involves weaving affect into the abstractions and theorisations of macropattern knowledge in ways that build meanings. The resulting signification in our macropatterned knowledge flags what is important, what is appropriate, and what is right (and also what is unimportant, inappropriate, and wrong). By adding valence to macropattern, signification can act as a balance to reductive abstraction, particularly when it might otherwise elide associated feelings and meanings. What may be perceived as significant in one context may not be in another. Signification is context-dependent (or paradigm-dependent) in terms of what rules, relationships, and meanings apply or matter. If we do not understand the context of application, if we have incorrectly identified the context, if we are unaware of there being contextual factors involved, or if we disregard them, then any signification afforded by macropattern knowledge will be misleading.

Formalisation is another broad macropattern characteristic. While macropattern knowledge generally involves abstraction, theorisation, generalisation, and signification, the remaining two epistemic factors, formalisation and systematisation are indicated in some but not all macropattern knowledge. Macropattern formalisation involves rendering one macropattern in terms of another macropattern or macropattern system. For instance, we may use language, logic, and mathematics to develop or test ideas, to conduct analyses, to generate proofs or exemplars, or to communicate our ideas once they are suitably developed. Sometimes our macropattern knowledge is well established and formalisation is simply a matter of rendering ideas into concrete forms. Sometimes we use formalisation to work through ideas. There is much truth in Didion's (2000) paraphrased observation that "I don't know what I'm thinking until I write it down".

Perhaps the most common approach to formalisation is to express potential macropattern ideas in words or symbols and cumulatively into sentences, equations, and from there to larger analytic structures. Alternatively, we might draw pictures or diagrams, a way of formalising macropattern ideas often favoured by architects, engineers, inventors, and designers as it can render nonlinear relationships and draw on spatial, textural, chromatic, and other characteristics not easily expressed in the linearities of text. Formalisation may involve creating pattern expressions or pattern representations, for instance, as mathematical equations, text-based theoretical statements or rules, or functional models. It may even involve the creation of pattern systems or pattern languages consisting of multiple expressions and/or representations. Whatever the purpose and medium of formalisation, it is, like abstraction, a reductive process. It is even cartographic in that, not only are less relevant details omitted, the details that are included are rendered in a unifying and consistent way (the same symbol being used for the same kind of entity). This consistency of formalisation can stabilise and normalise an otherwise complex construct, and it can make the pattern knowledge produced more tractable. Without formalisation, pattern knowledge could not translate to or form the basis of scientific knowledge, it could not be shared, taught, and otherwise socialised within communities and networks of enquiry, and it could not therefore have been the catalyst for so much of our modern world.

One last macropattern characteristic to consider is that of *systemisation*; translating pattern knowledge into practices, structures, customs, institutions, and organisations. For instance, our shared macropatterns reflect societal values (democracy, justice, power, hierarchy) and structures (such as fiscal, health, and military policies) and the organisations that support them. We can draw on the concept of doxa to make sense of this. In classical philosophy two kinds of knowledge were often contrasted: doxa (habitual and accepted knowledge) and episteme (knowledge produced by reasoning), with doxa considered the poorer if the more common of the two. The concept is captured in modern concepts of orthodoxy (official knowledge), and of unorthodoxy and heterodoxy (at variance with orthodoxy). Bourdieu (1977) redefined doxa to refer to knowledge we acquire through socialisation and that we then reproduce through participation in social activity. These concepts of macropattern systematisation are about conforming or following (or not) the macropattern knowledge of the collective (or of those in control of the collective). Doxa is about how much we defer our individual macropattern thinking to that of the collective and how much we defy it.

We might be more agentic in macropattern systematisation, for instance, by participating in and shaping doxa for the collective though political or other forms of leadership, or through creative and intellectual acts (such as research, critique, literature, and the dramatic arts), albeit with the ever-present normalising potential of engagement with others. Macropattern systematisation can involve our macropatterns being disseminated out into the communities and cultures we engage with, and it can involve our assimilation of shared macropattern thinking into our own patterned minds. Typically, as we are educated and socialised to particular forms of human activity long before we can make any material contribution to them, assimilation is likely the much more common and indeed pervasive aspect of macropattern systematisation, and indeed, based on any individual's circumstances, opportunities, and choices, they may contribute little or none of their own macropattern thinking to the collective.

Macropattern or Macropatterning?

I earlier noted apparent functional differences between the left and right hemispheres of the human brain, particularly with respect to analytical thinking and meaning making. McGilchrist (2019) described the left hemisphere as broadly adept at abstraction, analysis, conceptualisation, and categorisation, while the right hemisphere is where emotion, empathy, imagination, creativity, morality, and humour are processed. It might be argued on this basis that the left hemisphere is where much of our macropattern development takes place. However, McGilchrist also emphasised that while the left hemisphere focuses on parts, the right hemisphere focuses on wholes. This, linked with other suggestions that thinking that has meaning and significance for the individual concerned tends to be located in the right rather than the left hemisphere (Shermer 2011), would suggest that both hemispheres are involved in macropattern formation. The left hemisphere focusing on the analytic aspects of macropattern development, i.e., abstraction and theorising, the right hemisphere focusing on the holistic and associative aspects of macropattern development, i.e., generalisation and signification. This suggests a tentative model of macropattern knowledge (as the product of macropattern thinking) that crosses back and forth between hemispheres – see Figure 7.7.

However, neat as this model may seem, it seems a little oversimplified. For instance, I cannot readily think of any macropatterns that lack abstraction or theorisation, from which I could provisionally conclude that these two characteristics or processes are indeed essential to macropattern development. There might not be much development and the abstractions and theories involved might be crude, but I think that abstraction and theorisation must be there in some way or other. Of course, this might say more about my cognitive limitations than it does about pattern epistemologies. As to aggregation, as patterns have no edges and their distinctiveness is largely illusionary, who is to say whether patterns are really aggregated, i.e., subsumed, in macropattern thinking? Rather than assimilation, the reality may be that macropatterning involves creating evermore complex

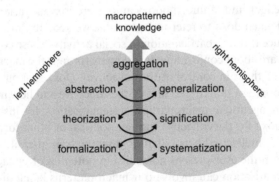

FIGURE 7.7   Macropattern knowledge can be understood as involving epistemic characteristics of aggregation, abstraction, theorisation, signification, formalisation, and system-atisation, which tend to occur in one or other hemisphere of the human brain. These processes would seem to interact in support of abductive cycles of knowledge development: aggregation with abstraction and generalisation, theorisation with signification, and formalisation with systematisation.

Diagram prepared by the author.

interconnections and associations between pattern nexuses. Aggregation may be more metaphorical or analogical, therefore, than it is real.

How about generalisability? It seems quite possible for macropatterns to be developed that do not apply outside their immediate context and are not expected to do so, such as patterns related to professional practice, particularly in subspecialties. Similarly, patterns that apply to a relationship with one person may have no currency or meaning outside of that relationship. Can macropatterns lack signification? Again, potentially, this is also possible. For instance, we may learn patterns of rules and regulations required to participate in a particular social context (school, work, home, etc.) and yet be indifferent to them, whether they are followed or breached. Similarly, those of us who struggle to learn a subject or skill may struggle not because the topic is difficult but because they simply do not care about it. What about formalisation or systematisation? Certainly, not all macropatterns need to be rendered linguistically or mathematically (or in any other symbolic way). Take social customs in high context cultures that are well understood by its participants without having been formally codified. However, that is not to say these patterns are not systematised as they are very present in social relations within these cultures. Similarly, patterns related to ideology or faith may not be formalised but can still be systematised. As a corollary, a scientist, philosopher, or inventor may engage in much formalisation of their pattern thinking but not seek to translate their ideas into practices or share them within their collectives.

So, of these characteristics of macropattern, I would argue that only aggregation, abstraction and theorisation are essential to macropattern, and even then, only in a broad sense. While the other four may often feature in macropattern knowledge and they may variously be essential for particular kinds of macropattern thinking, they are not intrinsic characteristics of macropatterns. Moreover, I should distinguish between the processes that generate macropattern knowledge and the knowledge that they generate. While macropatterning processes involve abstraction and theorisation, is the macropattern knowledge thus created abstract or theoretical? Abstraction is a relative concept after all; abstract compared to what? Whether such knowledge is theoretical depends on whether the knowledge articulates theory or is simply dependent on the theories that were used to develop it. Is macropattern knowledge generalised? Macropattern knowledge might be cast in general terms, but generalisability is again a relational construct and can only be assessed relative to whatever context or contexts it is to be applied to or in. What about signification in pattern knowledge? Signification is about feeling and value and as such requires an embodied mind; macropattern knowledge can be significant only while it is being thought. As to whether macropattern knowledge is formalised, that would seem to depend entirely on the processes that generated it. And finally, as to whether macropattern knowledge is systematised, that depends entirely on what is done with it.

It would seem, therefore, that macropattern knowledge is every bit as variant as the thinking that has generated it, even as we (or at least I) think about it as a common if somewhat abstract phenomenon. Indeed, this would seem to be another family resemblance, which in turn suggests that macropattern is itself a macropattern and the knowledge that we have or can suggest regarding it is subject to the same characteristics and limitations. Essentially, the epistemology of macropattern is itself macropatterned. Such circular reasoning is problematic I readily admit, and I think it is my terminology that is at fault here. Rather than talking about macropattern knowledge, I should have used the adjectival term "macropatterned knowledge" such that macropatterning is a characteristic rather than an identity. From this perspective, knowledge might be better understood as being differently macropatterned based on the admixture of abstraction, theorisation, signification, generalisation, formalisation, and systematisation that went into its creation and its articulation.

To illustrate this, let us compare language and mathematics. Language is our most fundamental and most extensively developed pattern system – so much so that language as an abstract concept refers to any symbolic system that can be used to assemble units of meaning that can then be shared between minds and recorded in various ways. Here I mean language as an everyday means of communication shared by a broad community, such as Swahili, Finnish, or English. Language depends on words (in thought, in speech, and in written form) and their combination into sentences and utterances. Words depend on perception-pattern relationships; the word "cat" refers both to the hairy thing in front of me and to my patterned memories of cats. Language is also closely tied up with our categorising capabilities and tendencies. Assembling words onto broader units of meaning depends on the grammar, syntaxes, and idioms of the language being used but most languages afford a good deal of flexibility in how they are used. Languages are also formative in that any given language enables thought in some respects and constrains it in others, and that different languages can allow you to think different things or think about things differently. Mathematics also has a symbol system – primarily made up of numbers, variables, and operators of various kinds, however, the things that these symbols refer to are typically more abstract and precise than the ideas that we use words to refer to.

I might argue, therefore, that mathematics is more abstract and theoretical than language, that mathematics is less generalised and less signifying, and that mathematics and language are formalised and systematised in very different ways. You may disagree of course and that is my point, macropatterning is relative and interpretive, it is not a concrete characteristic except, perhaps, when translated to pattern representations such as laws or computer code. All knowledge might, therefore, be understood as being relatively macropatterned in that some knowledge is more macropatterned than other knowledge, and that any assessment of degrees of macropatterning depends on which criteria are used and how they are assessed and by whom. There are also exclusive and combinatory factors to be considered in making these assessments. Some are exclusive in that there are things you can do in everyday language that you cannot do in mathematics (such as speechmaking and poetry) and there are things you can do in mathematics that you cannot do in everyday language (such as bookkeeping and calculus). This also implies combinatory uses in that different kinds of macropatterned knowledge used together can do more than any one on its own. For instance, we need mathematics and language to communicate the processes and findings of scientific inquiry.

It would seem that, the more macropatterned knowledge becomes, the more fractal its patterned epistemic basis (patterns of patterns, pattern epistemologies or pattern epistemologies) becomes. On one hand this recursiveness could be seen as a weakness in the argument I have been advancing, not that it is wrong necessarily (the alignment with evidence and other theory is strong after all) but that it is incomplete. For a start, patterns cannot be "turtles all the way down" (or up). If we were to delve into the parts of patterns and the parts of part at some point micropatterns would no longer be patterns, they would be neurological structures and biochemical behaviours. On the other hand, as much as we are adept pattern thinkers, there are likely to be functional limits to how large and complex our patterns can be. Macropatterning is not (and cannot be) simply a matter of endlessly sticking other patterns together, it requires abstraction and generalisation to remove unnecessary details to maintain an efficient pattern density. Indeed, we might consider or explore pattern density empirically, for instance, in terms of the numbers of contributing or active neurons contributing to a pattern, the amount of traffic activating a pattern creates, or the density or complexity of any representations we might create from such a pattern. However, is pattern density the same as macropatterning? Denser patterns could mean more emotionally rich or more abstract, except abstraction reduces pattern density – bifurcation of macropatterning between rich pattern memory and theoretical abstract macropatterns. Density might also be reflected in the numbers of interacting layers of patterning, the panarchic relationships between layers, and the number/extent/complexity of interacting pattern knowledge cycles that any given macropattern has.

Another apparent characteristic of macropatterned thinking and knowledge is decentring. Micropatterned knowledge (taking a liberal definition of what constitutes knowledge) is intrinsically woven up with self as it is the closest to direct experience. Mesopatterned knowledge is more abstract, but it is still embodied in that it is the basis of perception. The pursuit of macropatterned knowledge may elide or side-line the self, in great part because the individuality of the mind in which the thinking arose becomes less important. This is not to say that macropatterned knowledge must elide the self, only that much macropatterned thinking seems to do so. This implies that we both to set aside our subjectivity in the pursuit of objectivity and abstraction and that we have the capability to do so. The tendency for macropatterned thinkers to decentre and elide themselves from their own pattern thinking may be why we often struggle to understand macropatterned thought; it intrinsically encourages an illusion of objectivity. While this can support a degree of rigour and critical thinking in some regards, it also means that our macropatterned thinking is often not fully appraised or understood, which can make us vulnerable to errors and misunderstandings. The one constant is that macropatterning is at the centre of all this. We cannot have advanced critical, analytical, scientific, and cultural thinking without macropatterning. Indeed, we develop patterned epistemologies that shape how we think about and how we interact with the world in all areas of human activity. Put another way, while mesopatterned thought is the primary lens through which we perceive the world, macropatterned thought is the primary way we understand (or think we understand) the world (Godfrey-Smith 2003).

Drawing the macropattern characteristics I have described together; we can consider a series of parallel continua for defining pattern knowledge and its relative states of micro- and macropatterning – see Figure 7.8. In advancing this model I should also say that macropatterning is not a mono-dimensional construct, rather it may vary across multiple dimensions. For instance, my knowledge of my home is aggregated yet concrete, experiential yet significant.

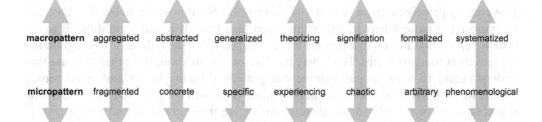

macropattern	aggregated	abstracted	generalized	theorizing	signification	formalized	systematized
micropattern	fragmented	concrete	specific	experiencing	chaotic	arbitrary	phenomenological

FIGURE 7.8 Pattern knowledge continua between micropatterned and macropatterned knowledge states.

Diagram prepared by the author.

Pattern Meta-epistemology

Before leaving the topic of epistemology, let us consider the epistemology of pattern theory and the knowledge that pattern theory can afford. The epistemology of pattern theory that I set out in the preceding chapters depended on a rather heterogeneous mix of theoretical and provisional empirical evidence. Moreover, some parts of pattern theory would seem to be more robust than others, at least in terms of evidence, alignment with other theory, and alignment within pattern theory. For instance, I called on my own experience of having a patterned mind (which was generated from within that mind) and of seeing patterned minds in others. Without robust empirical inquiry, I must admit the subjectivity of much of this and the potential for bias and error in the theory I have advanced, not least because much of the science (at the time of writing) remains exploratory and provisional. Not only are we unable to access each other's minds directly (we only do so indirectly from what we can perceive), we cannot consciously access much of our own minds. In many cases, phenomenological evidence, such as it is, is the best evidence we can bring to bear. This is not ideal, but it is the state of the art.

On the positive side, I have tried to connect many phenomena that have been explored more robustly to a theoretical whole. Indeed, I have argued that pattern thinking plays an intrinsic role in language, culture, science, art, social relations, and law, indeed in all aspects of human existence. Pattern allows us to deal with ambiguity and uncertainty. Pattern allows us to be our individual idiosyncratic and error-prone selves and yet still engage in collective activity.

In considering meta-epistemologies of pattern theory, we might also consider alternatives to pattern theory. For instance, what if we were just to agree that "pattern" as a term can mean anything we want it to mean? As I described in Chapters 1 and 2, this can create many more problems than it solves. I would argue, therefore, that a more critical and integrated theory of pattern is needed, and that includes more precision in the terminology that is used within that theory. That said, any meta-epistemology of pattern includes pattern as memory, pattern as thought, and pattern as reasoning. I do not want to criticise or erase science in these areas, and yet pattern is heavily implicated and would seem to afford a defensible theory of how much of our minds work.

Another aspect to consider is the meta-epistemology of pattern as illusion. Pattern thinking is what allows us to be mistaken, to be deceived, and to be inappropriately selective, inattentive, or ideologically blinkered. We think we see patterns around us, but these are just reflections of our own minds. There is regularity and there is a form in the sense of things exist outside our minds and they follow laws and tendencies of our universe. However, by perceiving them in terms of patterns (as we almost inevitably do), then are we imposing a pattern lens to understand a pattern lens? It would seem difficult to do otherwise, but that means that pattern theory is open to all the failings of pattern thinking as well as to its strengths. There may be little we can do about this other than use this understanding to analyse and critique our pattern thinking habits and conventions. Surely, we need some level of awareness of quite how patterned our minds are to stand any chance of being able to use them to maximise their strengths and mitigate their weaknesses.

Can pattern theory be turned to inappropriate ends? Of course. Any advances in science can be co-opted and used for bad things every bit as much as they can be used for good things, and often some of the bad things turn out to be good and many of the good things turn out to be bad. If we take a pattern theoretical approach to things, are we eliding other perspectives? I would suggest that pattern theory is a metatheory in the sense that it allows us to orient and understand other cognitive theory rather than being yet another competing theory of mind.

Pattern Axiology

Having considered ontology and epistemology of pattern, I now turn to its sibling, axiology. Axiology covers issues such as ethics and aesthetics, but its main focus is on values, and value systems. This begs the question, what are values? Do we mean feelings, preferences, principles, or beliefs? Do we mean personal values or something shared, such as the values of a community, profession, or society? Are these values used to guide decisions, to select between or prioritise options, or to make ethical or aesthetic judgements? Do we mean values that are normative (ideals) or descriptive (realistic)? These are all axiological questions, and there are many others.

Both Hartman (1967) and Edwards (2010) sought to outline a formal science of axiology. Much of their work was based on the idea that the closer something is to its ideal the more value it has and the further from the ideal the less value it has. This logic depended on how these ideals were defined and how instances were to be compared to ideals, and much of their formal axiological expositions wrestled with these issues. Edwards built on Hartman in distinguishing between systemic, extrinsic, and intrinsic values as the basis of any kind of axiological science; systemic values relate to ideas and concepts, extrinsic values relate to things in the world, and intrinsic values relate to people. Both Hartman and Edwards argued that intrinsic values are more important than extrinsic values, which in turn are more important than systemic values. That individuals are prepared to kill or be killed for ideas illustrates the fallacy in this argument. I would argue against such arbitrary structuration since values, like patterns, are intrinsically cognitive and experiential. We may share, encode, and otherwise abstract our values into structures and systems beyond our minds, but minds are required for and are central to any sense of value.

Hartman separated cognitive value thinking as something rational and, therefore, distinct from other kinds of value-oriented thinking, while Edwards took a more holistic perspective:

> We relate to the things we value not only cognitively, but also through feelings, emotions, attitudes, preferences, likings, and desires.
>
> *(Edwards 2010, p. 92)*

Although deliberate and analytical thinking about value may be more macropatterned than thinking based on feeling and emotion, these would seem to be matters of variation and degree rather than intrinsic differences, and I would argue, therefore, that all evaluative thinking is reflected in patterns and pattern thinking.

Noting the challenges of translating feelings to objective value statements, Edwards also considered differences between positive and negative feelings, and the different degrees of feeling we associate with certain values. While such valence is undoubtedly important, this model still separates value from feeling as if value were some kind of abstraction informed by (but quite separable from) feeling. If instead, as I have suggested, we consider emotions in pattern memory as cueing current evaluative thinking, and the results and experiences of current evaluative thinking as something that is woven back into pattern memory, then the entanglements between value, feeling, and pattern thinking become more apparent – see Figure 7.9.

As Edwards (2010) observed, we are constantly in a state of assessing and responding to values, both our own and those of others. Some of our values are things we feel; they reflect things that attract us and things that repel us, they reflect things we want and things that we want to avoid. Other values may be more abstract and, for instance, reflect an ideal or code of what is good and bad, or right and wrong, defined either internally (such as morals) or externally (such as ethics). Evaluations that are based on feelings and those that are based on codification may occur in isolation or together such that one can inform or lead to the other, or they can be bound closely together. Evaluations may be entirely affective, or they might be something we do by making comparisons, either between a thing and an ideal or code ("is this thing good?"), or between two or more things ("which of these is better (or best)?". Often, it would seem that we combine all these modalities. Values may also be more or less canonical or negotiable. For instance, assumed values are by definition not interrogated or evaluated, they are treated as given unassailable facts, typically based on socialisation or familiarity (or both).

Our own values and the values of others (as much as we are aware of them) can influence our thoughts and actions in they allow us to evaluate or assess what has value and what value it has, and they allow us to function in the many value-infused contexts in which those thoughts and actions take place and to which they refer. Over time we weave values

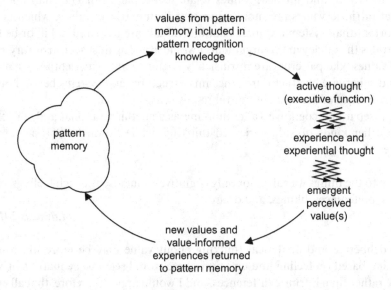

FIGURE 7.9 The pattern knowledge cycle rendered in terms of values. When a pattern is triggered and moved into active thought its value elements shape perceptions of value. These experiences can modify perceived values such that new or variant values may result, some of which are later returned to pattern memory.

Diagram prepared by the author.

into our sense of identity, purpose, and morality. We can also treat, analyse, or otherwise work with our values in non-evaluative and abstract ways (as I am doing right now) i.e., as the subject rather than the means or context of thought. When we think about others' values then we likely draw on our own values (again as part of means or context) as well as considering them more abstractly.

Given this, what characteristics of value do we hold in pattern memory or invoke to inform our current thinking? Rather than using a normative framework of systemic, extrinsic, and intrinsic values, we might start with our direct experiences of value. At its heart I would argue that value is experienced in terms of various combinations of feelings and knowledge. I do not see these as two ends of a continuum but rather as parallel factors – one experience of value (for instance, related to loyalty or integrity) may be laden with both feeling and knowledge while another (for instance, related to casual aesthetic reactions) may lack feeling and knowledge. Indeed, rather than value being made up of entwined parts of feeling and knowledge in some metaphorical double helix, what if we were to consider value to be the product of memories of feelings and knowledge? Knowledge without feeling would be highly abstract and impersonal, and, by definition, we would be disinterested in it. Feeling without knowledge may involve a limbic sense of value in terms of primal emotions linked, for instance, to danger, interest, and attraction (amongst others) but there would be no understanding, interpretation, or linking to past experiences. If value is the product of knowledge and feeling, then the more knowledge and feeling is involved then the greater would be the sense of value. However, this is more a definition of significance (the value of value) than value.

If value does not come by the pound or the yard, then what is it? Given that value seems to have different elements (feelings, facts, memories, symbols, understandings, etc.), that elements can be present in different proportions or emphases, that values can be developed, elaborated, and abstracted, and that values can be expressed and represented external to minds, then value would seem to be a particular kind of pattern nexus. Let me propose then that values are patterns that combine memories of experience and feeling. It would be very difficult to consistently and reliably define values with any greater precision, since, as patterns, they will always be more variant than any syntactic definition could capture. This would also mean that value, like other pattern phenomena, has open and potentially generative ontologies and epistemologies.

Although global definitions may be impossible, we might be able to say something more substantive about the regularities of value-as-pattern. For instance, do we use value in selecting matching pattern memory? In terms of pattern perception or other pattern recognition, when we are searching (or scanning or trawling) our pattern memories for matches to current stimuli, is this only a syntactic process matching the objectivities of experience to the objectivities of pattern? Or, as I might suggest, is pattern perception semantically rich such that value-in-pattern and value-as-pattern are part of this matching process? Pattern recall very often has an emotional aspect in that this match feels right (or wrong), or more important, and so on. In this way priority, probability, risk, and other affect-related characteristics that are central to value are tied to pattern knowledge. The affective component of any pattern may not be particularly dominant, but it will nevertheless be present in some form or other. If we were to remember things entirely without affect, then thinking becomes mechanistic, predictable, but lacking in humanity. Affect, therefore, is critical to our ability to function and to realise our human natures. Indeed, rather than affect being an occasional and parallel element alongside many others, it might be that affect is the most important functional part of a pattern thinking mind.

Rather than being an esoteric concern, affect is woven into all our thinking. Certainly, we use value from pattern memory to evaluate situations, objects, things, and ideas, as well as ourselves and other people. We do not simply identify things, we note whether they present threats or opportunities, whether they are consonant or dissonant relative to our ideals or norms for them or relative to our understanding of their usual contexts or presentations. We use values in making judgements about similar differences and different similarities. We use values in deciding how to act in or react to different situations, as well as in forming our identities, in forming our theories, in forming everything we think or do. We use values in developing and advancing goals, motives, and desires, and from these, our various fixed or fluid dispositions towards value and its uses.

Another consequence of value-as-pattern is that we can use a faceted approach to consider different characteristics of value patterns even as we acknowledge that each value pattern or pattern instance will vary in its composition. A faceted model of value is more akin to a repertoire that considers building blocks that come together in different ways, to different extents, and with different implications. Value-as-pattern implies elements and connections that outline value referent(s) and relationship(s). After all, value is constructed as a relationship between an individual mind and some other thing. Value is, therefore, relational and emergent both in the moment and over the duration of the relationship. Sometimes a referent might be very specific, such as a particular person or thing, and sometimes the referent might be relatively abstract (and macropatterned) such as ideas and other values. Whatever the relationship, value is realised in value relationships rather than being intrinsic to the value pattern alone. A second facet relates to elements that reflect the different kind(s) of value(s) involved in a value relationship. We tend to categorise values (or our understanding thereof) in terms of wants, needs, desires, principles, and preferences, to name but a few. In categorising them we are in effect weaving further connections between value patterns and the patterns of the things that are being valued. A third facet is related to value orientation. Although value pattern relationships tend to lie along a continuum of strength of feeling from strongly positive to strongly negative, orientation can also include the degree of normativity of a value relationship in terms of ideals, rules, and conventions (which are themselves grounded in patterns). A fourth facet reflects the logic(s) of evaluation applied in a value relationship. We process value relationships differently according to the norms, ideals, rules, and conventions associated with a particular value or its referent(s). In this way, value relationships may be logical and analytic (reflecting macropattern qualities of aggregation, abstraction, generalisation, theorisation, signification, formalisation, and systematisation) or affective and holistic, or some combination of the two. A fifth facet, but by no means the last, reflects the metavalue (the value of value). We typically have (or develop) a sense of value of the value relationship in terms of our confidence in that relationship, and its apparent significance, importance, and/or negotiability.

Although patterns can express and represent value patterns in personal, professional, legal, and other system contexts, at a more abstract level we can distinguish (categorise) this thinking in terms of ideologies, ethics, economics, and aesthetics. An ideology is a set of beliefs that they function as a system. Given that values are patterned then, ideologies too must be intrinsically patterned. This is reflected in Zizek's (2009) assertion that no one who has a mind is ever without ideology. Indeed, we typically have many active ideologies in respect of different things, and we are aware of many more. These can be assimilated or connected, for instance, through moral codes, a sense of identity, through personal philosophies or politics, or through loyalty and commitment to others, to a system (such as a profession or state), or to a particular religion. Again, as per Žižek, although ideology is often positioned or understood as a negative disposition, a weakness, or an unconscious or wilful bias, we cannot escape or erase ideology, we must confront our own ideologies as well as the ideologies of others we interact with. This would seem to involve a degree of macropattern development of ideology through aggregation, abstraction, generalisation, signification, theorisation, systematisation and so on, although, given the affective basis of value and belief, some factors, such as signification, may play a larger role than others. Moreover, because an ideology is a system of values and beliefs then it should be more properly understood as an implicit pattern system. From that perspective, there should be no surprise that parts of an ideological pattern system are often rendered as pattern expressions (for instance, in actions and arrangements that reflect an ideological position) and pattern representations (for instance, in codifying an ideology in terms of rules and regulations). Sharing, subscribing to, or being subject to a collective or dominant ideology (for instance, ideologies that reflect social, political, or economic value systems), individual codes and beliefs may come into conflict with those of the dominant group, particularly if they are imposed on individuals, or individuals may subsume or align their value systems to those of the collective. Pattern also guides our thinking regarding ideology, both in terms of our own ideology by organising systematising, providing category and organisation to our values and beliefs, but also allowing comparison, analysis, and critique of other ideologies. Indeed, our understanding of others' beliefs, ideas, and underlying motives would be impossible without engaging in shared pattern thinking to some degree: pattern thinking might even be argued to be the basis of theories of mind.

Ethics is primarily about codifying values in ways that we can use to guide or appraise our attitudes and actions to hold ourselves or others accountable for them, or to be held accountable by others. To that end, ethics is both normative and analytic in its application to the protection and stewardship of integrity, responsibility, and obligation with respect to a system of ethical values. Although there are some distinctions between ideology and ethics, at a pattern level there would seem to be no apparent differences. Ethics can be understood as being woven out of our patterns of memory and thought and have patterned affordances and limitations that are not appreciably different to other implicit or explicit pattern systems. This is true for other axiological macrostructures such as economics and aesthetics. Similarly, economics is about making or justifying value calculations and assessments, which in turn are often based on symbolic signification of value in terms of money, time, effort, or disruption. These economic calculations and assessments can help us to make decisions, or analyse decisions, made under conditions of scarcity or contention. Aesthetics can also reflect individual and collective assertions of value, although not necessarily in terms of codes or rules, but rather in terms of experiences and abstractions thereof, in service of assessing and evaluating quality, goodness and badness, beauty and ugliness, what is pleasing and what is displeasing. Different as they are in terms of purpose and scope, ideology, ethics, economics, and aesthetics are all patterned phenomena and behave as such.

Clearly, we readily attach values to our patterns, although that value cannot be said to be intrinsic to the pattern but rather to the pattern holder, observer, or user. Patterns that reflect or shape social structures and conventions have values wrapped around them as well as being woven through them. We also readily perceive patterns of values, ethics, and aesthetics, whether it be in philosophies or creeds, in professionalism or in lawfulness, or in systems of cultural expression (literature, poetry, music, dance, art, cuisine and so on). Patterns can please us, and they can disgust us, we can see ourselves in patterns "as in a mirror darkly", and we can see others in a similar light. Despite this, it is hard to make any global observation about pattern and axiology other than to note that there are a great many actual and possible pattern axiologies but no one overarching pattern axiology, unless it be the profound utility of patterns and pattern thinking in general.

Humans have evolved a disposition and a capacity for pattern thinking, we school our children to be pattern thinkers, and participation in human society is virtually impossible without pattern thinking. Given, therefore, that pattern and all its aspects (pattern memory, pattern perception, pattern thinking, etc.) form the basis for much of the way we experience and think about things, asking whether it is valuable would be like asking whether blood is valuable. Not only can we not live without it, having it allows us to be us, to be our idiosyncratic selves in a world that is complex, changing, ambiguous, and deeply nuanced.

Chapter Summary

In this chapter, I asked whether patterns exist given their intrinsic cognitive basis, and I explored issues and implications from the perspective of debates around the "mind-body problem". I explored how pattern related to theories of instances, types, and tokens, and I considered ontologies of pattern expressions, representations, systems, and languages. I considered how patterns work as a way of considering their ontologies, and I considered the ontologies that patterns can give us, and in doing so I outlined some of the qualities of a robust pattern thinker. Drawing these points together I argued that pattern has a minimalist ontology in that, rather than having necessary concrete features, its general ontology reflects a range of interacting characteristics.

Turning to pattern epistemology, I described the epistemologies of different pattern phenomena. I started by defining the idea of pattern knowledge and the different meanings it can have. I expanded on this to describe a pattern knowledge cycle moving from pattern memory to active thought and back again. I also noted how pattern knowledge diverges from traditional models of justified true belief by being based on intersecting degrees of satisficing, utility, and confidence. I then considered epistemologies of pattern aggregation and association, and in so doing expanded on theories of layered pattern integration to give an alternative definition of pattern as: a confluence of layers of connected and interacting nexuses of memories that functions in pattern-like ways. I also considered epistemologies of pattern meaning and signification, interconnected epistemologies of pattern and category, and epistemologies of pattern fuzziness. I also paid particular attention to macropattern epistemologies, describing their basis in terms of intersecting processes and outcomes of aggregation, abstraction, theorisation, signification, formalisation, and systematisation.

In considering pattern axiology, I outlined five core facets of pattern axiology given that value is realised in relationships: referents and relationships, kinds of values, value orientation, logics of evaluation, and metavalue. I then considered axiological pattern systems in the broad areas of ideology, ethics, economics, and aesthetics, noting that while they have functional differences, they can all be understood as pattern system phenomena. I closed the section observing that value plays an essential role in pattern theory.

In writing this chapter, I generally found it easier to describe what patterns and pattern thinking do and how they work than to describe what they are. This would seem to reflect an ongoing tension between the functional and existential aspects of pattern theory. That said, I do not think this as an intrinsic flaw or limitation. Rather, it illustrates that pattern is not a single fixed thought or concept, but a shimmering dynamic system of memory, interpretation, and consciousness and that pattern theory is no more than a reflection of this.

References

Adamson P. *Philosophy in the Hellenistic and Roman Worlds*. Oxford, UK: Oxford University Press: 2015.

Aurenhammer F, Klein R. Voronoi Diagrams. In: Sack JR, Urrutia J (eds.) *Handbook of Computational Geometry*. Amsterdam, NL: Elsevier: 1999.

Bateson G. *Mind and Nature: A Necessary Unity*. New York, NY: EP Dutton: 1979.

Bourdieu P. *Outline of a Theory of Practice*. Cambridge, UK: Cambridge University Press: 1977.

Cheng E. *The Joy of Abstraction: An Exploration of Math, Category Theory, and Life*. Cambridge, UK: Cambridge University Press: 2022.

Chi MTH, Ohlsson S. Complex Declarative Learning. In: Holyoak KJ, Morrison RG (eds.). *The Cambridge Handbook of Thinking and Reasoning*. Cambridge UK: Cambridge University Press: 2005.

de Grefte J. Knowledge as Justified True Belief. *Erkenn*. 2023; 88: 531–549.

Diesing P. *Patterns of Discovery in the Social Sciences*. Chicago IL: Aldine: 1971.

Edwards RB. *The Essentials of Formal Axiology*. Lanham MD: University Press of America: 2010.

Ellaway RH, Bates J. Exploring Patterns and Pattern Languages of Medical Education. *Medical Education*. 2015: 49: 1189– 1196.

Falguera, L, Martínez-Vidal C, Rosen G. Abstract Objects. In: Zalta EN (ed.) *The Stanford Encyclopedia of Philosophy*: 2022: https://plato.stanford.edu/archives/sum2022/entries/abstract-objects

Godfrey-Smith P. *Theory and Reality: An Introduction to the Philosophy of Science*. Chicago, IL: University of Chicago Press: 2003.

Grenander U, Miller M. *Pattern Theory: From Representation to Inference*. Oxford University Press: 2007.

Guyer P, Horstmann R-P. Idealism. In: Zalta EN, Nodelman U. (eds.) *The Stanford Encyclopedia of Philosophy* (Spring2023 Edition). https://plato.stanford.edu/archives/spr2023/entries/idealism/ accessed 30 April 2023.

Hartman RS. *The Structure of Value: Foundations of Scientific Axiology*. Carbondale, IL: Southern University Press: 1967.

Hartmann S, Sprenger J. Bayesian Epistemology. In: Berneckert S, Pritchard D (eds.): *The Routledge Companion to Epistemology*. New York, NY: Routledge: 2011.

Ison MJ, Quiroga RQ, Fried I. Rapid Encoding of New Memories by Individual Neurons in the Human Brain. *Neuron*. 2015; 87(1): 220–230.

Jacquette D. *The Philosophy of Mind: The Metaphysics of Consciousness*. Bloomsbury: 2009.

Didion J. Why I write. In: Sternburg J (ed.). *The Writer on Her Work*. New York, NY: WW Norton: 2000: pp 17–26.

Lakoff G, Johnson M. *Metaphors We Live By*. Chicago, IL: University of Chicago Press: 1981.

McGilchrist I. *Ways of Attending: How Our Divided Brain Constructs the World*. Abingdon, UK: Routledge: 2019.

Medin DL, Rips LJ. Concepts and Categories: Memory, Meaning, and Metaphysics. In: Holyoak KJ, Morrison RG (eds.). *The Cambridge Handbook of Thinking and Reasoning*. Cambridge UK: Cambridge University Press: 2005.

Merton RK. *Social Theory and Social Structure*. Glencoe, IL: The Free Press: 1949.

Miller EK, Wallis JD. Executive Function and Higher-Order Cognition: Definition and Neural Substrates. In: Squire LR (ed.) *Encyclopedia of Neuroscience* (vol 4). Oxford, UK: Academic Press: 2009.

Nikolenko VN, Rizaeva NA, Beeraka NM, Oganesyan MV, Kudryashova VA, Dubovets AA, Borminskaya ID, Bulygin KV, Sinelnikov MY, Aliev G. The Mystery of Claustral Neural Circuits and Recent Updates on its Role in Neurodegenerative Pathology. Behav Brain Funct. 2021; 17; 8.

Otten LJ, Henson RNA, Rugg MD, Depth of Processing Effects on Neural Correlates of Memory Encoding: Relationship between Findings from Across- and Within-task Comparisons. Brain. 2001; 124(2): 399–412.

Pinker S. *Rationality: What it is, why it seems scarce, why it matters.* New York, NY: Viking: 2021.

Pompa L. Family Resemblance. The Philosophical Quarterly. 1967; 17 (66): 63–69.

Rodriguez-Pereyra G. Nominalism in Metaphysics. In: Zalta EN (ed.) *The Stanford Encyclopedia of Philosophy* (Summer2019 Edition). https://plato.stanford.edu/archives/sum2019/entries/nominalism-metaphysics/ accessed 30 April 2023.

Root-Bernstein RS. On Paradigms and Revolutions in Science and Art: The Challenge of Interpretation. *Art Journal.* 1984; 44(2): 109–118.

Rosch E, Mervis CB. Family Resemblances: Studies in the Internal Structure of Categories. *Cognitive Psychology.* 1975; 7: 573–605.

Rutishauser U, Mamelak AN, Schuman EM. Single-Trial Learning of Novel Stimuli by Individual Neurons of the Human Hippocampus-Amygdala Complex. *Neuron.* 2006; 49(6): 805–813.

Rutishauser U, Reddy L, Mormann F, Sarnthein J. The Architecture of Human Memory: Insights from Human Single-Neuron Recordings. *Journal of Neuroscience.* 2021; 41(5): 883–890.

Shermer M. *The Believing Brain.* New York, NY: St Martin's Press: 2011.

Titelbaum MG. *Fundamentals of Bayesian Epistemology 1: Introducing Credences.* Oxford, UK: Oxford University Press: 2022.

Tononi G. Consciousness as Integrated Information: A Provisional Manifesto. *Biological Bulletin.* 2008; 215(3): 216–242.

Voss JL, Bridge DJ, Cohen NJ, Walker JA. A Closer Look at the Hippocampus and Memory. *Trends Cogn Sci.* 2017; 21(8): 577–588.

Wagner AD, Koutstaal W, Schacter DL. When Encoding Yields Remembering: Insights from Event-Related Neuroimaging. *Philosophical Transactions of the Royal Society of London. Series B: Biological Sciences.* 1999; 354 (1387): 1307–1324.

Westphal J. *The Mind-Body Problem.* Cambridge MA: MIT Press: 2016.

Wetzel L. *Types and Tokens: On Abstract Objects.* Cambridge, MA: MIT Press: 2009.

Wetzel L. Types and Tokens. In: Zalta EN (ed.) *The Stanford Encyclopedia of Philosophy* (2018). https://plato.stanford.edu/archives/fall2018/entries/types-tokens

Wittgenstein L. *Philosophical Investigations.* Chicago, IL: Wiley: 2009.

Yadav N, Noble C, Niemeyer JE, Terceros A, Victor J, Liston C, Rajasethupathy P. Prefrontal Feature Representations Drive Memory Recall. *Nature.* 2002; 608, 153–160.

Zai A, Brown B. *Deep Reinforcement Learning in Action.* Shelter Island, NY: Manning: 2020.

Zizek S. *The Sublime Object of Ideology.* Verso: 2009.

8

THE BASIS OF PATTERN INQUIRY

In which I consider pattern theory as the basis for deliberate scientific inquiry

There is little point in arguing that the idea of patterns should be introduced to scientific thought; the concept has long been present in scientific discourse. Patterns may be the goal of inquiry, patterns may be used as part of the processes of inquiry, and patterns may be "found" as a result of inquiry (Lawson 2005). There are patterns of conceptual, experimental, and review-based approaches to inquiry (Hanson 1958) as well as patterns of positivist, postpositivist, constructivist, and realist philosophies of inquiry (Schultz 1987). There have been descriptions of patterns in methodologies; some of them broad, such as those that typify quantitative and qualitative approaches, and some of them more detailed in their patternedness, such as in the different kinds of grounded theory or phenomenology (Glaser 2014). There have been descriptions of patterns in study designs, particularly in the logics and sequencing of different study designs and the constraints thereof, such that, although there are many kinds of trials, cohort studies, and naturalistic studies, they have many common pattern characteristics. There have been descriptions of patterns in methods, including those in instrument development and testing, in configurations of research contexts, and in approaches to data collection, recruitment, ethics, and data curation. There have been descriptions of patterns in analysis, including the various analytical algorithms we use (such as mathematical processing or qualitative coding), many of which involve seeking and describing "patterns in data" (Kantowski 1977). There have been descriptions of patterns in appraising scientific inquiry including patterns of peer review and patterns of critical appraisal, such as those used in evidence-based medicine. There have also been many descriptions of patterns in the ways that science is communicated, such as IMRAD reporting and the various ways of formatting citations and references (Siler & Strang 2017). As much as pattern is a recurring concept in science, quite what patterns are in the context of scientific inquiry is ambiguous at best (as I described in Chapter 2). To illustrate this, I have outlined some of the different uses of the term pattern in a single well-cited methodological textbook – see Table 8.1. Note that this is not intended as a criticism of this book or its

DOI: 10.4324/9781003543565-8

TABLE 8.1 Quotes from Creswell & Creswell's 2018 textbook *Research Design: Qualitative, Quantitative, and Mixed Methods Approaches* (5th ed.), illustrating the variety of applications and framings the concept of pattern can be put to in the context of structured inquiry

Pattern is …	Quotes from Creswell & Creswell (2018)
… what we study	"Ethnography is a design of enquiry coming from anthropology and sociology in which the researcher studies the shared patterns of behaviors, language, and actions of an intact cultural group" (p. 13).
… a step in an analytical process equivalent to thematic or category development	"Researchers make interpretations of the statistical results, or they interpret the themes or patterns that emerge from the data" (p. 16).
	"The literature does not guide and direct the study but becomes an aid once patterns or categories have been identified" (p. 27).
… theory	"Qualitative inquirers use different terms for theories, such as patterns, theoretical lens, or naturalistic generalizations, to describe the broad explanations used or developed in their studies" (p. 49).
… a quality of theory or a type of theory	"… rather than the deductive form found in quantitative studies, these pattern theories or generalizations represent interconnected thoughts or parts linked to a whole" (p. 64).
… different from but equivalent to a theory as the endpoint of inquiry	"… rather than starting with the theory (as in postpositivism), inquirers generate or inductively develop a theory or pattern of meaning" (p. 8).
	"The development of themes of categories into patterns, theories, or generalizations suggests varied endpoints for qualitative studies" (p. 63).

authors (I have used it in my own teaching), rather it is an illustration of a widespread tendency to use the term pattern indiscriminately.

As I have argued, many if not most of the references to patterns in scientific discourses refer to regularities (phenomena that both recur and that create and sustain conventions through their recurrence) rather than patterns as cognitive phenomena. I might accept an argument that methodologies are similar to pattern representations (at least when they are about logics and reasons) and methods are similar to pattern expressions (when they are about actions and interactions), but neither of them are patterns. If investigators were to claim that had found patterns in their phenomenon of interest, then I might counter that these cannot be patterns because patterns are cognitive. Alternatively, I might give a more nuanced (and constructive) answer that, they have recognised patterns by matching their perceptions to their pattern memories or they have discovered patterns by adapting existing patterns or constructing new patterns to reflect their perceptions of apparent regularities. Either way, patterns are the source of understanding, meaning, and significance in these studies, they are not *in* the phenomenon under investigation, nor are they *in* the data collected from the phenomenon, no matter how much that seems to be the case. That said, the data that researchers collect are intrinsically patterned by the pattern thinking that goes into study design. Whether a researcher uses surveys, questionnaires, interview questions, sensors, or measures, they all impose regularities on the data they capture. That subsequent analyses of these data yield patterns should not be surprising. Indeed, does searching for patterns give us anything more than echoes of our own minds in the context of scientific inquiry? A pattern theoretical perspective on scientific inquiry offers a way out of this apparent paradox.

Trainee scientists are expected to internalise the shared pattern thinking of their chosen discipline. Its patterns should be developed, nuanced, connected, and elaborated in each trainee's mind to form a personally coherent cognitive pattern system that can support the necessary cognitive functions and capacities required of a scientist. That said, the need to develop robust pattern thinking is not exclusive to scientific training, it is a fundamental part of participating in any profession (such as medicine, engineering, legal practice, the military, and the emergency services) or any other skill-based undertaking (such as sports, music, joinery, or cookery). It is perhaps tautological to assert that expert pattern thinking is required whenever expertise is required, but given our inattention to pattern thinking in general, it is still an important point to make. Researchers' patterns and pattern thinking habits change according to the field of expertise involved, but all human activity beyond the autonomic has its shared patterns.

We can again draw on Bourdieu's (1977) concepts of a "field", its "doxa", and participant "habitus" to help us to understand these dependencies. Doxa (the truths and rules of a particular field that define what things are done and how they are done) can be understood as shared pattern thinking that individuals are expected to assimilate to participate in a field. A discipline's pattern repertoire is not just its knowledge of the world, it also includes its practices, actions, conventions, philosophies, roles, values, and ideologies (Devlin 1994). Habitus can be understood as an individual's pattern thinking developed in response to a field's doxa. Not only does this illustrate the inevitably individual understanding and alignment individuals have with respect to the doxa of their field(s), it also illustrates the necessity of an individual's broader pattern thinking in participating in a field. After all:

> … what [is it] that really makes the productive scholar. That he has learned the methods? The person who never produces anything new has also done that. It is imagination that is the decisive function of a scholar.
>
> *(Gadamer 1977, p. 12)*

I invite you to consider (and hopefully explore) what this often disregarded cognitive and epistemic meta-phenomenon means for the development of knowledge and understanding in your field(s) of inquiry. For instance, what if we consider the development of scientific traditions as reactions to and a means to overcome the limitations of earlier non-scientific pattern thinking based on magical, mythic, and superstitious thinking? We might conclude that science is one philosophy of pattern thought among many.

This chapter draws the threads of pattern theory together to develop a conceptual and practical basis for *pattern inquiry*, the study of patterned phenomena using patterned theory and methods. I start by considering patterned theory. I then outline principles and concerns in pattern inquiry in terms of the purposes of pattern inquiry, what pattern inquiry is focused on, the dispositions of researchers engaged in pattern inquiry, the processes of pattern inquiry, and the products of pattern inquiry. I consider a range of methodological precedents for pattern inquiry, and I consider three paradigms of pattern analysis (typological, topological, and pattern language).

Patterned Theory

I start my exploration of pattern inquiry with a consideration of theory based on the Kantian interlocking of theory and inquiry: theory guides inquiry, inquiry guides theory. This would suggest that there is no pattern inquiry without pattern theory and no pattern theory without pattern inquiry. However, since the goal of this volume is to build a body of pattern theory, I will refer instead to patterned theory; theory that is based on or reflects pattern theoretical concepts.

We may perceive regularity in the form of repetition, but our minds rapidly and instinctively try to connect this to pattern memory such that we understand it in terms of regulation. Our theories of regularities (what they are, what they mean etc.) become manifest in our perceptions of instances of them. Indeed, given that theories can account for the nature of order and regularity, including the causes and likely future disposition of particular regularities, it might be more useful to understand theory in terms of deepening and enriching our understanding of regularities.

Macropatterns are intimately bound up with theory, both because they become theories and because theorisation is a sign of macropattern development (alongside aggregation, abstraction, signification, formalisation, and systematisation). We might ask, therefore, whether all pattern thinking is theoretical. We need relatively little abstraction or conceptualisation for something to be perceived, albeit indistinctly, as a bear or a rock. But is this a theory? Not as such, but pattern recognition is, I think, where theory begins. It suggests coherence, however tentative, in the midst of incoherence. Even very young and inexperienced humans can establish identity and map consequences of that identity onto something they perceive, even though their pattern repertoires and habits of pattern thinking are still developing.

To say that all patterns are theories dilutes the meaning of theory. Rather, I would argue that patterns have the potential to contribute to theorisation (and abstraction, aggregation, signification, formalisation, and systematisation) as part of macropattern development. It might be argued that if all theories are macropatterns then all macropatterns are theories, but that does not hold either. There are clearly many macropatterns that are not theories (at least in scientific sense), such as customs, conventions, identities, and values (although some of them may imply theory or depend to some extent on theory). As to whether all theories are macropatterns that depends on what you consider a theory is. For instance, it would be hard to say that a theory is *in* a pattern in someone's mind if the definition (or expectation) of a theory is that is shared in some way. We might, therefore, be better asking what roles different pattern phenomena play in our development and use of theory. In that case, patterns and pattern thinking can certainly be seen as contributing to the development and articulation of theory, while pattern expressions and pattern representations are ways of externalising and sharing theory.

Does activating a pattern (pulling on it, triggering it) return theoretical knowledge to our active thoughts or do we recall the ingredients of theory from pattern memory and then assemble them into theoretical knowledge in real time? It would seem to be both and neither. *Both* because all pattern recall is subject to elaboration and adaptation even though learning involves laying down memory of our thoughts and ideas such that we do not have to recreate them every time we need them. I also say *neither* because this suggests a fundamental distinction between pattern memory and pattern thought, which we should be cautious of. Acknowledging that this edges into the somewhat contentious matter of consciousness, it seems likely that there is a continuous connection and interaction between pattern thought and pattern memory such that patterns are only relatively active or inactive and pattern thinking is continuous with all thought.

A pattern theoretical perspective on theory acknowledges the macropattern basis for the development of theories in general, it situates all theory as subject to the strengths and limitations of pattern thinking and the pattern knowledge it produces, and it notes the uses, strengths, and limitations of pattern expressions and pattern representations in articulating and sharing theory. Beyond that, a patterned perspective on theory might be expressed in many ways. It could focus on developing and testing theories of pattern phenomena in general (as I am attempting to outline in this book). It could focus on theories of particular patterns or pattern phenomena, or it could focus on theories created out of patterns and other pattern phenomena. However, coming back to a Kantian duality perhaps patterned theory cannot be meaningfully developed without attending to its development and application both of which are based in acts of inquiry.

Pattern Inquiry

All inquiry is shaped by pattern thinking, it cannot be otherwise. So too are the theories and frameworks by which we conduct and evaluate acts of inquiry. Rather than go over the patterned basis for all human thought again, I will focus on how pattern theory might be used to shape different stages of the research process. For example, methodologies are the broad conceptual orientations to inquiry around which studies can be both built and appraised. A pattern theoretical approach to methodology might focus on those that support the examination or generation of pattern phenomena or those that use pattern theoretical concepts as a lens. Methods are specific action-based ways of solving problems in acts of inquiry. Pattern methods are activities that engage pattern theory in some way or other, such as asking pattern-focused questions, sampling or gathering data that reflect patterned phenomena, or using pattern theoretical concepts to analyse data or to communicate findings. It would seem, therefore, that although a blending of pattern theory and modes of inquiry might be productive, it creates so many possibilities that it is hard to establish a sense of coherence. We need more (macropatterned) structure to be able to move forward.

Whether talking about methodologies or theories, there are three broad dimensions of inquiry to consider: what, how, and why – see Figure 8.1. *What* methodologies are concerned with what it is that inquiry is exploring and as such are primarily ontological in focus. For instance, the various schools of phenomenology have much to say about the nature of human experience but relatively little about the process of inquiry. *How* methodologies (such as statistics, case study, and grounded theory) have much to say about processes of inquiry but have little to say about what they are applied to or why. *Why* methodologies (such as participatory action research and appreciative inquiry) tend to focus on ethics and values and are, therefore, primarily axiological in nature. This divergence can be problematic for junior researchers who feel they need to commit to one particular theoretical or methodological position only to find that it can only meet some of their needs. Typically, researchers need to combine theories and methodologies so that what, how, and why questions can all be answered – see Figure 8.1.

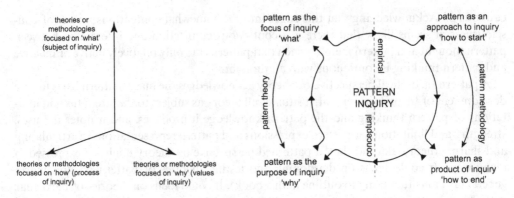

FIGURE 8.1 Building from (left) an orthogonal sense of different theories and methodologies that focus or load on dimensions of what is the focus of inquiry, how is inquiry conducted, and why engage in inquiry, the scope of pattern inquiry can be outlined (right). Note that these are not solitudes; any given act of inquiry may involve more than one of these characteristics.

Diagram prepared by the author.

I can expand on this to consider five aspects of pattern inquiry (purpose, focus, disposition, process, and product), which can be combined with pattern theoretical principles I described in earlier chapters. I have selected 22 core pattern theoretical principles for this purpose – see Table 8.2.

Purposes of Pattern Inquiry

We can start with the *Cognition* principle by asking why investigators might be interested in patterns as cognitive phenomena. As a species, not only are humans adept pattern thinkers, language, art, culture, science, and philosophy are all products of individual and collective pattern thinking. Understanding how patterned minds work would seem to be a good reason for engaging in pattern inquiry. Pattern inquiry could advance investigators' understanding of their own minds as well as the minds of others, and it could explore how individual pattern cognition translates to shared pattern thinking. Alternatively, inquiry could focus on disorders of or damage to patterned minds, or on ways to change, extend, or improve patterned minds. It may also provide opportunities to explore philosophical and existential questions related to the cognitive basis of pattern thinking.

Purposes of inquiry related to the *Constitution* principle reflect the implications of pattern memory being constituted in our interconnected memories of experiences, thoughts, and feelings. This constructivist perspective emphasises the idiosyncratic, individual, and embodied nature of memory and understanding. Investigators might want to understand how different kinds of memories can shape thinking (individually and collectively) or how dependent different individuals are on different kinds of pattern memories. A constitutive basis for pattern inquiry might be expanded using the *Logics* principle in that pattern memory can also include meanings, functions, causalities, and significances.

TABLE 8.2 Core pattern theoretical principles used to outline the nature of pattern inquiry

Principle	Description
Cognition	Patterns are cognitive phenomena.
Constitution	Patterns consist of memories of experiences, thoughts, and feelings.
Logics	Patterns can include logics in terms of meanings, functions, causalities, and significances.
Uniqueness	Patterns are unique to the mind in which they are formed.
Awareness	We can be aware of our own pattern thinking, but only to a limited extent.
Dynamism	Patterns are dynamic, they grow and shrink, fuse and fracture, and they can reflect phenomena and experiences that are temporal and ephemeral.
Continuum	Patterns have no edges – there is a continuum of pattern in our minds.
Categorization	We weave categories, concepts, symbols, and language into our patterns to give us the sense of differences and distinctions.
Formation	Patterns are formed by pattern thinking.
Dependency	Pattern thinking depends on patterns such that pattern thinking can be both enabled and constrained by the patterns it draws on.
Modalities	There are different modalities of pattern thinking including pattern perception, pattern recognition, pattern development, and pattern elaboration.
Triggering	Patterns can be triggered, pulled upon, and otherwise activated such that they are available to and used in active memory and thought.
Elaboration	Patterns in active memory and thought can be elaborated, rebuilt, adapted.
Weaving	Patterns in active thought can be woven back into pattern memory.
Micro-Meso-Macro	There are differences between micropatterns, mesopatterns, and macropatterns, in terms of the ways in which they are built, adapted, and used.
Layering	Mesopatterns and macropatterns are aggregative and emergent across layers of pattern memory and are therefore layered, scaled, panarchic.
Similarity and Difference	By working with similar differences and different similarities, patterns afford fuzzy and fluid ontologies and epistemologies.
Generation	Patterns are generative, they allow us to work with possibilities as well as actualities, and in so doing they enable imagination and creativity.
Limitations	Patterns and pattern thinking can be constrained, limited, repressed, and damaged.
Sharing	Patterns can be shared but only in imperfect, complex, and constructivist ways.
Latency	The patterns and pattern thinking of others cannot be directly experienced by anyone else.
Expressing and Representing	Patterns can be shared through creating pattern expressions, pattern representations, pattern systems, and pattern languages.
Medium of Expression	Pattern expressions and representations can be communicated discursively, expressively, performatively, and through artefacts.

Purposes of inquiry related to the *Formation* principle might explore how pattern memory forms, while the *Dynamism* principle might involve inquiry into the ways in which patterns grow, shrink, fuse, and fracture. Inquiry drawing on the *Elaboration* principle might explore how it is that pattern memory can be elaborated, rebuilt, and adapted based on new experiences and thinking, while inquiry based on the *Weaving* principle might explore the ways in which this happens. The fluid nature of pattern memory might also be explored drawing on the *Continuum* principle that patterns have no edges, they are nexuses on a continuum of changing pattern memory in our minds. Inquiry based on the *Categorisation* principle might explore the ways in which we weave categories, concepts, symbols, and language into our patterns and pattern thinking. Inquiry based on the *Similarity and Difference* principle might explore the nature and function of fuzzy and fluid ontologies and epistemologies. Inquiry based on the *Generation* principle might explore ways in which patterns and pattern thinking allow us to work with possibilities and probabilities. The *Layering* principle could translate to inquiry into how pattern memories are layered and aggregated to form ever more coherent macropatterns. The *Micro-Meso-Macro* principle might be used to guide inquiry into the formation and use of micropatterns, mesopatterns, and macropatterns.

The *Triggering* principle might be used to guide inquiry into how it is that patterns can be triggered, pulled upon, and otherwise activated in ways that make them available to and used in active memory and thought. The *Modalities* principle could guide inquiry into the similarities and differences between different modalities of pattern thinking such as pattern perception, pattern recognition, pattern development, and pattern elaboration, while the *Limitations* principle could guide inquiry that explores how these modalities can be constrained, limited, repressed, or damaged.

The *Dependency* principle could be used in designing studies to explore ways in which pattern thinking can be enabled and/or constrained by the patterns it draws on, while the *Uniqueness* principle could guide inquiry into the uniqueness of patterns in individual minds and the resulting subjectivities. The *Awareness* principle could be used to explore how aware we are of our own pattern thinking, which has implications for our understanding of consciousness, free will, and for the reliability and rigour of scientific inquiry.

The *Sharing* principle can be used in complementary or dialectical ways with *Uniqueness, Awareness, and Latency* principles to explore the strengths, weaknesses, and varieties of pattern expressions, which are also reflected in the principles of *Expressing and Representing* and *Medium of Expression*. Pattern inquiry of this kind might also explore the ways in which we collaborate through shared pattern thinking.

Clearly, these many possible purposes of pattern inquiry span the development of knowledge and understanding about minds, persons, groups, societies, and cultures. Researchers may seek better understand of themselves or each other, better understanding of how minds work individually and collectively, better understanding of the strengths and limitations of pattern thinking including how reliable it is, how malleable and manipulable, and how robust or fragile it can be.

Foci of Pattern Inquiry

What questions might investigators explore starting with the pattern theoretical principle of *Cognition*? They could explore the pattern basis of different cognitive phenomena (such as behaviours, skills, attitudes, beliefs, knowledge, and competence) or the products thereof in ways that are naturalistic ("how do we think?") or normative "(how can we improve our thinking?"). Pattern inquiry might focus on the pattern thinking of research subjects or on the interplay between the pattern thinking of subjects and investigators.

The pattern theoretical principle of *Constitution* could lead to questions about how sensoria are translated into perception, thought, and memory. This includes memories of feelings, emotions, and the embodied nature of experience, and the wicked problems of consciousness and learning. Other questions may focus on the emergence of patterns from parts, on topologies of patterns, on different constituent pattern parts and the role of different parts and clusters of parts in pattern thinking. Given that patterns are intrinsically made up of memories of past experiences, thoughts, and feelings, pattern inquiry could be somewhat forensic ("what were the causes of pattern formation?") or morphogenetic ("what is the origin of the form of pattern memory?").

The *Logics* principle notes that patterns can include understanding of what things mean, how they function, what their causal relationships are (or seem to be), and the significance of all these things. Research questions might focus on how we use these logics, how we develop the capacity to cast them into pattern memory, and how we use them to support other cognitive processes. Investigators might also seek to better understand how logics work in terms of their representations in memory and how they are activated and used. Are these logics simply recalled or reactivated or are they dynamically reconstructed from pattern elements on recall? Investigators might also seek to explore how different logics interact, for instance, whether understandings, meanings, causalities, significances, likelihoods, etc. are in any way distinct or whether these are imposed conventions (patterns) that are applied to experiences and interpretations of these cognitive capabilities.

The pattern theoretical principle of *Uniqueness* focuses investigators' attention on the individuality of patterns and pattern thinking and the need to consider similar differences and different similarities rather than identities. Research questions could focus on the individuality of patterns in different observers' minds that ostensibly refer to the same external phenomenon or how much similarity or alignment there is in patterns and pattern thinking across minds, or to what extent pattern formation is unique or variant, something that could substantiate constructivist theories of learning. Another direction might be to explore individual pattern signatures of individual minds, and similarities in pattern signatures of minds that share experiences, practices, or cultures. Studies might even explore whether patterns are identifiable as a biometric signature of the mind in which they formed.

Drawing on the pattern theoretical principle of *Awareness*, pattern inquiry could focus on the ways in, degrees of, and limitations to which we can know our own patterns and pattern thinking. This would seem to be a foundational issue both as the focus of research and as a mindfulness practice for researchers. If self-knowledge of our own patterns and pattern thinking is possible then to what extents, in what ways, and with what limitations can this happen? How do our perceptions of our pattern perceptions compare to the deeper and pre/unconscious aspects of our minds? Do we see patterns at all clearly or realistically? A pattern theoretical approach could be used to explore reflexivity, bias, paradigm, value, and ideology. Moreover, starting from the pattern theoretical principle of *Awareness*, investigators may query or otherwise seek to research their own pattern thinking. However, as this is typically subjective, care should be taken in interpreting and building knowledge claims on such reflection.

Starting with the pattern theoretical principle of *Dynamism*, investigators may focus on the pattern dynamics of individual minds or those of multiple minds. Investigators might explore pattern dynamics in specific contexts, for instance, in training scientists (or anyone else). Research questions might focus on how patterns grow, shrink, shift, adapt, come together, and break apart in individual minds, or they might consider how it is that patterns can be shared across a community or culture grow, shrink, shift, adapt, come together, and break apart across multiple minds. Investigators might also explore how we relearn or reform patterns that have been lost, for instance, through degeneration or injury, or through neglect or coercion.

The pattern theoretical principle of *Continuum* could lead to studies that seek to identify specific patterns or pattern thinking processes in the mind or brain. For instance, Metzger et al. (2023) reported working with a stroke patient who, although she could still form language in her mind, was unable to speak her thoughts. The researchers implanted an imaging array on the surface of her cortex over the speech centre. The patient was given a series of training phrases to say in her head and the data from the array were fed into a neural network which linked regularities in the data to the intended phrase, gradually training the computer to match the patient's neural activity to phonemes and then to words and sentences. Although it took some practice, after a few weeks the computer was able to turn the patient's thoughts into words (rendered first as text then as text-to-speech) with a reasonable level of accuracy. From a pattern inquiry perspective, this kind of approach might allow us to explore how neural activity relates to pattern-in-mind, at least to some extent. Investigators might also explore the nature of pattern continua, or they might explore how the connections and associations in our minds work and whether are there really differences between pattern connections (inside pattern) and pattern associations (between patterns). Not only could this research provide a more substantive neurological foundation for much of pattern theory (or perhaps refute it), it could also map connections and translations between brain and mind, a topic that has long challenged and perturbed scientists in many disciplines.

The pattern theoretical principle of *Categories* could be used by investigators to explore how we know or decide whether things exist and what kinds of things they are. Are categories, as I have suggested, pattern indexes, a symbolic layering of macropatterning to render our intuitive and unregulated pattern thinking coherent and organised? Do categories create an illusion of distinctiveness, coherence, and identity in an otherwise chaotically continuous universe? There are many extant methodologies and philosophies of category and categorisation, and to that end pattern inquiry may afford new perspectives on these debates. Inquiry into pattern phenomena may also focus on the use of categories, concepts, symbols, and language as guides to participants' patterns and pattern thinking (*Categories*), for instance, by understanding them as parts of pattern matrices such that they reflect its otherwise inaccessible structure (like bones in an X-ray, or radar or sonar echoes of distant objects) rather than being the entire phenomena in and of itself.

There is so much potential for exploration in regard of the pattern theoretical principle of *Formation* (as I noted earlier) that I will simply suggest that the principle affords many jumping off points for inquiry into the relationships between pattern memory and pattern thinking. Can one exist without the other? Are all patterns in a state of potential excitation and is pattern thought distinct from or woven into these patterns? Do we indeed have patterns of pattern thinking and as such are our thought processes themselves guided by pattern?

The pattern theoretical principle of *Dependency* could translate to questions that explore the differences between naïve, experienced, and expert pattern thinkers. Investigators might explore what it means to have a paucity of patterns, poor pattern formation, skewed pattern formation, or other debilitating circumstances, and might we consider it a human right to have the opportunity to build a rich pattern memory? The pattern theoretical principle of *Dependency* might also focus researchers' attention on how disruptions or significant changes to pattern disrupts participants' other patterns and pattern thinking, while Formation might shift the focus of inquiry to enabling and constraining nature of how specific experience translates to laying down patterns, such as in childhood or in training.

The principle of *Modalities* could be used to focus inquiry on participants' pattern perception, recognition, development, or elaboration. There are many possible intersections with ongoing research into perception, pattern recognition (albeit with different understandings what this means), and cognition. To that end I would refer to my suggestions for future research in earlier chapters rather than reiterating them here. I have also already considered many reasons for researching these phenomena and directions that research might take based on the pattern theoretical principle of *Triggering*. However, I would stress the importance of exploring how our patterns are searched, matched, and triggered or activated. For instance, earlier, I used metaphors, such as grabbing a part of a pattern and pulling in the rest like a fishing net, but what actually happens, and how do our perceptions of this in our minds differ from the reality?

Patterns are adapted or elaborated every time they are activated. Starting from the pattern theoretical principle of *Elaboration* questions could explore whether this is always true or whether patterns reach some kind of saturated state beyond which there is little change. Are recall and elaboration also how (or part of how) we solidify long-term pattern memory? Does elaboration involve disassembling patterns or does this already happen when patterns are triggered? Similarly, there are different sides to the pattern theoretical principle of *Weaving* of pattern memory. Investigators might focus on exploring how pattern nexuses are formed, on how triggered patterns are subsequently elaborated as a result of triggering and conscious thought, or on whether formation and elaboration are two different processes or variations on the same process.

There is much that can be explored in and around the pattern theoretical principle of *Micro-Meso-Macro*. In the context of research, macropatterns and macropatterning seem particularly important because they form the basis for all scientific theory and empiricism. A *Micro-Meso-Macro* perspective also affords a novel perspective on processes of empiricism – how sensory experience is transformed into data, and then progressively, into knowledge and wisdom.

The pattern theoretical principle of *Layering* again affords many opportunities for inquiry. Despite a history of reductive and graph/set models being used in pattern-focused research, I do not think that a simple directed acyclic graph (or some other nodal topology) can easily represent the ways in which individual neurons come together to create pattern elements let alone patterns. It seems likely that it takes many layers of neuronal learning to represent even the most basic mesopatterns and that they are emergent across skeins of neural activity rather than directly keyed to individual neurons (this one is "cat" than one is "dog"). Investigators might, therefore, focus on the bundling of memory to create pattern elements, and it might also focus on how memory emerges across these intertwined bundles of neurological connectivity and interactivity. Notably, in the Metzger et al. (2023)

study I mentioned earlier the resolution of the embedded array was still very coarse relative to the neurons it overlaid; at best it was able to detect broad regional variations given that neuronal density is in the region of several 100,000s per square millimetre of cortex (Young et al. 2013). Panarchy also raises the dynamic nature and consequences of changes to one aspect or layer of a pattern can lead to adaptive cycles across other aspects or layers of the pattern. This in turn raises questions of pattern fragility or resilience to such changes. Not only can ecological systems theory be adapted to consider these kind of pattern characteristics (adaptive cycles, panarchy, hysteresis, etc.) pattern theory may provide reciprocal benefits.

There are many possible directions investigators might take based on the pattern theoretical principle of *Similarity and Difference*. For instance, this principle might be used to consider differences in the causes and nature of fuzziness: is this a matter of observer uncertainty or perceptual ambiguity, is it about intrinsic variability and uncertainty or is it about fuzziness as the result of insufficient pattern distinctiveness (is it pattern A or B when neither is a good fit) and, therefore, a need to elaborate or develop new patterns? Second, exploring similarity and difference is often a primary focus of many kinds of empirical inquiry. Approaching similarity and difference from a pattern theoretical perspective can recentre such considerations from intrinsic properties and observations to patterned and patterning perceptions and the strengths and limitations thereof. This principle also affords empirical starting points to exploring Bohm's theory of the importance of distinguishing similar differences and different similarities in regularities, and to that end, the principle may also link pattern-focused inquiry to theories of fuzziness and precision in science, logic, and mathematics.

One of the broad generic goals of research is to understand a phenomenon so well that its future behaviours can be predicted. A pattern theoretical approach might focus on how we adapt and elaborate our patterns to accommodate new variations as well as the points at which such elaboration falters and new patterns are calved off or developed de novo to capture the new. Research based on the pattern theoretical principle of *Generation* also creates pattern representations, systems, and languages as concrete generative epistemologies for representing phenomena that may take on different configurations to those observed but that are still based on the same underlying pattern building blocks. This principle creates opportunities to explore questions that consider what is possible as well as what is observed in a phenomenon, and how both extant and potential configurations can be expressed and modelled. On a more positive note, if investigators are interested in exploring what can be done with pattern inquiry, then the *Generation* principle might focus their attention on how participants' patterns and pattern thinking are used generatively, for instance, in support of imagination and creativity.

Limitations is a reciprocal principle to the previous one given its focus on constrained pattern phenomena. A consideration of limitations can focus investigators' attention on the limitations of individual minds involved in research, the limitations researchers face in working collaboratively, and the constraining influence of pattern thinking and pattern expressions on research activity. These issues can be considered as the focus of research (how other researchers were constrained in acts of inquiry) as well as reflexively (how investigators understand their own constraints in undertaking a particular study). A third direction might be to focus on the limitations and constraints of pattern thinkers in other contexts, for instance, in socially, culturally, or economically limited or constrained circumstances. If an investigator's focus is on the principle of *Limitations* then they might consider how participants' patterns and pattern thinking can be constrained, limited, repressed, damaged, or punished. Alternatively, the principle of *Continuum* might focus on the tensions between experiences of pattern distinctiveness and the experiences of pattern continuity.

The pattern theoretical principle of *Sharing* can serve as a starting point for considering the means, practices, limitations, and complexities of sharing ideas and knowledge that are pattern-based (as they always are) particularly when patterns are unique to individual minds (*Uniqueness*), when they are dynamic and changing (*Dynamism*), and they coalesce and develop in unique ways (*Formation*). How do the patterns that form the basis of ideas and knowledge differ between minds, in terms of formation, change, and exchange? Postpositivist perspectives on science consider both the imprecision of instruments, measurements, samples, and other procedural aspects of empirical inquiry, as well as the biases, paradigmatic asymmetries, and human fallibility of investigators. A pattern theoretical perspective on imperfection could focus investigator attention on the latter rather than the former, particularly in terms of variance, historicity, and situatedness rather than error. If investigators cannot interact with shared pattern directly then they can either sample individual experiences of shared pattern thinking and extrapolate across a collective, or they can explore the artefacts of pattern thinking at a collective level. For instance, an investigator might consider the use of language and the exchange or articulation of ideas and values as reflecting shared pattern thinking such that linguistic and discourse analyses may provide valuable insights. Indeed, they might interpret shared patterns and pattern thinking in terms of an archaeology of its remains and other echoes.

Starting from the *Latency* principle, investigators could explore whether the focus of a specific act of inquiry is on core latent cognitive phenomena or on the non-latent products or reflections of pattern phenomena. Although *Latency* relates back to the uniqueness of patterns, this principle can be used to expand on this to consider both pattern practices (either the focus of research or the product of research) and pattern thinking (the practices of research). Generically, investigators might ask questions about how people can reliably and meaningfully explore, record, model, or otherwise understand represent patterns and pattern thinking. The *Latency* principle also circles attention back to theory and metatheory related to latent constructs and latent variables and the ways in which and extents to which they can be explored meaningfully.

Given that macropatterns and their expression as representations, systems, and languages are created as products of research (see later in this chapter), there is a meta-research or metascholarship aspect to the pattern theoretical principle of *Expressing and Representing* that focuses attention on how and why pattern systems and pattern languages are developed, and how they are used, in which case there may be interesting overlaps with theories of the diffusion of innovations and of communities of practice. Starting from the pattern theoretical principle of *Medium of Expression*, there are at least two broad questions; what are the most effective ways of communicating pattern expressions and representations, and what are the ways in which pattern expressions and representations have been communicated? Much of this overlaps with linguistics, communication theory, and cognitive psychology, and more practically with graphic and instructional design.

Dispositions of Pattern Inquiry

Pattern theory can also be translated to investigator dispositions. The pattern theoretical principle of *Cognition* suggests that the nature and quality of inquiry (whatever it is focused on) is dependent on investigators' pattern thinking, which should be taken into consideration in study designs and in generating and critiquing the products of inquiry. More specifically, the pattern theoretical principle of *Constitution* suggests that investigators need to attend to the conscious and deliberate macropatterning of their experiences, perceptions, and thought processes, particularly as they will inevitably create pattern knowledge from their studies. The pattern theoretical principle of *Uniqueness* emphasises that the pattern thinking individual investigators bring to bear is unique to them. Who investigators are and how they think matter, and this draws in principles of reflexivity, transparency, and accountability. The pattern theoretical principle of *Awareness* suggests that investigators should be mindful of their own pattern thinking and seek to focus it, account for it, and critique it within the context of structured inquiry. The pattern theoretical principle of *Dynamism* suggests that investigators' pattern thinking should be expected to change as the result of the research process, which in turn raises questions regarding how investigators think at specific points in time and in particular contexts, and what the implications are for executing a study. For instance, results are typically presented as a series of abstract knowledge claims, but they could also include an account of the changes to the investigators' (and perhaps the participants') pattern thinking.

The pattern theoretical principle of *Logics* might be applied in different ways. For instance, investigators could productively reflect on how their own pattern logics contribute to their understanding of the phenomena under investigation, to identifying or deducing meanings or causality, or to attributing likelihood and significance. A second application could be to explore how, in what ways, and to what extent investigators defer to external proxies for their pattern logics. As an example of this, the use of measurement and statistical modelling can afford a more objective alternative to internal assessments of quantities and probabilities, while discursive multi-coder analysis of qualitative texts can (or at least it is often assumed it can) diffuse individual coder bias.

The associative and continuous nature of pattern thinking can both enable inquiry and disrupt it. The pattern theoretical principle of *Continuum* suggests that investigators should be attentive to the strengths and weaknesses of their pattern thinking (and that of others) in the context of inquiry. The pattern theoretical principle of *Categories* suggests that investigators should be deliberate and mindful in their use of categories, concepts, symbols, and language in the context of inquiry, noting their strengths and limitations, their pattern provenance, and their patterning proclivities. The pattern theoretical principles of *Formation* and *Dependency* suggest that investigators should be deliberate and mindful in how they develop and refine their patterns and pattern thinking, and that they should be attentive to how this shapes the execution and reporting of their studies. For instance, when they are analysing their data, investigators could pay particular attention to their feelings and how they are shaping their thinking, particularly if they have a sense of growing completeness or symmetry in developing findings or theories.

The pattern theoretical principles of *Modalities* and *Elaboration* suggest that investigators should be deliberate and mindful in their use of pattern perception, pattern recognition, pattern development, and pattern elaboration, and that they should seek to refine these capabilities where possible. This includes being mindful of how and when their own patterns are activated in the context of inquiry (theory, method, analysis, synthesis), both deliberately and unintentionally (*Triggering*), and how and what gets written back into pattern memory (*Weaving*). Investigators should be mindful of the macropatterns they draw on in the context of inquiry (*Micro-Meso-Macro*). They should also be mindful of the macropatterning processes of aggregation, abstraction, theorisation, signification, formalisation, and systematisation that they draw on in acts of inquiry, as well as the strengths and weaknesses of their macropatterns, and how they engage with the macropatterns of their chosen field, methodology, and paradigm. Investigators should also be mindful of layering, scaling, and panarchy in their own pattern thinking and the ways in which this may strengthen or weaken it (*Layering*). For instance, how are theory or methodology disrupted when an investigator's pattern understanding shifts?

Investigators should be mindful of how they work with similar differences and different similarities in their pattern thinking (*Similarity and Difference*). They should also understand and make good use of the fuzzy and fluid epistemologies and ontologies afforded by a pattern theoretical perspective, including the generative possibilities of their pattern thinking (*Generation*). For instance, investigators should be mindful of how they work with possibilities as well as actualities, in particular how patterns and pattern thinking is the basis of imagination and creativity.

Investigators should always be mindful of investigator bias, prejudice, enculturation, blind spots related to their pattern thinking and the pattern thinking of the collective (*Limitations*), and of the limitations of collaboration in inquiry that result from divergent pattern thinking. Investigators should be mindful of the confidence they have in the pattern thinking and pattern knowledge they have and that they engage with. They should also be mindful of the pattern expressions of their scholarly communities, paradigms, and discourses. Collaborating investigators should seek to make their pattern thinking clear to their colleagues and to those the investigators with whom they want to share their research (*Latency*). Investigators should be mindful of how they have developed their research knowledge, skills, and values from interacting with macropattern artefacts and the macropatterns of others (*Expressing and Representing*). They should also be mindful how they engage others through their macropattern thinking and macropattern artefacts, and of the different ways they can communicate macropattern ideas and concepts (*Medium of Expression*).

There is a lot here for investigators to accommodate, although much of this reflects existing thinking about rigour in scientific inquiry. What a pattern theoretical approach adds is a more delineated range of issues to attend to and a direct link between investigators' pattern thinking and the values and aspirations of scientific inquiry. How this translates to training, codes of conduct, and the identities of researchers I will leave for another time.

Processes of Pattern Inquiry

The pattern theoretical principle of *Cognition* suggests that pattern methodologies will often need to approach pattern phenomena in terms of non-latent characteristics to explore latent pattern characteristics. This may involve investigators analysing artefacts, behaviours, discourses, symbols, and /or systems to explore the shared or individual pattern thinking that has shaped them. What kinds of subjects, procedures, and data are involved will depend on the specific research questions involved and the methodological position taken, as will the processes of analysis and synthesis. I would note that, practically speaking, exploration of *Cognition* would also involve exploration of *Latency* given the need to use proxies or other indicators of pattern thinking. This in turn suggests the *Limitations* principle given that research may as interested in dysfunctions of pattern thinking if not more than in ideal examples.

I would also observe that *Latency* is a common concern in statistics when latent variables are explored in terms of observable variables, which in turn is based on assumptions that the two are somehow causally related. However, as Boorsboom et al. (2003) argued, investigators need to consider the assumptions associated with their use of latent and observable variables. On one hand investigators might take the latent variable (most likely some form or aspect of pattern thinking) as their fundamental real construct and any observed variables as no more than a means to illustrate the central latent variable ("entity realism"). On the other hand, investigators might be more interested in the observed variables with the latent variable simply being a way of connecting or grouping the observed variables ("theory realism").

The principle of *Constitution* focuses on pattern elements and their connectivity, and to that end, although the questions or tasks involved may differ, investigators would need to select methods that allow them to explore their participants' pattern memories. This might involve activities such as journaling, phenomenological interviewing, or think-alouds. Although subjects, procedures, and data depend on the methodology and study design, analysing the constitution of specific patterns could involve investigators making models of pattern elements and the connections between them (as in General Pattern Theory), but may be both illusory and overly reductive. Investigators should examine their assumptions regarding the discreteness and accessibility of individual pattern elements and connections (*Continuum* principle). Because of the *Latency* principle, *Constitution*-focused inquiry would also need to involve *Categorisation* in terms of how participants' use of categories, concepts, symbols, and language create a sense of differences and distinctions. From a practical perspective, investigators might explore *Constitution* in terms of individual or group expressions of patterns and pattern thinking (with parallels in linguistics, discourse analyses, and narrative reviews) or they might explore individual or group awareness and understanding of the patterns and pattern thinking they use.

The *Uniqueness* principle focuses attention on similarities and differences between minds. Investigators might use a variety of imaging and task related analyses to explore the neurological basis of patterns, or they might, as before, use a variety of psychological and social science methodologies to explore participant awareness or interpretations of their own patterns. In terms of analysis whether investigators are interested on individuality, commonality, or both will determine what they look for in data. As with *Constitution*, any answers that are predicated on distinct patterns should be examined more closely given that differences between two or more individual's patterns for ostensibly the same experience may reflect the subjectivity of experience or arbitrary perceptions of pattern boundaries (*Categorisation*) rather than variant or similar qualities of the patterns themselves.

The pattern theoretical principle of *Logics* parallels exploring pattern phenomena from the perspective of the *Awareness* principle in that both of them involve (or at least are likely to involve) deeply subjective accounts and as such methodologies such as phenomenology, phenomenography, and autoethnography might prove useful. Investigators might seek to explore issues of fidelity, but who is to say what a 'true' pattern is given that it is not directly accessible? Rather, they might explore individual thresholds of awareness of pattern thinking and factors that influence individual awareness. Alternatively, investigators might explore awareness of continua of pattern thought or of processes of pattern thinking (*Continuum, Modalities*), or they might explore awareness of pattern activation (*Triggering*), awareness of how patterns change in active thought (*Elaboration*), or awareness of how patterns are returned to pattern memory (*Weaving*).

Practically, the pattern theoretical principle of *Dynamism* presents both opportunities and challenges. Opportunities because dynamic phenomena can be easier to detect and track than static ones, and because how patterns change may be particularly illuminating. Challenges because tracking change can be difficult if they are very fast or slow, and challenging when, if everything is in motion, then everything becomes relative and fluid such that absolutes become elusive. From an empirical perspective, change (*Dynamism*) would also require consideration of *Dependency* in terms of the disruptions or perturbations in patterns and pattern thinking that result from change. In terms of methodologies, investigators might use varieties of realist inquiry such as Pawson's program theoretical approach or Archer's morphogenetic approach, or they might use more experiential methodologies, such as phenomenology if they were more interested in participants' experiences of changing patterns and pattern thinking.

The pattern theoretical principle of *Formation* could be considered using the lenses afforded by the principles of *Awareness, Dependency,* and *Constitution,* but, as an education scientist, I admit I have a particular interested in how pattern formation can be guided and focused to achieve particular ends, and as such the many methodologies used in educational research could also be employed in exploring pattern formation in terms of learning. This might be realised in study designs that explore (variously) what is learned, by whom, in what circumstances, to what extent, in what ways, and with what consequences. However, the problem of *Latency* is still with us as we can explore learning obliquely through behaviour, performance, or autobiography, but we cannot access it directly.

The pattern theoretical principle of *Micro-Meso-Macro* might also be realised in many ways, but, given that macropatterns are generally the most accessible and substantial kinds of patterns we hold, and they reflect the way that our minds contribute to collective undertakings such as culture and science, investigators may wish to focus their inquiry on them. Practically speaking, any kind of sociocultural analysis can be interpreted in terms of individual and shared macropatterns and the systems and languages we build around them. A discourse analysis is in effect drawing out shared and contested macropatterns from discursive artefacts (typically texts). Systematic reviews, in synthesising knowledge, are in fact extracting the macropatterns expressed by researchers in the primary materials they review and recasting them into new or adapted macropatterns of the reviewers. Alternatively, research that focuses on macropatterning processes and experiences thereof may employ experiential methodologies or it might focus on modelling the associative and emergent characteristics of macropatterns. To this end, the pattern theoretical principle of *Micro-Meso-Macro* is entangled with the pattern theoretical principle of *Layering* given that macropatterns are emergent across layers of pattern memory and are, therefore, layered, scaled, panarchic, all of which afford yet more methodological options and opportunities.

Investigators might consider applying the principle of *Similarity and Difference* to research topics such as reasoning and recognition (*Modalities*) or to how patterns in working memory are changed and adapted according to new experiences and understanding (*Elaboration*). For these kinds of approaches investigators might use the kinds of methodologies I already outlined regarding experience, behaviour, and performance. Alternatively, they might consider the principle of *Similarity and Difference* in terms of the fuzzy and fluid ontologies and epistemologies patterns can afford. From this perspective, investigators might use methodologies that explore pattern reasoning, understanding, and categorisation and how these might change according to different kinds of pattern triggering or elaboration. For instance, a study may explore how different pattern interventions (experiences, knowledge) change participants" pattern reasoning in predetermined tasks. Alternatively, a study might explore confidence in pattern recognition or association and how these things change according to context and task.

The pattern theoretical principle of *Generation* reflects the ways that patterns allow us to work with possibilities and by doing so enable imagination and creativity. Studies might teach or otherwise provide a new set of patterns to participants and then get them to apply them in novel circumstances, for instance, in problem-solving activities of various kinds. Alternatively, participants might be provided with familiar resources and then be challenged to solve problems or otherwise be innovative in using familiar things in unfamiliar ways. Methodologies may involve observing participants undertaking these tasks, getting participants to think aloud during or debrief at the end of an activity, analysing the products of these creative processes, or following participants over time to explore how their learning and creativity develop from these activities. As before, methodologies may also simply focus on the experiential and awareness aspects of imagination and creativity.

The pattern theoretical principle of *Expressing and Representing* might be combined with those of *Sharing* and *Medium of Expression* to consider studies that explore how patterns (typically macropatterns) are shared through expressions, representations, systems, and languages. This might involve case studies of pattern system or pattern language development (as I provide in the next chapter) to identify good practices or to compare different approaches, contexts, and needs in developing pattern representations. More naturalistic methodologies (such as grounded theory or ethnography of various kinds) might be used to explore how, why, and with what results collectives engage in pattern sharing of various kinds. Indeed, these approaches might also be focused on exploring the limitations and complications of pattern sharing, such as the ways in which it can be co-opted and manipulated or used to manipulate others. I would also note connections with the *Generation* principle in terms of how collectives work with their pattern systems and languages, and connections with *Similarity and Difference* in terms of the lumping and splitting of pattern thought to create distinct representations and the implications thereof. Note that I take up the matter of pattern expressions, systems, and languages in the following section.

When it comes to the practicalities of pattern inquiry there can be many intersecting pattern principles in play. This in turn suggests that pattern inquiry may often involve (although not necessarily that it *must* involve) mixed and multiple methods. An exposition of these issues is beyond the scope of this volume, so I will simply note that methodological approaches in pattern inquiry need to be combined with care and attention to issues of convergent and/or divergent inference and to commensurability, alignment of constructs, and the nature of syntheses that cross methodological boundaries.

Products of Pattern Inquiry

As much as the four previous dimensions of pattern inquiry afford a rich palette of possible directions and approaches, that inquiry can generate or elaborate patterns in the minds of investigators and that those patterns can be shared with others through pattern expressions, systems, and languages is what transforms focused and thoughtful inquiry into the collective project of science. The pattern theoretical principle of *Cognition* should remind investigators that, although patterns may be developed or elaborated as a result of inquiry, they are only ever in the minds of the investigators. These patterns can be externalised as pattern expressions and pattern representations, systems, and languages as part of the research process, but they are at best reflections of these patterns. This raises issues related to the principle of *Constitution* in that there is of necessity a translation from a pattern's nexus of experiences to an external representation. Pattern representations are abstractions of patterns rather than a complete rendering of all a pattern's elements and connections. To that end, given that all pattern expressions are of macropatterns, we can also draw on the *Micro-Meso-Macro* principle that focuses attention on the processes of macropattern building in creating representations: aggregation, abstraction, theorisation, signification, formalisation, and systematisation. Indeed, these might be used as a conceptual framework for the development of pattern expressions, systems, and languages. Central to this is the pattern theoretical principle of *Logics* given that this is where explanation, understanding, meaning, causality, likelihood, and prediction lie.

The *Continuum* principle raises the issue of the perimeters and scope of the patterns being represented, particularly as, however perimeters are drawn, there are likely to be many associations that transcend these perimeters. The pattern theoretical principle of *Categories* requires attention to the categories, concepts, symbols, and language the author of a pattern representation uses, as well as to the artefactual tendencies of using one pattern system to express another. Related to *Continuum*, the principle of *Similarity and Difference* draws attention to issues of how authors go about lumping and splitting representations, in particular when one representation is better split into several representations or when several representations are productively merged into one. The principle of *Dependency* is also important as pattern systems can include some sense of the relationships and dependencies between their constituent patterns while pattern languages absolutely require this in the form of grammars and syntaxes. Those building pattern representations, systems, languages should consider the principle of *Layering* to deepen the strength and utility of what they produce, while also anticipating, within reason, how layering, scaling, and panarchy may weaken the uptake or understanding of the pattern representations, systems, and languages that they create.

Creating pattern representations also raises issues of bias and subjectivity in that what seems important or significant (and what seems unimportant and insignificant) to an author trying to represent a particular macropattern may be quite subjective. Indeed, a pattern expression may say as much about the mind of the investigator(s) who developed it as it does about the pattern it refers to. This raises concerns related to the principle of *Uniqueness*; how to express pattern thinking that represents the phenomena rather than the patterned mind of the investigator(s) as well as how to express pattern thinking in way that others can understand in the way intended by the investigators.

Some pattern theoretical principles can be interpreted in terms of caution and limitation in developing pattern representations, systems, and languages. The principle of *Dynamism* reminds us (qua Alexander and others) that pattern expressions and representations are snapshots, samples of pattern thinking that may continue to change and evolve after they have been created. On the one hand the representative artefacts of pattern thinking cannot track the dynamic nature of the patterns they represent, while on the other representations of pattern thinking may long survive the minds that created them. The principle of *Sharing* reminds us that those building pattern representations, systems, languages should be mindful that whatever pattern representations they build and however well they build them, those who subsequently engage with them will interpret and use them in ways that differ from the authors' expectations and goals. The principle of *Sharing* also serves as a reminder that authors of pattern representation should be mindful of the conventions for communicating their pattern ideas (whether explicitly or tacitly) and the ways that these conventions might enable or constrain such expressions as they choose to create. Care is needed in not overstating or understating, or overinterpreting or underinterpreting what aspects of that pattern thinking are apparent.

Reflexivity and caution are needed in developing pattern expressions, systems, and languages, not least because, as the pattern principle of *Awareness* reminds us, there are limits to which we are or can be aware of our own pattern thinking and its influence on our work. The principle of *Formation* reminds investigators to be mindful of whether expressions are intended to reflect or represent patterns, pattern thinking, the relationships between them, the enabling and constraining nature of those relationships, or some admixture of these factors. Care should be taken to distinguish between pattern representations that reflect researchers' pre-existing pattern thinking and the patterns they developed or elaborated from particular acts of inquiry and discovery. The principles of *Modalities, Triggering*, and *Elaboration* can draw investigators' attention to how other researchers' use of pattern perception, pattern recognition, pattern development, and pattern elaboration are or might be used in building pattern representations – these issues alone could spawn many research programs. For instance, Dearden & Finlay proposed a pattern language research agenda related to human-computer interface design that included researching different ways of developing pattern languages, exploring how to express or organise languages such that they can be used effectively, exploring the values that are expressed through pattern languages, and exploring how to use pattern languages more effectively in training and practice (Dearden & Finlay 2006).

Other principles can help investigators to focus on how pattern representations, systems, and languages might be received or used by others. For instance, the principles of *Formation, Weaving, Triggering* and *Elaboration* can be used to consider how users of pattern representations interact with them and how they weave them into their own patterns and pattern thinking. The principles of *Limitations* and *Latency* can serve as reminders to those building pattern representations, systems, and languages that they should be mindful of how their pattern expressions can be constrained or limited in development, and how, once disseminated, they can be further constrained, limited, repressed, damaged, or even punished. The principle of *Generation* might productively be used to explore how pattern representations, systems, languages are used in creative and imaginative ways.

Going deeper to consider progression of inquiry regarding building pattern representations and from that to possibly create systems of systems, and languages of languages, the pattern theoretical principle of *Expressing and Representing* might be used to how pattern expressions are used to develop pattern representations, how pattern representations are used to develop pattern systems, how pattern systems are developed into pattern languages, and how systems and languages of pattern systems and pattern languages might also be developed and used.

Methodological Precedents

Although many existing methodologies might be used in a pattern theoretical context, there are some that would seem to be more closely aligned with pattern inquiry than others. In this section, I describe a range of pattern-oriented methodologies and methods. Note that they all predate the pattern theoretical approach I have set out in this volume and as such are not formal or explicit examples of pattern inquiry, but rather examples from which pattern inquiry might borrow or learn.

Typological Pattern Analysis

A typological analysis involves analysing pattern phenomena in terms of categories, themes, classifications, or symbols. These may be predefined (such as in a template analysis [King 1998]) or may be abductively developed as part of the analysis (such as in thematic analysis [Braun & Clarke 2006]). Typologically, GPT researchers modelled patterns in terms of six fundamental components: generators, bonds, configurations, regularities, transformations, and images. "Generators" (also called "building blocks" or "blocks") were the unique objects from which patterns were built (Tarnopolsky & Grenander 2006). "Bonds" (also called "connectors") connected generators indicating the direction and nature of connection. When generators bonded together, they created "configurations". As bonds and generators in a configuration changed over time then configurations also changed, emerged, or disappeared. Regularities were the rules of a pattern that included both local regularity (allowable bonds between generators to create configurations) and transformations (allowable configurations). A sixth component of patterns in GPT was "image" which referred to the potential disconnect between perceptions of a pattern configuration and its reality, and in particular the computational problems of recognising objects whose appearance changes according to the relative positions of the object and the observer and to variations in observer conditions (lighting, distraction, movement, etc.). In a later (and more discursive) work, Tarnopolsky & Grenander (2006) argued that (independent of the modelling techniques employed) a GPT pattern model had four essential aspects: "building blocks, rules to connect them, general architecture, and changes that are allowed." "Blocks" were the vertices in a graph that represented both the forms, structures, capabilities, characteristics, and behaviours of the pattern, and those things with which the pattern has relationships. "Rules" were the edges in a graph, and they qualified the direction, nature, behaviour, and strength of the relationships between blocks. Architecture was the cumulative form of all the blocks and rules (the vertices and edges) in a pattern model. A pattern model's architecture was, therefore, emergent from its blocks and rules. Changes were the topological capacities and limitations of the model.

A typological approach was also used by Stevens (1974) who outlined recurring "natural patterns" including (variously) symmetry, shape, geometry, turbulence, spirals, and meanders, branching and trees, bubbles and films, and packing and cracking. Bell (2012), who drew on Stevens' typology of basic pattern forms, focused on types of patterns in the landscape based on their apparent qualities, origins, and functions, which included: "… position and orientation, shape, interval, texture and density, colour, visual force, interlock, enclosure, rhythm, balance, scale and proportion, asymmetry, hierarchy, transformation, similarity and continuity". Bell's approach depended on developing a suitably macropatterned repertoire of kinds of regularity and then engaging in systematic pattern recognition in matching visual perceptions to pattern archetypes, and when matches were found then these were taken to be pattern instances. This abstraction of Bell's methodology is illustrated in Figure 8.2. Importantly, he acknowledged the intrinsic subjective and contingent nature of this approach:

> We look at a scene with a purpose in mind: to enjoy a natural prospect, to find a route through it, to evaluate it as a good place to live or to build a factory. This affects where we look, how we look and what visual cues we seek out, such as easy walking places, good house sites, level ground or water supply. Thus, some people will see some objects and patterns and not others; yet all may believe that they perceive the reality of the scene in objective terms.
>
> *(Bell 2012, p. 59)*

Although I disagree with Bell what and where patterns are, the principle of developing a macropattern repertoire, a cognitive pattern system or language in fact, and then systematically applying it and based on experience appraising and elaborating it would seem to afford a robust approach to conducting fieldwork. The implications are that you need to configure a suitable pattern repertoire before you can apply it as a research tool, that the

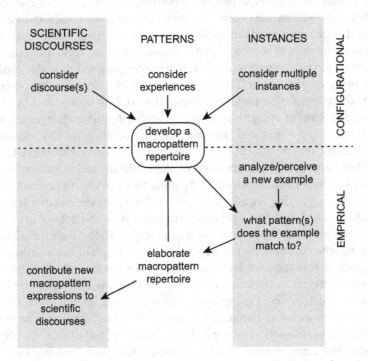

FIGURE 8.2 An abstract logic map of the typological pattern methodology from Bell (2012).

Diagram prepared by the author.

triggering of particular patterns can fill in meanings, understandings, and significances, and that, by triggering certain patterns, then they are elaborated (and presumably refined).

Having raised the idea of archetypes, another typological methodological approach we might draw on is archetype analysis in which patterns are defined as abstractions of recurring regularities in or of phenomena, and patterns that recur are defined as "archetypes". Oberlack et al. (2019) argued that identifying an archetype required multiple instances of a phenomenon across multiple contexts, and that no definition of an archetype was or could be exclusive of other definitions or perspectives. This reflects the principle outlined by Fiske that archetype analysis is a middle-range approach which can lead to multiple valid but variant models of an archetype rather than one general one (Fiske 1986). We can take two pattern inquiry principles from this. The first is the need for multiple examples of using a particular pattern or pattern system, in part because a pattern can be triggered by pulling on different pattern elements, and in part because each perceived instance can elaborate the pattern. The second is a more positional one in considering to what extent and in what ways particular approaches to pattern inquiry are middle ranged. It may be that, given that patterns afford aggregative but fuzzy epistemologies, pattern inquiry is inescapably middle ranged, an issue I will return to later.

Another approach that might be used is categorical or concept analysis. I do not mean by this the well-established use category theory in mathematics (a "general mathematical theory of structures and of systems of structures" [Marquis 2023]), although this might conceivably be useful in some way or other (see my later comments on set theory). Rather I mean categorical or concept analysis as naturalistic ways of exploring and mapping the categories and concepts we use. For example, Walker & Avant's (2019) outline methodology for concept analysis might serve in analysing both constitution and variation in particular patterns. Their approach included developing a model case of the "normal" use or operation of the concept along with complementary cases that were borderline (mostly the concept but not quite), related (similar to the concept but not the identical), and contrary (definitely not the concept). There were also steps that involve identifying antecedents (what is required for the concept to exist), consequences (what must follow directly from the concept), and empirical referents (phenomena that reflect the concept). Individually and collectively these kinds of techniques might be usefully applied to pattern analysis.

Topological Pattern Analysis

A topological analysis focuses on the ways in which parts are connected or otherwise associated in phenomena of interest. Examples include mathematical graph theory (Gross et al. 2018) which has been used to compare network topologies to identify shared regularities (Maheswaran et al. 2009) and actor-network theory (Latour 2005). Many GPT researchers rendered and then analysed patterns topologically using mathematical graph and set theoretical models (Tarnopolsky & Grenander 2006). This involved hybrid methods that combined deductive logic (creating ideal topological or algorithmic models of patterns) and inductive logic (defining patterns in terms of sets of all their legitimate variations). Of course, this was predicated on elements or nodes in networks being amenable to identification and representation, something that seems particularly elusive in the case of patterns. At best this approach might be used to capture abstractions of macropatterns. Moreover, as per the pattern theoretical principles of *Uniqueness, Awareness, Dynamism, Continuum, Latency,* and *Layering* (amongst others), patterns are not static networks of static elements, they are fuzzy, dynamic, and emergent and as such a graph approach may be somewhat limited. It is not surprising therefore that Fiske (1986) suggested that smaller patterns are easier to work with than larger ones. However, "small" in this case refers to the numbers of elements and connections, not the micropattern or macropattern qualities of a particular pattern. The patterns these approaches are able to reflect are all, I would argue, macropatterns. Moreover, if, from a reductive GPT perspective, the larger and more complex the pattern the harder it is to work with, then it is likely to be of little help in exploring the larger more complex (and more interesting) patterns of understandings, meanings, theories, conventions, and systems.

Set theory can also be used to represent clustering, identity, and similarities and differences, which are all associated with patterns. In addressing this, I differentiate between naïve set theory which is natural language and formal axiomatic set theory that is rendered using symbolic notation and logical operators and rules (Halmos 2017). In Table 8.3 I have outlined a number of broad naïve set theoretical principles and I consider some of their equivalences in a set theoretical space. As with graph theory, set theory would seem to have limited application to the kinds of complex and dynamic patterns researchers are likely to be interested in. The subset of "fuzzy set theory" (Zadeh 1965) may be better aligned, but again this is a matter for another time. I would note though that exploring the parallels in topological axioms between sets and patterns is illustrative and there may be other ways of using set theory in the context of pattern inquiry – I will leave that question for others to explore.

TABLE 8.3 Comparing set theoretical principles with those of patterns

Set Theoretical Principle	Equivalence in Pattern Theory
Sets are equal only if they have the same elements	Patterns in different minds are equivalent (although they cannot be identical) if they relate to the same phenomena.
The identity of a set is determined by membership	The identity of a pattern is defined by much more than membership (its constituent elements) and it does not have well-defined edges by which any absolute identity can be established or maintained.
Order does not matter in a set, only membership	Symbolically sets are linear (one symbol after another) but pattern elements can be associated with many other elements, and patterns with other patterns such that relationships matter a great deal.
The union of any sets is itself a set	The union of any patterns is itself a pattern.
Any subset is itself a set	I am unsure whether a subset of a pattern is always a pattern. This is true of a pattern of patterns such that its constituent patterns are patterns.
There can be sets of sets	There can be patterns of patterns – indeed this is the basis of moving from micropatterns to mesopatterns and from mesopatterns to macropatterns.
A set can have just one member	Can a pattern have just one element? I think not as patterns are intrinsically aggregative and associative.
A set can have no members	Can a pattern have no elements? Again no, patterns emerge from clustered memories.
Sets can contain themselves (the basis of Russell's paradox)	Can a pattern contain itself? Again, no. A pattern of itself is itself, and the principle of merging similarities into common patterns would prevent a pattern turning itself inside out by referring to itself.

Pattern Language Analysis

A third methodological approach that has precedents for pattern inquiry is that of pattern system and pattern language development. As I noted earlier, a pattern language is a system of related pattern representations with a common grammar and syntax that sets out how patterns in the language are expressed and how they are combined. However, although many pattern languages have been developed, there has been relatively little published on the methodological approaches used. For instance, although Alexander et al. did not disclose their methodology in advancing their pattern language of architecture in its original publication, Alexander (1999) did later describe some of the processes they had used. These included identifying things that the developers perceived as having had profound impacts in the domain of practice (patterns needed to be significant nexuses of thought and practice) or that had contributed to some greater and more coherent whole (patterns needed to establish coherence). Alexander also noted that pattern development was inescapably abductive, fluid, and discursive, and based on consensus building.

The work of Alexander and colleagues has long been the commonly accepted point of origin of pattern languages, and there have been many who followed their example, often mirroring their inattention to method and methodology. However, there have been some researchers who have been more fulsome in describing their approach, notably Iba & Isaku (2016) who outlined a three-stage approach based on "pattern mining" (collecting candidate patterns and pattern elements), "pattern writing" (abstracting and grouping similar patterns), and "pattern symbolising" (codifying the patterns). Leitner (2015) too focused on collecting examples and observations, discussing them, and coming to a consensus on the patterns that can represent them.

Pauwels et al. (2010) described the development of a pattern language for human-computer interaction design that was to be used in support of a single software application (rather than for a domain of practice). The authors described four stages in developing their pattern language. The first involved mining (collecting) existing design solutions in the form of screenshots and user interactions. The second involved interviewing users and other stakeholders to "analyse problems and contexts". The third step involved the iterative workshopping of new design solutions and building consensus within the project team. The final step involved testing draft patterns in practice and only those that passed this testing were included in the pattern language, its uniting aspect being the software application context.

Mor & Winters (2008) described the development of a pattern language for instructional design. Their process started with the project team developing a pattern typology, which was used as the basis of "pattern elicitation workshops" at a series of international conferences where participants provided case studies related to the typology and the "best" cases were developed to be "seed patterns" that consisted of a description of the context, the problem, and a "skeletal" solution. At the end the draft patterns were presented back to the whole group for feedback. The product seems to have been the typology with related cases and seed patterns.

Güven (2021) described the development of a pattern language of decision analysis. Patterns were developed by a small team using a workshop format and consensus methods. The main focus of their paper was on testing the draft patterns by getting students to use them in an assignment, then interviewing some of the students, and finally recording workshops where the patterns were used. Other than stating that data were analysed "using qualitative methods" few other methodological details were provided.

Felstead & Thwaites (2023) described the development of a pattern language for cohousing solutions that was developed using a hybrid approach combining principles of grounded theory inquiry. This seemed more methodologically robust than some of the other examples I reviewed, particularly in their attention to theoretical sampling and theoretical saturation. Nevertheless, their process was similar to that used by Iba and colleagues: problem-solution mining, pattern clustering, pattern writing, pattern cataloguing, and language testing. They described the value of their pattern language as a "participatory research tool" the sharing of ideas, and the development of creative solutions to design problems.

Setting aside the deficits in methodological orientation and detail (which are common to much of the pattern language literature), they generally focus on collecting examples or cases, refining them using workshops and other consensus techniques, and only a few engaged in any kind of validation or testing process. However, all three of these stages could have drawn on established methodologies. For instance, in gathering examples and pattern candidates, there are well-established systematic review and document analysis methods for ensuring suitably exhaustive searches of the literature and other sources, and there are well established methods for sampling (random, purposive, theoretical, convenience) that could also add rigour to these kinds of processes. There are also many ways of conducting task analyses and workflow analyses that might also be useful. In terms of revising, editing, and consensus building, there are again established consensus methodologies that might be used including "Delphi" techniques, "nominal group process" approaches, and "consensus development panels", all of which have well-documented procedural characteristics (Waggoner et al. 2016). Consensus methods are not the only way in which pattern representations might be built, indeed there is a wide variety of knowledge of techniques that could be used including knowledge syntheses, template analyses, realist inquiry, and the progressive coding used in certain kinds of grounded theory (open, axial, selective). In terms of testing and validating pattern languages, Pauwels et al. (2010) used conformance and field testing, Mor & Winters (2008) seemed to stop at the consensus stage, and Güven (2021) used a variety of piloting techniques, Felstead & Thwaites (2023) used "serious gaming" activities to pilot their pattern language. There are many other approaches that might be used including usability testing, various evaluation approaches, and theoretical saturation.

A particular concern I have with this body of work is the absence of theory in methodologies for developing pattern languages. Felstead & Thwaites perhaps came the closest to theoretical engagement. But, despite their allusions to grounded theory, they did not build theory other than their pattern language. Although they did not expressly state their pattern language was in effect their theoretical product, they did claim that:

> As a theoretical research tool, the grounded pattern methodology identifies common problems and solutions within complex, "live" environments and provides a clear link between empirical evidence, patterns, and broader concepts.
>
> *(Felstead & Thwaites 2023)*

A more rigorous approach might reasonably consider theoretical positions regarding the sources of knowledge and expertise they use, the issues, topics, or domains they are working in, the processes they engage, and the knowledge they generate – see Figure 8.3.

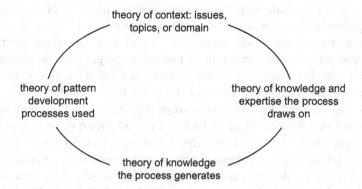

theory of context: issues,
topics, or domain

theory of pattern
development
processes used

theory of knowledge and
expertise the process
draws on

theory of knowledge
the process generates

FIGURE 8.3   An outline cycle of theoretical considerations for pattern representation, pattern system, and pattern language development.

Diagram prepared by the author.

In terms of theoretical positions towards context or domain for pattern language development, investigators might consider the ontological characteristics of the domain: what kind of domain it is, what domains it is similar too, and what its functions, behaviours, and consequences are. For instance, if investigators were to develop a pattern language for writing an academic book, they might start with considering the purposes of academic writing in general and how that compares to the specific challenges of writing academic books. Investigators might also consider the different kinds of academic texts that might be written, the kinds of academic writing found in a particular field or discipline, and how academic writing is both similar and different from other kinds of writing.

In terms of theoretical positions towards the knowledge and expertise pattern language development draws on, investigators might consider what possible sources there are available to them, what the authority and utility of different sources might be, what the consequences of combining sources might be, and how different epistemological positions and assumptions might be adopted or avoided as part of the development process. Normative pattern language methodologies require experts in the domain to be both the source and the developer of patterns, while naturalistic methodologies require the perspectives of participants, but they do not need to be experts *per se*, and the analysts and developers are likely to be expert pattern researchers but not necessarily expert in the domain of interest.

In terms of theoretical positions towards the processes used in developing a pattern language, investigators might consider using the pattern theoretical principles of macropattern development (aggregation, abstraction, signification, theorisation, formalisation, and systematisation) as a framework for pattern development. They might also note that naturalistic pattern languages may include negatives, conflicts, waste, irrationality, harms, and other kinds of problems that would be very unlikely to be considered appropriate for a normative pattern language. Investigators might also reframe both the concept of and product of validation in terms of evidence arguments rather than binary transformations by drawing on the validity theories of Messick and/or Kane.

In terms of theoretical positions towards the knowledge they generate, investigators might again consider the differences between normative and naturalistic pattern languages, noting that there may be quite different standards and directions of knowledge production from those used in naturalistic pattern language development. For instance, normative pattern languages have a primary focus on practical problem solving, not just in terms of developing pattern representations but also in terms of how these representations are subsequently supposed to be used. Naturalistic pattern languages may use problem solving as an analytical frame, but they could use others (such as a realist framing on context, mechanisms, and outcomes) and there is no expectation that the use of the resulting pattern representations will be used for problem solving. For example, a pattern representation template for individual patterns might be based on tabulating or connecting pattern elements that represent variously: experiences, knowledge, feelings, meanings, and predictions. As another example, a pattern representation template for macropatterns might involve cataloguing levels and forms of aggregation, abstraction, and formalisation. Investigators might instead adapt related frameworks such as Lorand's model of perimeters, variations, and ordering principles, although this might be better applied to analysing perceptions of pattern instances than of the patterns those perceptions draw on.

Given that a great many methodologies might be used in pattern inquiry, we might instead ask what does *not* constitute a pattern theoretical methodology? A strict reading would exclude anything that does not engage pattern theoretical concepts regarding the cognitive basis of pattern and its duality with pattern thinking. However, this seems unnecessarily exclusive so instead I would suggest that pragmatism and adaptation allow for many approaches to be applied in a pattern theoretical space. Moreover, the thinking, reasoning, and creativity that goes into developing these approaches and into generating knowledge from them is patterned and the products are also patterned such that they all depend on interpretation, often in accordance with a disciplinary or paradigmatic norms (shared pattern thinking). There is a suggestion of hermeneutics again but also a dialectical sense of interpreting interpretations in terms of patterns and pattern thinking of the interpreters and of the shared pattern thinking of the collective in which their work is conducted and disseminated.

Chapter Summary

The ideas set out in this chapter reflect my original intended focus for this book, that of outlining approaches to scientific inquiry informed by, drawing upon, or responding to patterns, patterned thinking, patterned knowledge, and other pattern phenomena. My long detour in building pattern theory was necessary in developing a solid and defensible basis for pattern theory and pattern inquiry. If inquiry is to respond to the realities of patterned thinking and patterned minds, it must take pattern theory into consideration, and it also needs to build on the opportunities and dimensions patterned minds afford us in engaging in inquiry. After all, patterns are not something I (or anyone else) invented, we have always been using patterns to think, and they are the basis for all science, philosophy, technology, culture, and society. Pattern thinking and patterned thinking is already omnipresent in every branch and paradigm of science. I have woven together and connect (in patternlike ways) a wide range of ideas and theories (some of them my own but many of them from others) into a coherent whole that I hope reflects the pattern nature of mind.

I have also argued that pattern theory has much to offer to scientific inquiry. One contribution is based on the argument that all researchers (and their audiences) are always thinking in and through patterns. Pattern is not something we might add to inquiry but rather something that is already and has always been central to it. The challenge to scientists, therefore, is to make more conscious and deliberate use of pattern rather than thinking about it as something new or foreign to current thinking. A second contribution is that pattern theory can be used to critique of much existing scientific thinking exactly because much of what we take as justified true belief does not stand up to scrutiny from a pattern theoretical stance. Pattern theory in this way also has much to contribute to the philosophy of science, not least that a reading of pattern theory in the context of scientific inquiry may serve both as a caution regarding knowledge claims and as the basis of making different kinds of knowledge claims. After all:

Science is a social enterprise that requires repetition or replication of results by other competent scientists, other minds. Without other minds [philosophically presupposed], there is no science. The hope of completely divorcing science from philosophy is utterly naïve.

(Edwards 2010, p. 82)

References

Alexander C. The Origins of Pattern Theory: The Future of the Theory, and the Generation of a Living World. *IEEE Software* 1999; 16(5): 71–82.

Bell S. *Landscape: Pattern, Perception and Process*. London, UK: Routledge: 2012.

Borsboom D, Mellenbergh GJ, Heerden JV. The Theoretical Status of Latent Variables. *The Psychological Review* 2003; 110(2): 203–219.

Bourdieu P. *Outline of a Theory of Practice*. Cambridge, UK: Cambridge University Press: 1977.

Braun V, Clarke V. Using Thematic Analysis in Psychology. *Qualitative Research in Psychology* 2006; 3: 77–101.

Creswell JJ, Creswell JD. *Research Design: Qualitative, Quantitative, and Mixed Methods Approaches* (5th ed.) Thousand Oaks, CA: SAGE: 2018.

Dearden A, Finlay J. Pattern Languages in HCI: A Critical Review. *Human Computer Interaction* 2006; 21(1): 49–102.

Devlin K. *Mathematics: The Science of Patterns: The Search for Order in Life*. Mind and the Universe: 1994: pp 4.

Edwards RB. *The Essentials of Formal Axiology*. Lanham, MD: University Press of America: 2010: pp 82.

Felstead A, Thwaites. A grounded Pattern Language: Testing a Methodology for Exploring Cohousing Residents Involvement in Shared Outdoor Spaces. *CoDesign*, 2023. DOI: 10.1080/15710882.2023.2289028

Fiske DW Specificity of Method and Knowledge. In: Fiske DW, Shweder RA (eds). *Metatheory in Social Science: Pluralisms and Subjectivities*. Chicago IL; University of Chicago Press: 1986.

Gadamer H-G. *Philosophical Hermeneutics*. Berkeley, CA: University of California Press: 1977.

Glaser BG. Choosing Grounded Theory. *The Grounded Theory Review* 2014; 13(2): 3–19.

Gross JL, Yellen J, Anderson M. *Graph Theory and Its Applications* (3rd ed.). Boca Raton, FL: CRC Press: 2018.

Güven AÖ. Synthesizing decision analysis: A pattern language approach. PhD Thesis, Stanford University, 2021 - https://www.proquest.com/openview/7583d92e1040f621573ddb65a189ff7b/1?pq-origsite=gscholar&cbl=18750&diss=y

Halmos PR. *Naive Set Theory*. Mineola, NY: Dover: 2017.

Hanson NR. *Patterns of Discovery*. Cambridge UK: Cambridge University Press: 1958.

Iba T, Isaku T. A Pattern Language for Creating Pattern Languages: 364 Patterns for Pattern Mining, Writing, and Symbolizing. PLoP '16: *Proceedings of the 23rd Conference on Pattern Languages of Programs*. 2016; 11: 1–63.

Kantowski MG. Processes Involved in Mathematical Problem Solving. *Journal for Research in Mathematics Education* 1977 May 1; 8(3): 163–180.

King N. Template Analysis. In: Symon G, Cassell C (eds.) *Qualitative Methods and Analysis in Organizational Research*. London, UK: Sage: 1998.

Latour B. *Reassembling the Social: An Introduction to Actor-Network-Theory*. Oxford, UK: Oxford University Press: 2005.

Lawson, A.E. What is the Role of Induction and Deduction in Reasoning and Scientific Inquiry?. *Journal of Research in Science Teaching* 2005; 42: 716–740.

Leitner H. *Pattern Theory: Introduction and Perspectives on the Tracks of Christopher Alexander*. CreateSpace: 2015.

Maheswaran R, Craigs C, Read S, Bath PA, Willett P. A Graph-theory Method for Pattern Identification in Geographical Epidemiology – A Preliminary Application to Deprivation and Mortality. *International Journal of Health Geographics* 2009; 8: 28.

Marquis J-P. 2023Category Theory. In: Zalta EN, Nodelman U. (eds.) *The Stanford Encyclopedia of Philosophy*. Online at: https://plato.stanford.edu/archives/fall2023/entries/category-theory/ - accessed 16th Nov 2023.

Metzger SL, Littlejohn KT, Silva AB, Moses DA, Seaton MP, Wang R, Dougherty ME, Liu JR, Wu P, Berger MA, Zhuravleva I, Tu-Chan A, Ganguly K, Anumanchipalli GK, Chang EF. A High-performance Neuroprosthesis for Speech Decoding and Avatar Control. *Nature* 2023; 620(7976): 1037–1046.

Mor Y, Winters N. Participatory Design in Open Education: A Workshop Model for Developing a Pattern Language. *Journal of Interactive Media in Education* 2008; 1: 1–16. https://eric.ed.gov/?id= EJ840809

Oberlack C, Sietz D, Bonanomi EB, et al. *Archetype Analysis in Sustainability Research: Meanings, Motivations, and Evidence-based Policy Making*. Ecology and Society. 2019; 24(2). https://www.jstor.org/stable/26796959

Pauwels SL, Hubscher C, Bargas-Avila JA, Opwis K. Building an Interaction Design Pattern Language: A Case Study. *Computers in Human Behavior* 2010; 26(3): 452–463.

Schultz PR. Toward Holistic Inquiry in Nursing: A Proposal for Synthesis of Patterns and Methods. *Scholarly Inquiry for Nursing Practice* 1987; 1(2): 135–146.

Siler K, Strang D. Peer Review and Scholarly Originality: Let 1,000 Flowers Bloom, but Don't Step on Any. *Science, Technology & Human Values* 2017; 42(1): 29–61.

Stevens PS. *Patterns in Nature*. London, UK; Penguin: 1974.

Tarnopolsky Y, Grenander U. *History as Points and Lines*. 2006. Unpublished book manuscript: online at https://www.dam.brown.edu/ptg/REPORTS/pointsandlines.pdf

Waggoner J, Carline JD, Durning SJ. Is There a Consensus on Consensus Methodology? Descriptions and Recommendations for Future Consensus Research. *Academic Medicine* 2016; 91(5): 663–668.

Walker OL, Avant, CK. Concept Analysis. In: Walker OL, Avant, CK. *Strategies for Theory Construction in Nursing* (6th ed.). New York, NY: Pearson: 2019.

Young NA, Collins CE, Kaas JH. Cell and Neuron Densities in the Primary Motor Cortex of Primates. *Frontiers in Neural Circuits* 2013; 7: 30. DOI: 10.3389/fncir.2013.00030

Zadeh LA. Fuzzy Sets. *Information and Control* 1965; 8(3): 338–353.

9

THE PRACTICALITIES OF PATTERN INQUIRY

In which I consider a worked example of pattern inquiry to explore the strengths and limitations of such an approach.

While the principles, practices, and precedents outlined in the previous chapter can provide a foundation for pattern inquiry, much of the research effort to date (mostly in and around the development of normative pattern languages) has been somewhat patchy in terms of rigour, methodological and theoretical grounding, transparency, and consistency. Robust methods and methodologies are needed if pattern inquiry is to have any place in the methodological canon. To that end, I provide an extended worked example, followed by a number of contrasting shorter examples as vehicles to explore practical issues in pursuing pattern inquiry.

Worked Example: Developing a Pattern System

The following worked example describes the development of a medical school admissions pattern system. I report this in the form of a descriptive case study, starting with the case context. My background is in medical education research and in 2019 I was asked to join a pan-Canadian think-tank organised by the Association of Faculties of Medicine of Canada (AFMC) to explore options for enhancing Canadian admissions processes to improve diversity in the physician workforce. This "Future of Admissions in Canada Think Tank" (FACTT) provided a valuable opportunity to begin to develop a pattern system for medical school admissions. I had already explored some of the challenges associated with the design and operation of admissions systems (Ellaway et al. 2019; Ellaway et al. 2018) and had found the research literature skewed in that most academic articles focused on the same few issues; what applicant information was gathered, how it was used, and how decisions were made. There were many other characteristics of admissions systems that were rarely considered and many problems and issues that could have benefitted from more attention. I hoped a pattern system could map the many topics and concerns that constituted medical school admissions as a way of broadening scholarly focus in this area.

DOI: 10.4324/9781003543565-9

Having worked with many medical schools and medical education scholars and leaders around the world, I was very aware that approaches to admissions varied significantly according to context. Some jurisdictions had national admissions exams and even admissions lotteries, different schools used different combinations of file reviews, interviews, and assessments used, and the resourcing, values, and sequencing of admissions processes also varied between systems. It was clear, therefore, that no one set of "best practices" would be able to be inclusive of or useful to all medical schools or systems. Rather, in thinking about the need for a model or guide for medical school admissions, I drew on some of my other earlier work in developing medical education information standards (Ellaway & Smothers 2013) to consider models that could be inclusive of many different implementations while still relating them to common concepts, processes, and structures.

Working with the FACTT with its representatives from most of Canada's (then) 17 medical schools along with several interested regulatory organisations, it was clear that, even there, there were many similar differences and different similarities in participant perspectives as to what admissions was for, what it should focus on, and what it should produce. In developing a pattern system, I wanted to acknowledge these differences in specific approaches at the same time as working with the common pattern thinking applied to underlying issues and problems in medical school admissions systems. I saw this as a way of mapping out these common concerns and dimensions of medical school admissions systems, and I hoped that the resulting pattern system could be used for describing and modelling medical school admissions systems more broadly in ways that could be used in appraising and evaluating them, guiding their development, and guiding scholarly inquiry. To that end, my study questions was: what is the nature of the shared pattern thinking of the medical education community regarding medical school admissions?

Developing a Methodology

As I described in the previous chapter, although there are some methodological precedents for pattern language development, they tend to be based on organic interpretations of consensus methods and do not appear to have been substantially grounded in mainstream methodological theory in the social sciences. Consensus methods can provide a degree of authenticity and utility, but they are limited in their sampling of expert or practitioner opinions and their products are typically provisional. Although I wanted to include some kind of consensus process in the study, I also wanted to expand the sampling of the shared pattern thinking of the collective in other ways. Medical education has a (fairly) robust research knowledgebase which reflects practices and thinking of medical educators around the world. I therefore decided to include a systematic literature review as part of the study, albeit one that sought to identify recurring macropatterns rather than appraise or synthesise findings. This would widen the geographical scope to include the contexts of whatever authors had published on admissions, although like most fields, the medical education literature is dominated by certain countries and regions (the USA and other higher income anglophone countries, the Global North, etc.). Moreover, although my goal was to create a naturalistic pattern system that represented the actual shared pattern thinking (rather than a normative focus on "best practices"), I anticipated that this would also require a completist approach to identify and include extant but neglected patterns to establish a comprehensive map of the medical school admissions landscape. I also wanted to focus on patterns that related to broad functional problems rather than specific operational ones.

Combining consensus and review methodologies meant that this was a mixed methods study, which raised the issue of how the two sub-studies would be combined. Using Greene's (2007) typology of stances, I took a purist stance in that the methods shared a common mental model, i.e., that there was an implied set of shared macropatterns that could be interpreted from the data. Using Greene's (2007) typology of inferential logics for mixed methods research my approach used a complementarity logic (results from different methods add breadth and depth). However, the design, although sequential, did not follow the convergent, explanatory, or exploratory sequential designs proposed by Creswell & Creswell (2018) as while the consensus and review stages involved different methods and different data, they involved the same abductive synthesis processes.

The first phase involved workshopping ideas with stakeholders which reflected many of the ideas proposed by Iba & Isaku (2016) in getting participants thinking about patterns and identifying possible candidate patterns, and thereby contributing to provisional pattern representations. Given the exploratory and unformed nature of understanding at the start of the study, more formal consensus methods such as Delphi (in its various forms) was not a suitable approach.

The second phase was grounded in principles of literature review and knowledge synthesis in that I sampled the wide ranging (or so I hoped) perspectives of authors and their institutions and their admissions systems as a way of exploring the broader shared understanding and priorities of the community for whom they were writing. Methodologically this meant using review techniques including bibliographic searches, sampling, inclusion and exclusion, and extraction. As I anticipated that peer-reviewed articles would focus on quality assurance and quality improvement issues (as much of the literature in applied fields such as medical education tends to do) and on issues of immediate interest or concern to practitioners, I augmented the peer-reviewed literature with grey literature sources from books on medical education and from admissions materials posted on medical school websites. I drew on template analysis for the analysis in that I had a pre-defined coding structure (the pattern system) and focused on elaborating and developing this coding structure rather than extracting or synthesising the findings of the articles I reviewed (King 2012). The execution of the study design was abductive in that I undertook multiple rounds of searching, extracting, and evaluating to test and develop the pattern system. This also involved adapting the methodology such that both the pattern system and the methodology were developed in tandem. In terms of contextual sensitivity, not only were the sources authentically grounded in admissions practices (and by implication its shared pattern thinking), I, as the analyst, was an experienced medical education researcher with a grounding (not surprisingly) in pattern theory. I will come back to a critical evaluation of my knowledge claims after describing the execution of the study.

Workshopping

I started this phase with the assumption that individuals involved in running and managing admissions systems in medical schools would have practical experience and perspectives that reflected shared if variant pattern thinking. In grounding this assumption, I drew on Wenger's theories of "communities of practice":

Over time, the joint pursuit of an enterprise creates resources for negotiating meaning ... they gain their coherence, not in, and of themselves as specific activities, symbols, or artefacts, but from the fact that they belong to the practice of a community pursuing an enterprise.

(Wenger 1998, p. 82)

Rather than symbols or artefacts, I focused on the implied shared pattern thinking of the group that supported its function as a community of practice. The patterns that I sought may not have been explicit or even common to every person or system, but there was, I assumed, a degree of shared pattern thinking across a group who had a shared practice and who interacted on a regular basis. I was also mindful that this shared pattern thinking would be implied in many different expressions (narratives, comments, personal theories, observations, questions) and that I needed to make sure that no one individual's patterns dominated the representations I created from this process.

The phase 1 process involved two workshop sessions (one online and one face-to-face), that involved members of the FACTT group working in small groups to identify generic problems and functions in admissions systems and then outlining what they involved. Returning to a larger group format, the putative patterns were discussed and refined leading to a draft pattern system, which was then shared back with interested workshop participants. Both during and then after the workshops I abstracted middle-range pattern representations (akin to Pawson's demi-regularities) from participant interactions using a template that drew on the work of both Alexander (1977) and Iba & Isaku (2016) with elements for the problem, the solution, the context, and examples. Some additional written feedback on these middle-range pattern representations was provided by a few interested members of the group following the workshops. The main product of this process was an outline set of pattern representations for medical school admissions, roughly equivalent to an initial programme theory in realist inquiry. I might have undertaken more rounds of workshopping or involved more individuals, but the FACTT project came to an end as the vehicle for pursuing the pattern system in this way.

Pattern Analysis

Throughout this process (in both stages of the study) I engaged in seven interlinked and abductive tasks that can be understood as making up the core of pattern system analysis: deductive and inductive pattern identification, lumping and splitting, resolving gaps, overlaps, perimeters and scope, and pursuing coherence and parsimony – see Figure 9.1.

Deductive shared pattern identification focused on identifying the implied pattern thinking that was necessary to yield both the extant and the implied structures and processes in the domain of interest. More specifically, this involved exploring what broad shared patterns had been and continued to be necessary to yield current medical school admissions systems. There were parallels in this with Archer's (1979) critical realist interpretations of morphogenesis (the origin of form) in that I considered shared pattern thinking to be a primary (although not an exclusive) causal basis of subsequent structures and practices, and that the ongoing shared pattern thinking in the domain was directed, and to some extent constrained, by past and present structures and practices. For instance, there was an implied shared pattern in discussions of participants' motivations in participating in

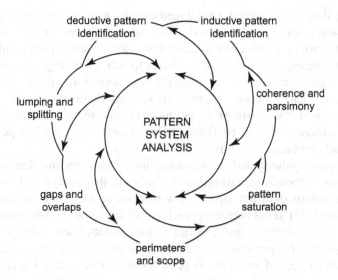

FIGURE 9.1   Interlinked aspects of pattern system analysis.

medical school admissions. There were many different reasons given (applicants wanted a career in medicine, raters wanted to make sure only the best candidates were admitted, and administrators wanted to make sure everything ran smoothly and efficiently) that I captured this in a single pattern representation "R03: Why individuals participate".

Inductive shared pattern identification focused on how actors in the domain articulated their pattern thinking in terms of what they said and did. Given that phase 1 data was drawn from workshop conversations and phase 2 data was drawn from academic articles, with no explicit fieldwork beyond this, inductive shared pattern thinking analysis involved looking for regularities in what was said across multiple voices. Practically, this involved taking notes, coding texts, building themes and microtheories, and exploring similar differences and different similarities in what was said. For instance, the pattern representation "R05: Participant experiences" reflected the personal experiential observations of individuals involved in admissions systems and comments about the experiences of others. It did not capture the specifics of the experiences, only that participating individuals had a wide range of experiences and that those experiences mattered in defining the shared pattern thinking in the domain.

Part of the analysis focused on identifying what was and was not involved in the domain of interest (perimeters and scope). Medical school admissions is a relatively discrete undertaking even within medical schools as it needs to its manage conflicts of interest and accountabilities with respect to its various stakeholders. As such, issues of perimeters and scope were not particularly challenging but they still required consideration in this study. For instance, when or how medical school admissions starts or ends may differ according to the percepions or understanding of different observers; applicants may begin preparing many years before they submit an application, while systems typically run annual admissions cycles that end with the selection of the class to be admitted. In part this part of the analysis involved developing definitional parameters in advance of the study and then adjusting them as the study progressed. To that end I worked the following definitional parameters: the domain of interest was all medical education admissions in the world; the construct of interest was a single medical school admissions system; each school, institution, or group thereof could have one or more admissions systems; each admissions system engaged multiple actors, activities, and resources to achieve their goals; each admissions system ran multiple admissions cycles, typically once a year; and there were many differences, both big and small, between admissions systems. Some of the pattern representations I identified also helped to define the perimeters of the domain in terms of those things the domain interacted with, in particular "G01: Context", "G02: Scope", and "G03: External oversight and accountability".

Another aspect of pattern analysis was attending to pattern saturation, something that built on concepts of theoretical saturation that propose that analysis can be concluded if new data or reanalysis does not lead to new theories or variations on existing theories (Glaser & Strauss 1967). It has been suggested that saturation should be more about testing findings in terms of constructs and theories as well as data (Low 2019). From a pattern analytical perspective, I approached saturation in a number of ways. First, patterns needed to have multiple sources of evidence to back them up (the lowest I accepted had nine sources). Second, I considered saturation in naming and developing the content of pattern representations in that I adjusted them as I came across new instances but that these adjustments were less and less significant as the study progressed. A third issue was noting that saturation in a pattern space is more pragmatic than absolute as there is an inescapable transience of pattern representations, systems, and languages. Saturation is, therefore, more a matter of judgement and of building a defensibly robust evidential basis for patterns and their representations than it is about finding now new patterns. I was also attentive to theoretical sampling in seeking different kinds of sources, using different searches, targeting different kinds of materials.

In defining patterns and their representations, I found that much work was needed to define how much of the domain they covered, much of which revolved around "lumping and splitting" them. On one hand, I sought parsimony in terms of the simplest and fewest pattern representations needed to cover the domain, while on the other I wanted to make sure there was sufficient detail and difference to allow the pattern system to cover all the main functions and issues in the domain. For instance, the representation "A02: Costs, resources, economics" lumped together what were provisionally separate patterns for resources and economics, while I split the early pattern for supports into two; "A05: Supports and opportunities in an admissions system" and "A06: External supports and opportunities". Again, this was a highly interpretive process and what seemed to me to be natural break points might not be perceived as such by others. This process was initially guided by those articles that had proposed frameworks for organising and/or appraising admissions systems (Edwards et al. 2001; Reiter & Eva 2011). However, as much as they suggested ways of lumping and splitting apparent shared patterns, I was mindful not be too influenced by these models, they were data and not pre-emptive findings. It was more important to focus on the aggregate implied shared pattern thinking across all the sources reviewed and to cycle back and forth through all six of the pattern analysis modes.

Checking for gaps and overlaps worked at two levels. One was at the domain or system level and involved checking for and resolving apparent gaps and overlaps between pattern representations. For instance, the two representations "G07: Intended and unintended impacts" and "D04: Quality of decisions made" originally covered much of the same ideas and functions. However, rather than merging them to address apparent redundancy, I found that I needed to be clearer in differentiating between them as G07 covered the outcomes and deliverables of an admissions system while D04 covered the objective and subjective measures and appraisals of the processes and the products of an admissions system. Analysing for gaps and overlaps between pattern representations reflected a completist pattern system philosophy. The other level was focused on gaps and overlaps in specific pattern representations, which involved identifying and removing duplications or redundancies in the examples and solutions each of them provided even as I expanded the range of examples and solutions to be more inclusive of different kinds of admissions systems. In conducting this part of the analysis, I was mindful of Ashby's (1956) concept of "requisite variety" in that a pattern system should account for variant admissions practices and for problems and malfunctions.

The last of the six aspects of pattern analysis involved checking and adjusting for coherence, consistency, and parsimony in the content of the pattern representations. A representation could span many pages and cover many examples and variations. However, my goal was to limit each representation to a single page and to be selective in the details provided such that they would cover a range of possibilities without including every possible variation. This also involved iterative alignment between pattern representations such that they employed a common style, common elements, and noted connections and dependencies between representations.

Perceiving patterns in the thinking of others involves finding or developing patterns in yourself. When we perceive patterns in data there is, it seems, a sympathetic resonance between what is perceived and what is understood from our patterned minds. Pattern is always in our minds, but it responds to, indeed it dances with our experiences of the world beyond and within ourselves. We need to understand that all of science and all the understanding it has afforded is patterned. This means that pattern analysis (like any methodology) is itself a form of pattern thinking.

Elaborating, Structuring, and Testing the Pattern System

The starting point for this phase was the draft pattern system developed in phase 1. Having decided to base this phase on a literature review, I needed to make a number of decisions as to what kind of review this would be, not least because there has been a proliferation of review methodologies over the last two decades. Some reviews focused on synthesising findings (systematic reviews, meta-analyses), some focused on the nature of the research effort (scoping reviews, metamethod), and some focused on particular constructs (realist reviews, critical reviews) (Gough et al. 2017), however, none of these was appropriate to my needs. I did not, therefore, focus on each article's methods or results as most reviews do. Rather, as my goal was to identify regularities in the implied pattern thinking that might be expressed at any point in the article, I was interested in what aspects of medical school admissions authors talked about and how they talked about it wherever this appeared in the article and whatever was said. I note some parallels in my approach with discourse analysis (Macleod et al. 2024) and metanarrative review (Wong et al. 2013) although I did not draw on either explicitly in developing the study. Rather, I drew on my experience conducting reviews of various kinds (Ellaway et al. 2016; Ellaway et al. 2018) to reflect established standards and practices in review methodologies.

I started by conducting pilot searches using PubMed (from the National Library of Medicine in the US – https://pubmed.ncbi.nlm.nih.gov/) to identify appropriate search terms and to collect a small initial sample of articles to use in developing and extraction instrument. Once I had developed a search strategy and an extraction instrument, I used PubMed to identify peer-reviewed articles, my university library and my personal library for book chapters, and Google searching for websites with substantive materials on admissions (although I found that many schools provided little information about their admissions systems). I applied no date range limitations to the PubMed searches, the books I drew on had been published in the last 20 years (although this was not a selection criterion), and the materials from the websites I looked at were whatever they contained on the day I searched them. Articles needed to be in English, articles in PubMed needed to have abstracts (either in PubMed or available from the publisher's website), only books available to me for review (either in print or in electronic formats) were included. I applied no filtering for type of study other than excluding letters and other short communications, and no other exclusions on methods or article type. All articles that had a focus on medical school admissions were included. I excluded noneducational articles dealing with other kinds of admissions such as admissions to hospital, articles where the topic of medical school admissions was a side-line issue, and articles where no abstract could be found. On this basis, I undertook six rounds of searching and pattern system development, all of which involved the six modes of pattern analysis I described earlier.

Round 1: Open search. Approach: PubMed search 1A on "medical school" and "admissions" and "system" sort by "best match" yielded 648 results. Screening the first 50 returns, 18 were relevant and so I discontinued that search. PubMed search 1B on "medical school" and "admissions" sort by "best match" yielded 4423 results. Screening the first 50 returns, 50 were relevant. However, reviewing more than 4,000 articles was beyond my means so I selected the first 100 from Search 1B as the basis of the first review corpus. Of the 100, 24 with no available abstract were removed, leaving 76 for review, there were no duplicates. I used the article abstracts as the basis of extraction. Results: The 21 pattern representations from phase 1 that I brought into this round were expanded to 26. I identified gaps in the issues articles had covered in stage 1: issues of scope, structure, streams, activities, who participants were and why they were participating, participant information, quality of participant information, resolving disputes, and interactions between participants. I found a preponderance of US articles (and a lot of Canadian) and articles published in last decade. I was also concerned that a large cluster of articles on equity and diversity topics from the USA might have skewed the sample and what it could tell me. I knew that there had been a large body of work published on admissions interviews and exams, and so I decided to switch to a more purposive sampling approach in round 2 focusing on issues not represented in round 1, a wider range of countries of origin, and a wider date range. Before moving on, I updated the draft pattern system by adding the new pattern representations and updating the extraction tool.

Round 2: Search adding "Structure". Approach: PubMed search using "structure of medical school admissions" gave 1,686 results. As before, I took the first 100 of the 1,686, of which 35 were not relevant to or about medical school admissions (typically either patient admissions, a brief mention of admissions, or admissions related to some later in- or post-programme issue) leaving 65, there were also 11 duplicates from R1 and 1 with no identifiable abstract, which left 53 for extraction. Abstracts were the basis of extraction. Results: I added 5 pattern representations giving a total of 31. There was also a better spread of countries, issues, and dates in the articles than in the previous one. Ordered by frequency over the round 2 extraction there were 15 pattern representations with >20 instances, 7 pattern representations with 10–20 instances, and 9 pattern representations with <10 instances. I decided to focus on the low-density pattern representations in the next round. Before moving on, I updated the draft pattern system by adding the new pattern representations and updating the extraction tool to reflect this.

Round 3: Low occurrence patterns. Approach: I undertook 5 PubMed searches based on low occurrence patterns from round 2 ("structure of medical school admissions", "streams in medical school admissions", "participant experiences", "how disputes are managed", and "costs, resources, economics"). I selected the first 100 selected from each search and put them through the relevance and duplicate checks I described earlier, which gave 46 articles for review. Results: I identified no new pattern representations, although I renamed one, I swapped the order of two others, and I moved some pattern representations between categories. Before moving on, I updated the draft pattern system by adding the new pattern representations and updating the extraction tool to reflect this.

Round 4: Re-reviewing articles from previous rounds. Approach: I re-reviewed the 129 articles from rounds 1 and 2 using the round 3 instrument and added this to the results from round 3. I used the article abstracts as the basis of extraction. Results: There were no new pattern representations and no changes to any existing pattern representations.

Round 5: Book Chapters. Approach: Five of the twelve textbooks on medical education I reviewed had chapters on selection and admissions. I reviewed the full text of these chapters using the instrument from round 3. Results: There were no new pattern representations and no changes to the existing pattern representations. Interestingly, although not unexpectedly, these longer texts with a focus on presenting a broad overview all had a higher density in terms of the number of pattern matches >0 – for chapters there was a mean of 24.5 (range 27–22) compared to a mean for articles of 15 (range 22–24). The loading was still on goals, values, outcomes + general and specific process + data/processing/ranking/decisions. Notably, some topics were not well covered, in particular "experiences", "interactions", "supports" (both internal and external), "pathways", and "disputes". However, two of the five chapters featured one dominant author and three featured another dominant author, so perhaps these were not as varied as I might have hoped.

Round 6: Policy documents and manuals/handbooks. Approach: The last set of resources I looked at were policy documents, manuals, and handbooks from medical school websites. I searched the websites of 25 schools in Canada, US, UK, Netherlands, South Africa, Australia, and New Zealand, of which only ten provided sufficient information to justify a data extraction. I reviewed both the admissions handbook and the admissions website where they differed in focus and language, this meant I reviewed two documents for five schools and one document for the other five schools. Institutions included: Universities of Calgary, Ottawa, California San Francisco, Edinburgh, and Warwick, Vanderbilt University, Flinders University, Stellenbosch University Imperial College London, and the American Association of Medical Colleges (AAMC). I reviewed the full text of these documents using the instrument from round 3. Results: I identified two new pattern representations and I made a small change in the ordering of existing pattern representations in the system. The utility of seeking different kinds of sources was illustrated by the patterns for policies and prospectus issues not coming up in the academic literature even though they were important problems that admissions systems needed to solve. See Table 9.1 for the list of pattern representations in the pattern system at the end of round 6.

Strengths and Limitations

As a sequential mixed methods methodology, the approach I took had the strengths of engaging multiple perspectives and sources (the perspectives of admissions practitioners and scholars both as FACTT participants and as authors of the 195 articles, chapters, and documents I reviewed), and different approaches to inquiry. By the end of phase 2, I had generated what I considered was an acceptably stable version of the pattern system. Nevertheless, I note several ways in which the study might have been improved or approached differently.

The phase 1 workshop methodology was a variation on a consensus methodology, and as such was similar to some of the examples of pattern language development I gave earlier, particularly in setting priorities and boundaries. Given that pattern analysis involves working with implicit and tacit knowledge, something more exploratory than consensus seeking was needed, an approach that was fluid and abductive in drawing out common patterns, rendering them as pattern representations, and connecting them to others as part of the pattern system. The question was not "what patterns are most important in this domain?"

TABLE 9.1 The titles of the 33 pattern representations in the medical school admissions pattern system

G01: Context	R01: Who participants are	D01: Reviewing and rating applicants
G02: Scope	R02: Participant qualities, skills, and characteristics	D02: Differentiating and ranking applicants
G03: External oversight and accountability	R03: Why individuals participate	D03: How decisions are made
G04: Values and goals	R04: What participants do	D04: Quality of decisions made
G05: How values and goals are defined	R05: Participant experiences	A01: Managing consensus and disputes
G06: How values and goals are applied	R06: Interactions between participants	A02: Costs, resources, economics
G07: Intended and unintended impacts	I01: Applicant information	A03: How quality is maintained
P01: Overall structure and approach	I02: Applicant information quality	A04: How accountability is maintained
P02: Specific activities and processes	I03: Gathering applicant information	A05: Supports and opportunities in an admissions system
P03: Streams and pathways	I04: Processing applicant information	A06: External supports and opportunities
P04: Policies, rules, and procedures	I05: Prospectus	
P05: How processes or rules vary		

Key to pattern representation IDs: G = global patterns, P = process patterns, R = participant patterns, I = information patterns, D = decision patterns, A = administrative patterns.

but "what are the explicit and implicit shared patterns of this domain?" This needed neither consensus nor good participant understanding of these implicit patterns, only experience as a participant in the domain of interest.

I could have added a third fieldwork-based phase where I disseminated the pattern system to stakeholder communities for their appraisal, evaluation, feedback, and application. To be meaningful rather than artefactual, this would need to be naturalistic and multidimensional, which would have been difficult to control or capture, and as such I considered it as falling outside of a formal methodology. Of course, at some point there might be some audit or evaluation of what stakeholders did with the pattern system but that too was out of scope for this current methodological case study.

The phase 2 review methodology was subject to the many limitations of secondary analyses: I could only review what existed and was identified in my searches; I could only review some of the many thousands of relevant articles; there were always issues of terminology and interpretation; and I was obliged to set thresholds and sufficiencies based on my own judgement and interpretation. Although I reviewed 195 articles, chapters, and websites, I could have included more articles or other sources at each stage, engaged more rounds, been more rigorous with my analysis (such as reviewing full texts rather than abstracts) and so on. I could have also extended the sampling further to include interviews or structured responses to the pattern system with individuals involved in admissions in various ways at different institutions. However, I would note the importance of pragmatism and parsimony in this kind of work. The effort involved was, I judged, proportionate to my aims of developing a broadly representative pattern system. This is also related to my decision to use four levels in my mapping instrument (primary focus, mentioned, implied, absent). This was proportionate to what an article abstract could tell me, but I accept that a more in-depth analysis could have employed more gradations and might have yielded more subtle findings.

Third, as with qualitative and secondary analyses in general, the study would likely have benefitted from more input to the extraction, analysis, and synthesis stages. I would argue my single analyst approach was appropriate in outlining this case as my goal was to illustrate and explore issues in pattern theoretical analysis. However, more "hands on" would be required in deepening the defensibility and likely applicability of the pattern system in practice. That said, prior to formalising concepts of pattern theory and pattern inquiry (the purpose of this book), the basis for working with patterns was at times hard to communicate to others, which was quite apparent in phase 1 as some FACTT participants displayed more difficulty and uncertainty in abstracting and translating their experiences and ideas to general patterns than others.

A fourth concern, and perhaps the most important conceptual one, was that of "lumping and splitting" in identifying, describing, and differentiating patterns in the pattern system (one of the six modes of pattern inquiry noted earlier). This is not an issue exclusive to pattern analysis though; it is a problem that is well acknowledged in qualitative research as a whole:

> Lumping gets to the essence of categorizing a phenomenon while splitting encourages careful scrutiny of social action represented in the data. But lumping may lead to a superficial analysis if the coder does not employ conceptual words and phrases, while fine-grained splitting of data may overwhelm the analyst when it comes to categorize the codes.
>
> *(Saldana 2021, p. 20)*

There are no absolute rules in this regard, rather it requires the analyst's attention to both the specifics of each pattern and to the balance and focus of the pattern system as a whole. As a principle (rather than a rule) I might argue that a pattern system is parsimonious in that it should have no more pattern representations than is absolutely needed to reflect the collective pattern thinking apparent in its target social systems and social realities. Thus, a simpler or less varied social system might be expected to require fewer pattern representations to represent it than one that is more complicated and varied. However, this is a principle, and the onus still falls on the analyst to decide when to lump problem-solutions together in one pattern representation and when to split them into different pattern representations.

Another principle to consider in lumping and splitting is equivalency in pattern representation abstraction. The degree of abstraction in pattern representations might be defined in advance of the analysis, but it is more likely to be developed abductively within the analysis itself such that the analyst seeks an appropriate level where the pattern representations are roughly equivalent even as they balance between parsimony (as few as possible) and requisite variety (enough to reflect the pattern thinking in target social system). In the case of this worked example, the level of the pattern representations reflected the thoughts of the initial workshop participants, the abstraction in the literature reviewed, and the principles of parsimony, variety, and levelling.

To that end, in advancing this particular pattern system (or any other), I do not say that the levelling I came up with was right or wrong, only that it reflected the balance point I found in my analyses. I fully accept that I could have added more detailed pattern representations under many if not all of the 33 pattern representations that I developed. For instance, under "quality of information" I might have added pattern representations for "validity", "reliability", and "trustworthiness". Under "specific processes" I might have added pattern representations for the many processes that schools employ such as "file review", "longlist", "interview", "shortlist", and "make an offer". In terms of "participants" I might have added pattern representations for "applicants", "reviewers", "interviewers", "committee members", and "administrators". The two points to note here are that any part of a pattern system represents a particular interpretive perspective on the domain of interest, and that different interpretive perspectives may be equally valid.

A fifth issue to consider is the way pattern representations are expressed. For this worked example pattern system of medical school admissions, I drew on Alexander's framing of patterns as general ways of solving particular kinds of problems, an approach that, as I have described, is common to the vast majority of pattern language work published since Alexander's in the late 1970s. However, I should note that positioning everything as a problem can be problematic. Sometimes, rather than considering problems, we might want to focus the development of pattern representations on opportunities or perimeters, or on requirements or regulations, or on actions or structures that reflect underlying shared pattern thinking. This is another issue deserving of more research. For now, in the context of this case study, I will simply say that framing pattern representations in terms of problems and solutions is an accessible and well-established approach, but it is not the only one to consider.

Although not a limitation per se, I should note that different patterns (as I interpreted them) attracted different levels of focus within and across articles and with different frequencies across all articles – see Figure 9.2. Some patterns were more often the primary

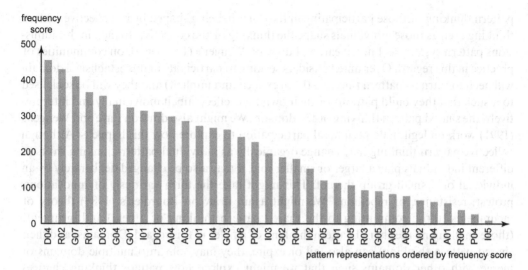

FIGURE 9.2   Frequency graph illustrating different cumulative scores for the different pattern representations ordered according to score. Note: scores calculated based on main focus = 3, mentioned = 2, implied =1, absent = 0.

Diagram prepared by the author.

focus of an article and were mentioned in many articles – the top 5 were: "D04: Quality of decisions made", "R02: Participant qualities, skills, and characteristics", "G07: Intended and unintended impacts", "R01: Who participants are", and "D03: How decisions are made". Some patterns were rarely mentioned or attended to – the bottom 5 were: "P04: Streams and pathways", "A01: Managing consensus and dispute", "R06: Interactions between participants", "P04: Policies, rules, and procedures", and "I05: Prospectus". Does this mean that "D04: Quality of decisions made" is the best or strongest pattern and "I05: Prospectus" is the worst or weakest pattern? Not at all. The scores reflect relative attention to certain topics, and it is not unreasonable that some topics are of greater importance or interest to scholars and educators than others. I would, however, say that higher scores reflect pattern representations that have a more normative provenance, and lower scores reflect pattern representations that have a more completist provenance, which emphasises the need to combine pattern system and pattern language modalities.

Another issue was the dynamic, socially constructed, and socially contingent nature of domains of practice. In my worked case study, I took the perspective that the pattern thinking in medical school admissions was largely stable and relatively homogeneous, but was that justified? After all, as a domain comes together its collective pattern thinking is the product of the individual pattern thinking of those forming the domain. We might draw on Tuckman's (1965) model of forming-storming-norming-performing in this regard such that we might want to explore how that shared pattern thinking coalesces from the interactions of the domain's pattern thinkers. The point at which a domain is established might be understood as the point at which its collective pattern thinking coalesces such that it takes on a momentum of its own.

Once a domain is established, although its collective or shared pattern thinking will likely continue to develop, this becomes a more ecological process with the individual pattern thinking of those participating in the domain being shaped by its collective pattern thinking even as those individuals shape the thinking of the collective. In fact, in the admissions pattern system, as I noted earlier, I drew on Wenger's (1998) work on communities of practice in this regard. Over time, outsiders seeking to participate in this established domain will need to learn its pattern thinking (both explicit and implicit) and they will be socialised to it such that they build patterns of their own that reflect, albeit individually and interpretively, the shared pattern thinking of the domain. We might also draw on Lave and Wenger's (1991) work on legitimate peripheral participation to explore how this happens. Although collective pattern thinking may change organically as individual pattern thinking shifts as different individuals play a larger or smaller role, it may also be changed deliberately by an individual or a small group thereof. This might take the form variously of innovations, protest, resistance, or rebellions. We might again draw on Bourdieu's (1984) theory of incumbents (those committing to the shared pattern thinking of a field) and insurgents (those challenging the shared pattern thinking of a field) in this regard. Domains (and their shared pattern thinking) can also fail or expire, they may split into multiple domains or merge with other domains such that we might explore how pattern thinking changes according to these broader social changes.

In summary, the knowledge claims made from pattern analysis have no more or less validity, reliability, objectivity, credibility, transferability, trustworthiness, or confirmability than those of other interpretive approaches or those based on broad latent constructs. Their utility is in the perspectives they can bring to bear, a point I will consider in describing the grammar and syntax that transforms a pattern system into a pattern language.

Exploring Grammar and Syntax

As I noted in Chapter 6, a pattern language's grammar consists of rules for how pattern representations must be combined, while a pattern language's syntax is about how pattern representations might be combined (style and idiom) such that a pattern language will have one grammar but may have many different syntaxes. My focus in developing a pattern system was on identifying and developing pattern representations that reflected a domain of practice. Indeed, it would seem that developing a pattern language first involves developing a pattern system. Representations, grammars, and syntaxes could be developed together in a single study but that does appear to be more onerous an undertaking.

Taking this sequential approach, I now consider whether and how my pattern system can be developed into a pattern language. There are two immediately apparent (at least to me) deductive grammars to consider, those that reflect temporal relationships and those that reflect logical relationships. Temporal grammar is about the necessary temporal ordering of pattern representations and the instances they reflect. That things happen in a particular order (and only make sense if they do occur in that order) reflects a strong pattern language grammar (as grammar should be) – see Figure 9.3. Logical syntax is about the necessary interdependence of pattern representations and the instances they reflect. Temporal relationships might be interpreted as one form of logic relationship but there are others. Figure 9.4 illustrates logical relationships between a selection of pattern representations.

The two examples I have given reflect a deductive rather than an inductive grammar based on my own interpretations and understanding of medical school admissions. For a

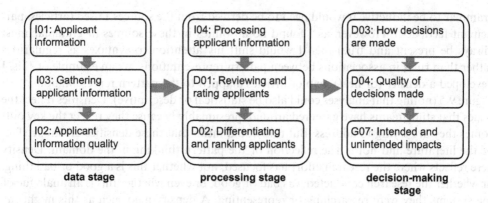

FIGURE 9.3   An example of temporal sequencing of pattern representations in the medical school admissions pattern system.

Diagram prepared by the author.

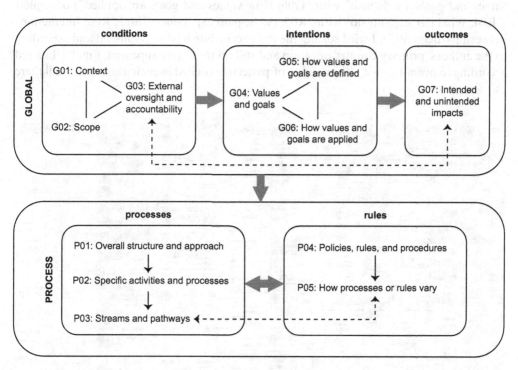

FIGURE 9.4   An example deductive and normative (principle rather than data-driven) functional mapping between pattern representations in the medical school admissions pattern system, suggesting a grammar for a pattern language.

Diagram prepared by the author.

grammar to be inductive it would need to be derived from the sources I used (articles, participant feedback). However, as I found no absolutes in these sources (a grammar must always be present and true) then I would shift my attention to syntaxes as regularities rather than rules in associations between pattern representations. As an example of this, I developed a density (or heat) map of articles mapped to the pattern representations – see Figure 9.5 (noting that analyses could also be statistical or descriptive). Densities reflect the issues that study teams have gathered around, presumably because they reflect the key outcomes they are trying to address. But this would mean that these densities are an artefact of the literature and not a true reflection of the pattern thinking in the domain. Density here reflects where the research effort has focused, not whether this is a good or bad thing, or whether the research conducted was bad or good, or even whether this is an analogue of the systems they were researching or representing. A density map such as this might say more about the sources reviewed in a study than it does about the pattern representations and the pattern system as a whole.

I found little regularity in the sources I looked at (rows) beyond couplets ("G05: How values and goals are defined" with "G06: How values and goals are applied") or triplets ("R04: What participants do" with "R05: Participant experiences" and "R06: Interactions between participants"). I tried sorting the rows according to date published and according to the authors' primary country of origin and still no regularity appeared. Finally, I sorted according to overall density (the number of patterns matched in each source) and still there

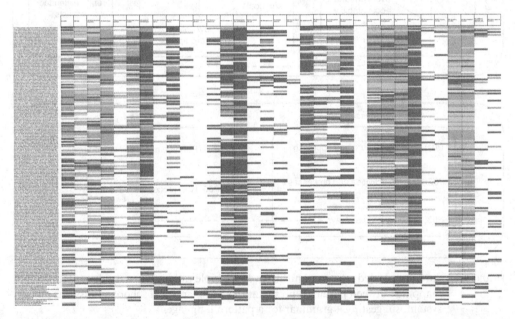

FIGURE 9.5 Density matrix showing sources as rows mapped to the pattern representations as columns. Where the pattern was the main focus of the article the intersecting cell is shown dark grey, if the pattern was mentioned then the cell is mid grey, if the pattern was implied then the cell is light grey, and if the pattern was absent then the cell is left blank.

Diagram prepared by the author.

were no apparent regularities. The only apparent row regularities were the differences between articles, chapters, and web sources. Articles were particularly focused on applicants, information, decision-making, the quality of the decisions, and outcomes. Chapters, with their extended word counts, showed a similar regularity to that of articles but with more detail. Web sources had very little to say about decision quality or outcomes but did consider logistical, rule, and support patterns that attracted little attention from either articles or chapters.

An alternative approach is to consider some kind of graph analysis of the dependencies between pattern representations within an article. If we model the pattern representations as vertices on a graph and the dependencies between them as edges to create a graph of the apparent connections between patterns described in a single article – see Figure 9.7. I used the online tool Polinode (https://www.polinode.com/) for the visualisation although there are many similar tools that might be used. Interestingly, only 13 of the 33 pattern representations were apparent in this article. This approach can be aggregated across all the articles I reviewed to generate a graph for the whole pattern language, such as that shown in Figure 9.7. I provide this for illustrative purposes only as a full graph analysis of this dataset could employ many techniques and approaches (such as path density, direction, and type) that are beyond the scope of this current volume. I will simply say that graph analysis would seem to afford many possibilities for syntax mapping and mining.

Despite these regularities, there were no apparent fixed rules of association between patterns and, therefore, no additional grammars, which may in turn reflect the divergent study perspectives, the divergent participant perspectives within an admissions system, and the different designs and values of different admissions systems. I conclude from this that the pattern system I developed can reasonably be extended to be a pattern language, albeit one that has a few strong grammars and a weak and diverse set of syntaxes.

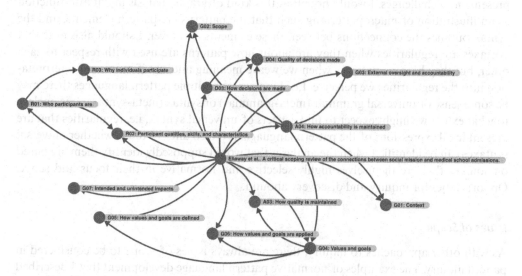

FIGURE 9.6 A directed graph rendering of the pattern associations in one academic article (Ellaway et al. 2018).

Diagram prepared by the author.

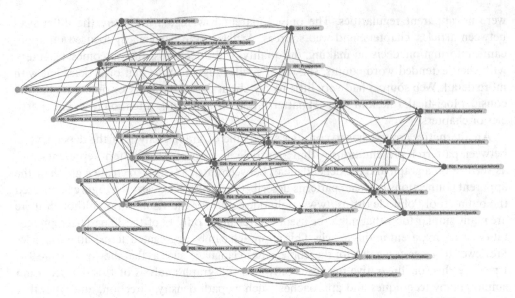

FIGURE 9.7 A directed graph rendering of all the sources in the pattern review of medical school admissions, illustrating the potential density of relationships between pattern representations in a pattern system or pattern language.

Diagram prepared by the author.

Cumulatively, this work could be aggregated to form a pattern atlas integrating many pattern systems or languages across a larger domain. We might eventually develop a pattern atlas of medical education, or of whatever other domain is involved. However, this would need common edges and connections between pattern languages and systems, which presents new challenges. I would note that this kind of graph analysis might also function as an illustration of macropatterning such that the patterns are pattern elements, and the syntax outlines the connections between these elements. However, I should also note that syntaxes are regularities when they are about how patterns are used with respect to each other, but they become patterns when we weave meaning and significance and interpretation into the regularities we perceive. Looking across multiple pattern languages there may be some sense of universal grammar (meta-grammar) or syntax (meta-syntax). It is unclear to what extent we might expect to find aspects of universal syntax, i.e. regularities that are present in all expressions of the pattern language. It is also unclear to me whether universal syntaxes can be identified when the analyses that would supposedly identify them are based on sources that are themselves highly selective and normative in their focus and scope. Opportunities for inquiry and discovery abound.

Issues of Scope

As with other approaches to inquiry, there are always issues of scope to be considered in pattern inquiry. The examples of normative pattern language development that I described in Chapter 6 and again earlier in this chapter had a relatively narrow scope in that they were developed as the product of a single study, they engaged a relatively small number of individuals, and there were no descriptions of revisions, iterations, or other longitudinal developments. I would note, however, that Alexander (1996) described their process as having gone on for a decade or so before they published their architectural pattern language, and that he had earlier (1979) emphasised that pattern languages should be living entities that continue to be revised and adjusted over time. One dimension of scope, therefore, is the timeframe for the study. Not only should this include when the study was conducted and how long it took to complete, there should also be some sense of any iterations, ongoing development, or revisions. After all, how current and representative are the products of any past study involving pattern inquiry? Thinking changes over time and all pattern languages and pattern systems, unless kept up to date, can quickly become dated. It is perhaps telling that so many of the normative pattern languages that have been developed do not seem to have been revised or developed beyond their first iteration. I would also note that a literature-based pattern system or pattern language approach can be a better way to capture thinking over an extended period of time (suggesting an approach to developing historical pattern systems or languages) than consensus, interview, or observation-based approaches – see Figure 9.8.

In recognising something we connect our current experiences to memories of past experiences and to our past thoughts that surrounded those experiences. Given that pattern is made of memory, it inescapably connects the past with the present, and it can be used to suggest possibilities that can help, within limits, to predict or at least suggest future events and outcomes. There is also a somewhat palimpsestic sense of pattern thinking as once we have developed some capability in pattern formation and application in the first few years of childhood then all subsequent pattern and pattern is based around these foundations. Understanding pattern means understanding its temporalities, the way our thoughts move through time and connect across instances of time.

Another dimension of scope to consider is place. Despite the many affordances of Internet technologies, most research continues to be focused on and conducted in relatively

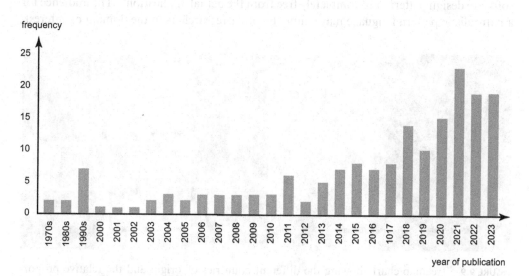

FIGURE 9.8 Numbers of articles and book chapters per year of publication reviewed in the medical school admissions study.

Diagram prepared by the author.

narrow locations such that generalisability and transferability are longstanding concerns in many fields and domains. There would seem to be little in pattern inquiry that sets it apart from the mainstream of social scientific thought in this regard. Studies conducted in a single location or with participants from a narrow range of contexts will perforce develop pattern representations of the thinking of that particular group but not necessarily much beyond. This was another reason for going to the research literature as these sources come from many contexts (although some contexts tend to be more represented than others) as illustrated in Figure 9.9 that shows that my review was dominated by articles from the United States, Canada, and the United Kingdom, in part reflecting an Anglophone bias in my sampling and in part the dominance of these countries in the medical education literature as a whole.

Just because a study is conducted in a particular location does not mean that it fully reflects that context; attention also needs to be paid to the sampling strategies involved. For instance, a normative pattern language based entirely on the pattern thinking of selected experts in a given field will be unlikely to be able to capture the pattern thinking of the rump of practitioners, but a naturalistic pattern language that draws on varying levels of experience as well as on varying perspectives may provide a fuller picture of the variance in pattern thinking. This is a limitation of any literature-based approach as those who conduct research and are able to get it published in peer-reviewed journals cannot be considered representative of everyone in an applied field. Noting the scope afforded by individual persons, I would note that another dimension of scope for pattern inquiry is its intended stakeholders and audiences. For instance, Gamma et al.'s audience was (and still is) trainee computer programmers whereas Alexander et al.'s audience was much broader than professional architects. Scheurer (2009), in contrasting Alexander's architectural design patterns with Gamma et al.'s software design patterns, noted that "Alexander's patterns largely refer to concrete application problems from the perspective of the end user" while "Gamma's software design patterns are completely free from the actual application". The audience for a naturalistic pattern language may simply be other researchers in the domain of interest.

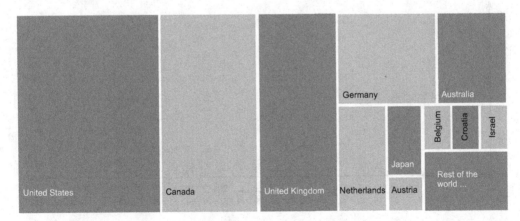

FIGURE 9.9   Treemap chart showing the different countries of origin and the relative proportions of articles, book chapters, websites, and handbooks reviewed in the admissions pattern language described earlier in this chapter.

Diagram prepared by the author.

Although time, place, and person capture many of the issues associated with the scope of pattern inquiry, I would also note Scheurer's (2009) observations that scale and scalability matter in architecture because "everything is based on the fundamental human dimension", whereas "software can be scaled at will, at least while it is being programmed, without any need to change the basic patterns". Clearly, scope in pattern inquiry can make a significant difference in what is synthesised and what is represented.

Alternative Approaches

My worked example is just one way of applying pattern theoretical principles to structured inquiry. I outlined a number of research foci and processes in the previous chapter, but I return to the topic here with a more grounded consideration of alternative approaches to pattern inquiry. In the worked example, I approached the task of sampling patterns and pattern thinking first by using an exploratory consensus approach and then more substantially conducting review-based pattern inquiry. However, I could have taken an approach that depended entirely on fieldwork. I might have explored the perspectives of admissions systems participants using interviews, "think alouds", or by observing participants going about their admissions activities. I might also have expanded the scope of the pattern thinking I was analysing to include the perspectives of applicants, accreditors, or patients. I could have conducted this sampling at one point in time or I might have sampled pattern thinking at different points in an admissions cycle or even across multiple admissions cycles. I might have compared the pattern thinking of different groups (such as reviewers, raters, committee members, administrators, and/or applicants) to compare and contrast their pattern thinking rather than presenting a single aggregate model of educators' pattern thinking. I might also have compared the apparent pattern thinking between different schools or between different regions or countries. There are clearly many more ways of framing a pattern inquiry study and many different methods and methodologies that might be employed in the service of exploring pattern thinking.

I took a naturalistic-completist perspective in my worked example but what if I had taken a normative or historical perspective (I set aside speculative perspectives in the context of research)? A normative perspective seeks to capture "best practices" or optimal solutions to the broad problems faced in a domain. In that case, I would have needed to attend to the quality of the evidence I drew on. Were the individuals I engaged in workshopping ideas experts in medical school admissions? Were the articles I drew on representative of expert voices or optimised perspectives on medical school admissions? Quality appraisal of articles is often carried out in systematic reviews, in part as an inclusion/exclusion criterion and in part as a way of describing the evidence base that the review drew on. Defining experts or expertise and then engaging this expertise in a study is a challenging and somewhat subjective undertaking, not least because what expertise is and who does (or does not) have it are often context dependent and subject to ideological positioning.

A historical perspective would need to frame both the historical and social context for inquiry. For instance, a study might consider "what was the apparent pattern thinking of those involved in medical school admissions in the United Kingdom between 1918 and 1939?" and then collect and appraise the available evidence accordingly. That said, although it has been long debated whether historical scholarship can be considered a science given that past events cannot be directly engaged or manipulated (Goodheart 2005), I would argue that robust scholarship can still be conducted by considering how pattern thinking is reflected in a domain's documents and artefacts. Of course, there would need to be extant sources to draw on to be able to conduct this kind of study. My worked example drew on a robust research literature, but it also meant that the pattern system that I developed reflected the pattern thinking of published academic authors rather than practitioners (although in medical education they are often the same people). An historical approach would not be feasible if there was no literature or document base to draw on, or if what was available did not fully reflect the past shared pattern thinking a researcher was interested in.

I also note that I did not say much about theory in the worked example even though much of my own research has been both theory-driven and theory-generating. I might argue that the act of engaging in pattern inquiry invokes pattern theory and that should suffice. However, that is not a position I am comfortable taking, so let me explore possible roles of theory in pattern inquiry in a little more detail. I noted in the previous chapter that there are at least three dimensions of inquiry to consider: what we inquire into, how we conduct the inquiry, and why we are engaging in inquiry. Theories of *what* (theories regarding what the phenomena of interest is, how it works, what we should pay attention to, how it can be explained, etc.) could be used to guide pattern inquiry every bit as much as they are used to guide other kinds of inquiry. If those theories involve or align with pattern inquiry then they should be even more useful in guiding inquiry. For instance, if an investigator was interested in meaning making in medical school admissions, then there is much theory (such as those offered by Frege, Quine, Searle, and Carnap) that could be used to guide their inquiry. Pattern theory is more aligned with some of these theories than others, such as Schiffer's (2006) work connecting meaning and belief, which might make them more amenable to pattern inquiry. Because they reflect macropatterned thinking (and most likely shared macropatterned thinking) theories of what could also be the focus of inquiry. For instance, an investigator might analyse various learning theories for their pattern implications or compare them on that basis as a way of developing or critiquing them.

Theories of *how* could be used to guide inquiry in terms of the methodologies used (which might be more or less aligned with pattern theory) or they too could be the focus of inquiry. For instance, an investigator might use constructivist grounded theory (Charmaz 2006) as a methodology or as the basis for analysing its patternlike qualities. Theories of how could also be used to appraise the products of other pattern inquiry. For instance, the framework of principles of pattern languages I adapted from Alexander (see Table 8.3) could be used to evaluate pattern representations, systems, or languages produced by other acts of inquiry (such as my admissions pattern system) by focusing on questions such as "are the pattern descriptions structured in coherent and meaningful ways?", "can the pattern descriptions provide usable solutions to problems that matter?", and "are the pattern representations flexible enough to be applied to different situations?"

Theories of *why* could again be used to guide inquiry or serve as the focus of inquiry. This is a particularly interesting aspect of pattern inquiry given the pattern basis of meaning, value, and significance. For instance, an investigator might ask why different methodological approaches might be used and what are the investigator, field, and paradigm pattern thinking implications of this? In the case of the admissions pattern system, what was it in my own pattern thinking that led me to design and conduct the study the way I did? After all, given my longstanding experience of working in and around medical school admissions, I took an emic perspective on this topic that an etic outsider would be unlikely to be able to bring to bear. My adding a review step might also be related to my experience in using these kinds of approaches, as might my particular framing of medical school admissions reflect my own experiences thereof.

I developed a pattern system and then provisionally expanded this to a pattern language. What if I had produced single pattern representations or other constructs that were not pattern systems or pattern languages? After all, research in the social sciences often builds theories, frameworks, or models of the phenomena of interest. I might argue that a pattern language is a kind of model or framework and that I have rendered it in different ways (such as in Figure 9.8). Nevertheless, there are other products I might have worked towards. For instance, if I had taken a normative approach, I might have ended with recommendations or even putative interventions to reshape thinking and practice to reflect these best practices whatever they might have been.

I would also note the work of my colleague Catherine Patocka as an illustration of alternative approaches to pattern inquiry. Her doctoral studies focused on developing a pattern system of feedback in medical education (Patocka et al. 2024) as a way of moving past the fragmented discourses and definitions of feedback different scholars in the field have pursued over the years. Catherine adapted Arksey & O'Malley's (2005) scoping review methodology as a literature-based way of developing her pattern system of 36 pattern representations. She followed this with a comparative case study methodology that mapped 11 models of feedback in medical education to the pattern system as a way of building validity evidence for the system. She also used this mapping process to identify regularities in groupings of pattern expressions that could then be used as starting points for translating the pattern system into a pattern language of feedback. She ended her programme of research by taking one of these regularities and exploring its expressions and characteristics as a syntax of her emerging pattern language.

Mapping a Landscape

Not only are my earlier worked example of a pattern language for medical school admissions and Catherine's pattern language for feedback in medical education examples of the application of pattern theory, they also contribute to methodological discussions around knowledge syntheses (such as scoping, systematic, and realist reviews). In great part this is because the approaches we have explored provide solutions to the problem of synthesizing heterogeneous literatures, particularly those with definitional divergence by mapping a common underlying semantic landscape. To that end, a pattern language can serve as a map of the pattern thinking in a domain, it presents a landscape within which many positions may be taken and many actors may situate themselves without having to occupy the same space. Indeed, they may wander through the landscape, or they may settle here or there according to their needs and interests. A map, as Korzybski observed, is not the territory; it is a representation of territory and as such omits some things while emphasising others. However, a map also emphasises that there is a territory, and a map is an invitation to explore and discover the territory it suggests.

Definitions can be understood as claiming a particular place in a domain's semantic landscape, with different definitions scattered like towns or villages across that landscape. Once you can see the map you can also begin to see the relationships between definitional points, and, as you do, you are no longer bound by single definitions but are instead free to visit other definitional points or to embrace the landscape as a whole. Definitions then are no longer prisons or castles but items of topographic interest. Should pattern systems be understood as a directory of key places in a pattern thinking domain? Are they, as I have suggested in my graph analyses, key nodes in a suggested landscape? Should pattern languages be understood as pattern thinking maps with their inclusion of nodes and the relationships between them?

While I accept that I may have stretched the cartographic metaphor, I do see a pattern theoretical stance as a form of emancipation for those caught in false dilemmas of choosing a definitional space based on assumptions of exclusivity, precision, and identity, particularly when structured and rigorous enquiry can, from a pattern theoretical perspective, be inclusive, fuzzy, and plural and not sacrifice the tenets of robust inquiry.

Pattern-Informed Inquiry

To this point, I have focused on approaches to inquiry that follow all five modalities of pattern inquiry (purpose, focus, disposition, process, and product), which in turn reflects pattern theoretical principles. Can researchers meaningfully engage in pattern inquiry that only employs some of these modalities or principles? In terms of dispositions to pattern inquiry that I outlined earlier, some (such as being mindful how investigators engage others through their macropattern thinking) are specific to pattern inquiry as they relate to particular methods and approaches. However, other dispositions (such as investigators attending to the conscious and deliberate macropatterning of their experiences, perceptions, and thought processes) are generally good reflexive practices for all researchers. Embracing these dispositions should lend substance and rigour to pattern inquiry, but inquiry could proceed without them, albeit in a somewhat reduced or weakened form.

Does it make sense to pursue inquiry that has a pattern theoretical framing (purpose) but that does not seek to explore explicitly patterned phenomena (focus)? I think not since, by adopting a pattern inquiry purpose, an investigator would already be conceptualising their phenomena of interest as being patterned. Conversely, exploring patterned phenomena without an orientation to pattern inquiry means an investigator would be disregarding the patterning or treating it atheoretically such that the phenomenon would not be understood as being patterned. Purpose and focus do seem intrinsically co-dependent.

What about having a pattern purpose and focus without using pattern methods? Given that pattern inquiry can make use of many extant methods and methodologies alongside those that are more specific to pattern inquiry, then yes, this is quite possible. Conversely, using specific pattern methods without a pattern theoretical purpose and focus does not seem viable as pattern methods would not be meaningful as such without a pattern theoretical framing. Can an investigator have a pattern purpose and focus but not generate pattern products? Again, I think not given that, from a pattern theoretical perspective at least, all findings and products of any kind of research are patterned. I should qualify this to note that inquiry with a pattern purpose and focus may not need to develop deliberate and stylised pattern products such as pattern representations, systems, and languages, only that what it does produce will be inescapably patterned. Can an investigator generate pattern products from inquiry that does not have a pattern purpose and focus? Yes, given that research products are inescapably patterned (see my earlier comments on periodic table and theoretical frameworks). However, developing explicitly patterned products requires both pattern methods and a pattern purpose and focus.

This illustrates a recurring tension between an explicit and an implicit focus on patterns and pattern thinking. Let me illustrate this with three examples from my own pattern-related work conducted before I began to develop the pattern theory I have set out in this volume.

The first example is Schrewe et al. (2018) which reported on a study exploring some of the clinical learning contexts encountered by family medicine residents undertaking rural and remote placements in Canada. We wanted to capture the uniqueness of different contexts while representing apparent regularities across and between sites. Data were drawn from a literature review on clinical learning environments and from a qualitative interview-based study that gathered family-medicine resident perspectives on their clinical learning environments. Our solution was to develop a simple pattern framework containing six broad pattern representations of the factors at play in different clinical learning environments: patients, practice, educational, physical, healthcare, culture. This framework was categorical as it allowed us to consider both ideal (normative) and actual (naturalistic) student experiences in different contexts, and it did not depend on problem-solution dyads.

The second example is Ellaway & Topps (2017), which reported on the development of a framework setting out broad functional domains of scholarly activity that could be used in teaching and professional development activities. The study involved outlining and differentiating domains of scholarly activity, both in terms of a broad focus on what a scholar does in general or what an article does, and a more detailed consideration of the nature of different scholarly approaches within a study or an article. Sources were a variety of texts on research practices, as well as discussions within the team and with students and other scholars. We used methods approximating to pattern analysis in that categories, scope, and collective coherence were developed abductively. However, this was based on task analysis rather than problem-solution dyads, and, as with the previous example, was categorical without focusing particularly on normative or naturalistic expressions of the patterns. The resulting system contained seven categorical patterns of scholarly activity: metascholarship, evaluation, translation, research, innovation, conceptual, and synthesis.

The third example is Roze des Ordons et al. (2019), which reported on a qualitative study that drew on aspects of activity theory, realist inquiry, phenomenology, and autoethnography to explore how clinicians conceptualized, observed, and expressed compassion in different clinical contexts. We only adopted pattern language methods at the synthesis stage of the study as a way of drawing together broad thematic findings as a series of integrated patterned facets. Findings were expressed as four primary patterns (activity, disposition, relation, and situation) and four orientational patterns (shared, individual, perceived, performed). We used the term "pattern language" to reflect the interactional and aggregative nature of the patterns in shaping compassionate practice in different settings.

That these studies did not incorporate the concepts of pattern theory or the practices of pattern inquiry that I have described in this volume does not invalidate them. Rather it situates the development of a pattern theoretical approach to inquiry as something that has been developing for some time and that will likely continue to evolve. These studies also suggest that there are many possible varieties and topologies of pattern inquiry, some fully embracing pattern theoretical principles, others only using some of them. This variety reflects intersections between pattern inquiry and other methodological stances and approaches, in reflects the pragmatic nature required in many research paradigms, and it illustrates the developmental nature of a pattern theoretical approach. Indeed, since pattern theory is only presented in this volume for the first time, the examples using this approach that I can refer to are necessarily limited.

Implied and Latent Pattern Thinking

I now turn to the issue of latent and observable (or manifest) pattern thinking in the context of pattern inquiry. From a pattern theoretical perspective, the broad goal in developing a pattern system is to capture the shared pattern thinking in a particular domain. However, as we cannot access this thinking directly, we need to work with those phenomena we can access and infer from them the pattern thinking that shaped or informed them. Translating this to a pattern theoretical space (and noting my earlier comments on the *Latency* principle), we should consider which is more fundamental to particular acts of inquiry; the implicit pattern thinking or the explicit expressions of that pattern thinking? An entity realist reading would focus on pattern thinking as being real and would take expressions as literal reflections of that pattern thinking. A theoretical realist reading on the other hand would focus on the reality of the expressions and would treat the ways that they imply pattern thinking as provisional and theoretical. I do not argue that one is superior to the other, only that they represent different empirical positions. In my case study, my focus on synthesising problem-solving approaches in the domain of medical school admissions was closer to theory realism particularly where it dealt with designs and actions. However, an entity realist perspective can also be present in pattern representations that are focused on values and principles. It would seem likely that some pattern systems might lean more to one or other perspective, while others might span both.

I would also note that there can be layers of latency to consider in exploring pattern thinking. First, there is the latency of the observable characteristics of pattern thinking with respect to the pattern thinking itself. Second, there is a latency of pattern thinking with respect to the patterns it draws on given the heuristic relationship between perception and the patterns we activate to establish it. Third, we can also consider the latency of the pattern thinking and its patterns with respect to the realities that they respond to or reflect given that our memories are not of the things we encounter but of our mental states and experiences related to those things. Given these intersecting latencies, if this were a forensic investigation, then the chain of evidence would be considered very patchy, which in turn suggests that the kinds of knowledge established from pattern analysis may be best understood in terms of the partialities and complexities of hermeneutics. Boorsboom et al. also argued that we need to be more deliberate in differentiating between within-subjects constructs (within individual subjects) and between-subjects constructs (shared across group of subjects):

> the between-subjects latent variable will not indicate the same process in each subject. Therefore, the causal agent (i.e., the position on the latent variable) that is posited within subjects on the basis of a between-subjects model does not refer to the same process in all subjects.
>
> *(Borsboom et al. 2003)*

We might expect there to be some degree of alignment between individual pattern thinking and the shared pattern thinking of the group with respect to the domain of interest (or any domain of common interest) without ever expecting or requiring identity. Our individual patterns must be unique at any point in time and uniquely dynamic over time. Any group pattern thinking we engage in will also change over time as the dynamic product of the pattern thinking of all the participating individuals. So, is our construct of interest individual pattern thinking in the context of a domain? If so, then we might be interested in exploring the similarities and differences between the domain patterns of individuals participating in the domain, for instance, as an indicator of training or socialisation to the domain. If, on the other hand, our construct of interest is collective pattern thinking then we would be less interested in individual divergence and more interested in aggregate and convergent pattern thinking. This was the position I took in my case study of medical school admissions, and it was also the focus of the many normative pattern languages I described in earlier chapters. There are, therefore, key conceptual differences to consider between individual pattern thinking and collective pattern thinking that have significant empirical implications.

It is axiomatic to pattern theory that patterns are the product of minds and, strictly speaking, can only be found in minds. When we examine the pattern thinking of a collective, we are working with something that is perhaps best understood as a convergence or gestalt of the minds involved in that collective. If we could image one (or all) of the patterns in individual minds that are part of a shared repertoire then we might somehow synthesise them, and in doing so reflect a more concrete sense of shared pattern thinking than inferential techniques based on observing the interactions in collectives, asking members of collectives to record their own understandings, or examining their shared artefacts and processes. Doing so is of course quite beyond the means of science at present, and it may always prove elusive. Even if we could image the patterns of a given mind, this would still not reflect the dynamic nature of patterns and pattern thinking of the individual and of the collective over time and based on the catalytic nature of engagement in a collective.

It may be, therefore, that developing gestalt models of the distributed, entangled, and dynamic shared pattern thinking of a collective is the best we can do, and that would mean that any such inquiry is inescapably dependent on inference and implication. If so, then there are several layers of implication to negotiate. The first is the ontological claim made with respect to the existence and tangibility of the implied shared pattern thinking that becomes the focus of a study. How do we define the collective? Does it have clear boundaries, or as is more likely, do we understand it in terms of degrees and varieties of inclusion and participation? Are all members involved in the same shared thinking or are there subgroups and allegiances to different kinds or aspects of particular kinds of pattern thinking? We might solve the problem of what constitutes a collective by defining it in terms of only those individuals who share the same patterns and pattern thinking, but then the patterns and pattern thinking that define the collective would first need to be defined. Rather than trying to anchor inquiry on what seem to be inescapably transient and emergent phenomena, we might work abductively by positing the existence of both a collective and its shared pattern thinking and then resolving boundaries and gestalts through iterative inquiry and model building. In doing so we would perforce be working with an implied collective and its implied pattern thinking. This fuzziness and uncertainty might well prove to be an inescapable characteristic of pattern inquiry.

Another layer of implication reflects expectations that the findings of a pattern inquiry study will generalise from the specifics of a study to other collectives and contexts. A pattern representation that is so thoroughly individual and idiosyncratic is likely to be of little use or interest other than as a reflection of a particular patterned mind. On the other hand, consulting every individual in a collective would likely be logistically prohibitive (and would still only capture a snapshot). So, following the principles of sampling theory, pattern inquiry should include sufficient pattern perspectives to account for individual idiosyncrasies and to reflect the characteristics of the collective (the population) to which it seeks to generalise. The shared pattern thinking of a collective that does not change much over time will likely have broader temporal generalisability than pattern thinking that is more fluid and adaptive. Like any review or other act of knowledge synthesis, the results of pattern inquiry will be out of date as soon as the study is completed. How quickly this happens and to what extent and in what ways will depend on many factors.

In terms of implied pattern representations, I noted in Chapter 6 that many if not most products of research and reasoning (models, theories, frameworks, etc.) are implied pattern representations in that they represent the macropattern thinking of the researchers or scholars involved. I would not claim that the authors of these products employed a pattern theoretical approach, humans have long created pattern representations without needing or using such a position. We might appraise the artefacts and products of research in terms of how explicit a pattern theoretical approach was in the work (including none whatsoever) but even there without explicit disclosure and orientation this depends on inferences and implications.

We cannot access our own patterns directly, at best we can have an impressionistic sense of their architecture and function. We are even less able to access other people's minds, and therefore the patterns that pattern inquiry works with are implied and inferred; pattern analysis is to a great extent a science of implication. In philosophy, implications are not causes, they are logical relationships, the product of one or more propositions – if A then B, if A and B then C. Implications can be understood as plausible and defensible but intangible relationships between things. When things become tangible then we might talk about causalities rather than implications. Patterns are, for now, latent constructs in the context of pattern inquiry. We work with intangible things such as ethics, morals, professionalism, integrity, competence, etc. all the time. We imply their existence inductively by unifying their expressions under a common category and we coordinate shared activity under that category through collective pattern building and pattern sharing. There are, therefore, implied gestalt patterns that result from the participation of many individuals each with their own versions of these patterns. These individual patterns will inevitably be imperfect and variant versions of the gestalt given that they are woven in individual minds with individual experiences of working with or otherwise interacting with the collective. We cannot observe these gestalt patterns directly, but we can observe them or interact with them in terms of the products and arrangements of participating members of a pattern-sharing collective. In other words, we can analyse the implied gestalt patterns by considering multiple individual pattern expressions and representations produced by participating minds.

It might seem, with so much implication caught up in acts of pattern inquiry, that its many uncertainties and subjectivities undermine its rigour and authority. Rather, I would suggest instead that the kinds of knowledge produced by such inquiry should be appraised with an understanding of their ontological bases and the limits to knowledge they imply. Noting my earlier observations regarding degrees of naturalism and different approaches to naturalism in pattern science, rigour and trustworthiness will depend on how hermeneutic concerns regarding interpretation or superimposition of analyst perspectives are addressed. It is not necessarily a good or bad thing that a pattern analysis reflects the analyst's pattern thinking, only that this is inevitable, recognised, and its implications understood. After all, Charmaz (2006) argued in the context of constructivist grounded theory that there is co-construction of meaning between analyst and participants.

Integrating Pattern Inquiry

Despite my focus in this chapter on outlining pattern theoretical approaches to scientific inquiry, I do not want to suggest that pattern inquiry is or should be decoupled from other approaches to inquiry. In this section, I outline some of the ways in which pattern theory and its application in pattern inquiry can contribute to other approaches to inquiry. For instance, pattern inquiry could be used in preparing for a study using some other methodology or paradigm. More specifically, pattern inquiry could be used to analyse apparent and implied collective pattern thinking regarding the topic or construct of interest. This could complement or replace a traditional literature review that situates the topic and briefs the researchers as to the main issues and concerns at play. Pattern inquiry could be used to augment a systematic review (in its many forms) by adding an examination of the implied shared patterns in a body of literature. Pattern inquiry could also be used in support of reflexivity as a way of examining the pattern thinking of those involved in the research, or it could be used to identify gaps and dissonances that may constitute limitations to a particular study.

Pattern inquiry might also be used as part of a mixed or multiple methods research study design. Mixed methods research combines methodologies and methods that share constructs or are expected to converge or complement each other, while multiple methods research uses methodologies and methods in parallel to explore different constructs with a broader and less coupled expectation of synthesis (Tashakkori & Teddlie 2010). For example, combining pattern inquiry with an ethnographic or ethnological approach could help to explain the rationale for the collective action and organisation under examination, while combining pattern inquiry with some variety of discourse analysis could help to provide a structural map of the discourses identified and/or examined.

Pattern inquiry might also play a productive role in a study that is nearing completion as a way of analysing the pattern thinking generated by a study in its findings, implications, recommendations, and suggestions the researchers share through scholarly communications. As an example, a pattern theoretical approach to peer review might focus on the apparent patterning of an article's arguments, constructs, and conclusions, both in terms of its strengths and its weaknesses. Another application might be to analyse the pattern thinking engendered in reaction to the findings of a study: how the findings are received, how they are interpreted, and what is done with the pattern thinking the study catalysed.

As we move further out from the locus of a given study, researchers might also use pattern inquiry to appraise and develop the training of new researchers, by making the shared pattern thinking of researchers and scholarly cultures that they participate in. Given that scientific paradigms (Kuhn 1962) are inescapably based on shared pattern thinking and the regularities and understandings produced by this thinking, pattern inquiry would seem to be a compelling way of analysing activity within and between different scientific paradigms. Indeed, there would seem to be little in the social sciences or in any other branch of the sciences where pattern analysis could not be used to develop a foundational understanding of how a field or discipline thinks and acts. Pattern analysis can be used to provide a foundation, a map, a basis for discussing or engaging in domains of scientific enquiry. Rather than suggesting that pattern theoretical lenses could add to these conversations, researchers might instead consider how robust scientific practice can be in the absence of such analyses. Mapping individual and shared patterns of thought and discourse in a community of practice, particularly a scientific one, could also serve as a way of comparing the foundations upon which different sciences stand.

Another issue to consider is how much pattern theory an investigator needs to include in acts of pattern inquiry. Do they need to accept or understand all aspects of pattern theory to use this position? As much as a full understanding and commitment might be helpful, I do not think it is strictly necessary. As an example (as I demonstrated in Chapter 2), many concepts in a given field have multiple meanings with the result that the conversations scholars think they are having are not conversations at all as they are not talking about the same things. Researchers can solve this problem by modelling the underlying structures of shared pattern thinking in a field, a domain, a discipline, or a community of practice. A researcher does not need to know, understand, or even agree with the spectrum of pattern theory that I have outlined to undertake this kind of work (although I think it would add depth and coherence) as they could still explore the underlying recurring structures and ideas that define what is said in a particular discursive space and how specific conversations about feedback relate to those structures. I call these things *patterns* and *pattern theory* provides a way to talk about these kinds of findings, but researchers do not need to use the P word to use the empirical logic of pattern theory.

Rigour and Integrity in Pattern Science

In Chapter 8, I discussed aspects of mindfulness required of those engaging in pattern inquiry. Mindfulness is required, at least in part, is because of the many ways in which pattern inquiry can falter or misstep. To some extent this is true of all science, hence the focus on rigour and reproducibility as markers of quality in science as a whole. However, quite what rigour involves varies according to paradigm. For instance, the National Institutes for Health (2023) in the United States defined "scientific rigor [as] the strict application of the scientific method to ensure unbiased and well-controlled experimental design, methodology, analysis, interpretation and reporting of results", which reflects a somewhat epidemiological, biomedical, and postpositivist worldview.

Qualitative criteria for rigour are typically captured in ideas of credibility, transferability, dependability, and confirmability (Guba & Lincoln 1989), which may include reflexivity (addressing investigator bias and positionality) and transparency (full disclosure of the research process) (Jacobs et al. 2012). Reflecting these varying perspectives, there are frameworks for assessing quality and rigour both in quantitative studies (Fawkes et al. 2015), in qualitative studies (O'Brien et al. 2014), and in the context of particular methodologies (Chiovitti & Piran 2003), all of which reflect general characteristics of credibility, dependability, confirmability, and transferability.

Credibility might be expressed or evaluated in terms of alignment between a study's problems, questions, contexts, theories, methodologies, and anticipated knowledge claims, dependability might be expressed or evaluated in terms of identifying and managing or mitigating bias, error, and imprecision, confirmability might be expressed or evaluated in terms of study reproducibility, transparency, and/or auditability, while transferability might be expressed or evaluated in terms of generalisability and contextual grounding. I shall take all these factors as given in considering what rigour means in the context of pattern inquiry.

What then are the particular challenges and issues regarding rigour in pattern inquiry? It could be argued that pattern inquiry is no less open to (and because of its hermeneutic interpretive stance possibly more open to) bias and error than other branches of science. All of the problems caused by insufficient training and insufficient resources (time, money, access), by mistakes and misunderstandings, and by inattention and compromise also apply to pattern inquiry and need to be attended to. Interestingly, pattern theory can explain the origins of bias and error when they result from failures in pattern perception, pattern thinking, and pattern reasoning. For instance, pareidolia is the condition of seeing patterns when they are not there. I would define *scientific pareidolia* as finding patterns when they are not there, either in the conduct of scientific inquiry or as the result of scientific inquiry. Clearly scientists (like all humans) sometimes see things that are not there and fail to see things that are there, an issue that, as Smith & Cordes (2020) argued, is exacerbated by the use of modern analytic techniques and the ever-growing volumes of data that they are applied to:

> ... the explosion in the number of things that are measured and recorded can provide us with useful information, but it has also magnified beyond belief the number of coincidental patterns and bogus statistical relationships waiting to deceive us ... The number of possible patterns is virtually unlimited and the power of big computers to find them is astonishing. Since there are a relatively few useful patterns, most of what is found is rubbish – spurious correlations that may amuse us, but should not be taken seriously. Don't be fooled by phantom patterns.
>
> *(Smith & Cordes 2020)*

This is not just an issue with computers and "Big Data". Not only have researchers often been found to be biased in terms of interpreting coincidence as correlation, and correlation as causation (Stanovich 2007), they are also prone to finding what they looked for because of the patterning effect their theories can have on their perceptions:

> ... discussions or interpretations of data are often the most important part of a research article ... [a] problem arises when these interpretations (i.e., theoretical propositions derived from the empirical propositions) are presented as facts or knowledge.
>
> *(Teo 2010)*

In responding to these issues, I return to the differences between regularities and patterns. Regularities are recurrences and/or apparent causes, patterns are meanings, interpretations, understandings. In pareidolia, humans perceive regularities in which they see patterns that reflect their triggered pattern memories. However, much of the analytic models used in "Big Data" look for regularities and, once found, identities and meanings are retrofitted to these regularities, often by human analysts. Does this mean that computers cannot identify patterns? Does pattern recognition need an embodied and limbic organic mind with all its idiosyncrasies and failings? I note again the work of Metzger et al. (2023) who were able to translate neural imaging into reasonably accurate synthesised speech. In this case regularities in inputs were filtered to selectively trigger words, sentences, and expressions from memory. I would suggest, therefore, that if computers can interpret regularities in terms of identities, meanings, and significances then they are close to identifying patterns, albeit pattern abstractions *in silico*. However, I would also argue that computers cannot approximate human pattern thinking (or that of other pattern thinking species) without being able to draw on the spectrum of human experience, in particular the embodied and limbic aspects of our conscious and unconscious minds. If a computer can neither feel nor emote then how can they ever reflect human pattern thinking? That said, computers could add other dimensions of experience to the detection and interpretation of regularities that are not available to humans. Computers could draw on extended sensoria (such as "seeing" wavelengths of light and sound beyond human perception), variant sensoria (such as "seeing" things that are very fast or very slow, very large or very small), and quite different sensoria (such as "seeing" ionisation or gravity), but could any of these replicate the experience of significance or meaning afforded by a patterned mind?

Leaving these issues for another time, a central function of pattern thinking in humans is to attribute identity, meaning, and significance to the things and situations they encounter, both within themselves and around them. When the attribution of meaning and significance to regularities is in error, we call this apophenia, particularly where doing so becomes problematic or it cannot be controlled. While *scientific apophenia* has previously been defined as "the tendency to find evidence of order where none exists" (Goldfarb & King 2016), this definition is more about pareidolia than apophenia, and I would rather describe scientific apophenia as the tendency to erroneously attribute meaning and significance to perceived regularities. Notably, Shermer (2008) reflected this tendency in his concept of "patternicity": finding "meaningful patterns in meaningless noise" and the implications thereof.

Scientific apophenia may follow pareidolia if it results from an erroneous pattern match, or the pattern match may be appropriate but the meaning or significance that follows from the match might be in error. In the latter case, we might attribute this to a broken or poorly constructed pattern, or we might consider it to be a failure in how those parts of a recalled pattern that are about significance and meaning were processed. For instance, if we are in error, is it because the pattern we match to our experiences does not contain indicators of meaning or significance, or is it because the pattern's indicators of meaning and significance are causing exaggerated or suppressed responses? There may be many reasons for overinterpretation, under-interpretation, or misinterpretation of perceived pattern matches, and a full exposition thereof is beyond the scope of this volume. I will simply say, therefore, that scientific apophenia can have many causes, many symptoms, and many impacts on research rigour and integrity.

On one hand these issues might be treated as a matter of risks and limitations, not just to pattern inquiry, but to all scientific disciplines and domains, and certainly not attending to these issues would likely be considered to be a lack of diligence. On the other hand, approaching these issues from a pattern theoretical perspective could present many opportunities for research into causes, impacts, and mitigations. A robust pattern inquiry approach to rigour and integrity is not, therefore, so much about limitations to pattern theory *per se* (although there are many limitations as I have described throughout this volume) but about how a pattern inquiry or pattern theoretical perspective can be used to examine issues of rigour and integrity in any paradigm or methodological framing of inquiry. For instance, our pattern thinking (both individual and collective) inevitably shapes our approach to inquiry in terms of theories (of what, how, and why), in terms of methodologies and study designs, and in terms of the specific methods and tools we use. That we shape inquiry in terms of our pattern thinking and then interpret what we find from the same patterned perspective is a significant concern in terms of objectivity and bias across science as a whole (Teo 2010).

Taking a pattern theoretical perspective can help us to understand the more fundamental cognitive limitations and strengths that all scientists face. Attending to how pattern thinking both shapes and forms the basis for interpretation may provide useful insights in the pursuit of rigour in scientific inquiry, and it may also provide a basis for more substantial approaches to reflexivity and professional identity formation. We should also be mindful of the patterned nature of the foundational narratives and beliefs of science, of the various approaches to and discourses in the philosophy of science, and of how various concepts of pattern have been used in scientific inquiry. In suggesting this it can bring us back to the issues I raised throughout Chapter 2 in terms of widespread use of pattern as a concept or frame of reference without any clear or consistent definition of what pattern is or might be.

Pattern theory can also focus attention on the heuristic character of pattern thinking, both in appreciating the agility and adaptability it affords and in attending to the limitations (but also opportunities) of working with provisional, abductive, and satisficing solutions rather than rigorous and definitive ones. Although trained and honed pattern thinking cannot solve all our weaknesses, it might go a long way to help us remove some of them, while being mindful of others such that we are better able to manage or mitigate them. For instance, a pattern theoretical position on theory requires us to be mindful that, while we may attend to the macro theories in a study (theories of what, how, and why), they are set against a background of the multitude of tacit pattern theories we hold that are critical to our stance and performance as scientists. These include theories of quality, behaviour, responsibility, and, yes, rigour and integrity. Our minds are pattern thinking engines, and as such every mind is biased and partial in its idiosyncratic pattern landscapes, the ways in which they have formed, what they contain, and how they are used. As a simple example of this partiality, we tend to remember pattern matches and convergences (particularly when accompanied by a positive emotive experience), and we tend to forget pattern misses and divergences (particularly if unaccompanied by a negative emotive experience).

Chapter Summary

Having set out principles of pattern inquiry in Chapter 8, in this chapter I focused on the practicalities of applying pattern theory to structured and scientific inquiry. Much of this was articulated around a worked example that described developing a pattern system of medical school admissions. I described the development of a methodology, I set out seven iterative and integrated principles of pattern analysis, I explored the strengths and limitations of the approach I took, and I considered possible grammars and syntaxes as a way of expanding the pattern system to be a pattern language. I reflected on issues of scope and on alternative approaches I might have taken, and I closed the example in describing how pattern inquiry can be used to map landscapes of shared pattern thinking. Developing ideas and issues raised in the worked example, I considered some of the implications of taking a pattern-informed approach to inquiry, and I explored issues in implied and latent pattern thinking. I explored ways in which pattern inquiry could be integrated into other methodologies and study designs, and I closed with a discussion of rigour and integrity in what we might call *pattern science*. As such, this chapter is a starting point for developing thinking and practice in pattern inquiry and should not be taken to be a comprehensive or definitive guide. Much work will be needed to expand and test the repertoire, and, more importantly, to develop rigour and meaningful applications of pattern inquiry in different research contexts.

References

Alexander C, Ishikawa S, Silverstein M. *A Pattern Language: Towns, Buildings, Construction*. New York, NY: Oxford University Press: 1977.

Alexander C. *A Timeless Way of Building*. New York, NY: Oxford University Press: 1979.

Alexander C. The Origins of Pattern Theory, the Future of the Theory, And the Generation of a Living World. 1996 *ACM Conference on Object-Oriented Programs, Systems, Languages and Applications (OOPSLA)*, San Jose, California. Transcript: https://www.patternlanguage.com/archive/ieee.html

Archer M. *Social Origins of Educational Systems*. London, UK: Sage: 1979.

Arksey H, O'Malley L. Scoping Studies: Towards a Methodological Framework. *International Journal of Social Research Methodology: Theory and Practice*. 2005; 8(1): 19–32.

Ashby WR. *An Introduction to Cybernetics*. London, UK: Chapman & Hall: 1956.

Borsboom D, Mellenbergh GJ, Heerden JV. The Theoretical Status of Latent Variables. *The Psychological Review* 2003; 110(2): 203–219.

Bourdieu P. *Distinction: A Social Critique of the Judgement of Taste*. London: Routledge: 1984.

Charmaz K. *Constructing Grounded Theory: A practical guide through Qualitative Analysis*. Thousand Oaks, California: Sage: 2006.

Chiovitti RF, Piran N. Rigour and Grounded Theory Research. *Journal of Advanced Nursing* 2003; 44(4): 427–435.

Creswell JW, Creswell JD. Mixed Methods Procedures. In: *Research Design: Qualitative, Quantitative, and Mixed Methods Approaches* (5th ed.). Thousand Oaks, CA: SAGE: 2018: 213–246.

Edwards JC, Elam CL, Wagoner NE. An Admission Model for Medical Schools. *Academic Medicine* 2001; 76(12): 1207–1212.

Ellaway R, Smothers V. *ANSI/MEDBIQ CI.10.1-2013 Curriculum Inventory Specifications*. Baltimore, MD: MedBiquitous: 2013.

Ellaway R, Topps D. *METRICS: A Pattern Language of Scholarship in Medical Education*. MedEdPublish. 2017, 6: 199. DOI: 10.15694/mep.2017.000199

Ellaway RH, Malhi R, Bajaj S, Walker I, Myhre D. A Critical Scoping Review of the Connections Between Social Mission and Medical School Admissions: BEME Guide No. 47. *Medical Teacher* 2018; 40(3): 219–226.

Ellaway RH, Malhi R, Woloshuk W, de Groot J, Doig C, Myhre D. An axiological case study of a medical school's admissions values. *Academic Medicine* 2019; 94(8): 1229–1236.

Ellaway RH, O'Gorman L, Strasser R, Marsh DC, Graves L, Fink P, Cervin C. A Critical Hybrid Realist-outcomes Systematic Review of Relationships between Medical Education Programmes and Communities: BEME Guide No. 35. *Medical Teacher* 2016; 38(3): 229–245.

Fawkes C, Ward E, Carnes D. What evidence is good evidence? A Masterclass in critical appraisal. *International Journal of Osteopathic Medicine*. 2015; 18(2); 116–129.

Glaser BG, Strauss AL. *The Discovery of Grounded Theory: Strategies for Qualitative Research*. Chicago, IL: Aldine: 1967.

Goldfarb B, King AA. Scientific Apophenia in Strategic Management Research: Significance Tests & Mistaken Inference. *Strategic Management Journal* 2016; 37(1): 167–176.

Goodheart E. Is History a Science? *Philosophy and Literature* 2005; 29(2): 477–488.

Gough D, Oliver S, Thomas J. *An Introduction to Systematic Reviews*. Thousand Oaks, CA: Sage: 2017.

Greene JC. *Mixed Methods in Social Inquiry*. San Francisco, CA: Jossey-Bass: 2007.

Guba EG, Lincoln YS. *Fourth Generation Evaluation*. Newbury Park, CA: Sage: 1989.

Iba T, Isaku T. A Pattern Language for Creating Pattern Languages: 364 Patterns for Pattern Mining, Writing, and Symbolizing. PLoP '16: *Proceedings of the 23rd Conference on Pattern Languages of Programs*. 2016; 11: 1–63.

Jacobs A, Büthe T, Arjona A, Arriola L, Bellin E et al. The Qualitative Transparency Deliberations: Insights and Implications. *Perspectives on Politics* 2012; 19(1): 171–208.

King N. Doing template analysis. In: Symon G, Cassell C (eds.), *Qualitative Organizational Research*. London, UK: Sage: 2012: pp 426–450.

Kuhn TS. *The Structure of Scientific Revolutions*. Chicago, IL: University of Chicago Press: 1962.

Lave J, Wenger E. *Situated Learning: Legitimate Peripheral Participation*. Cambridge, UK: Cambridge University Press: 1991.

Low J. A Pragmatic Definition of the Concept of Theoretical Saturation, *Sociological Focus*. 2019; 52: 2: 131–139. DOI: 10.1080/00380237.2018.1544514

MacLeod A, Ellaway RH, Cleland J. A meta-study analysing the discourses of discourse analysis in health professions education. *Medical Education* 2024; 1–13. DOI: 10.1111/medu.15309

Metzger SL, Littlejohn KT, Silva AB, Moses DA, Seaton MP, Wang R, Dougherty ME, Liu JR, Wu P, Berger MA, Zhuravleva I, Tu-Chan A, Ganguly K, Anumanchipalli GK, Chang EF. A High-performance Neuroprosthesis for Speech Decoding and Avatar Control. *Nature* 2023; 620(7976): 1037–1046.

National Institutes of Health. *Enhancing Reproducibility through Rigor and Transparency*. Bethesda, MD: National Institutes of Health: January 26, 2023: https://grants.nih.gov/policy/reproducibility/index.htm accessed 20 Nov 2023

O'Brien BC, Harris IB, Beckman TJ, Reed DA, Cook DA. Standards for Reporting Qualitative Research: A Synthesis of Recommendations. *Academic Medicine* 2014; 89(9): 1245–1251.

Patocka C, Cooke L, Ma IMY, Ellaway RH. Navigating Discourses of Feedback: Developing a Pattern System of Feedback. *Advances in Health Sciences Education* 2024. Accepted, in press.

Reiter H, Eva K. Selecting for Medicine. In Dornan et al. *Medical Education Theory and Practice*. Edinburgh, UK: Elsevier: 2011: 283–296.

Roze des Ordons AL, MacIsaac L, Everson J, Hui J, Ellaway RH. A Pattern Language of Compassion in Intensive Care and Palliative Care Contexts. *BMC Palliative Care*. 2019; 18(1): 15.

Saldana J. *The Coding Manual for Qualitative Researchers*. Thousand Oaks, CA: SAGE: 2021.

Scheurer F. Architectural Algorithms and the Renaissance of the Design Pattern. In: Gleiniger A, Vrachliotis G. *Pattern: Ornament, Structure, and Behaviour. Context Architecture*. Basel, CH: Birkhauser: 2009: pp 41–56.

Schiffer S. Two Perspectives on Knowledge of Language. *Philosophical Issues*. 2006; 16: 275–287. http://www.jstor.org/stable/27749869

Schrewe B, Ellaway RH, Watling C, Bates J. The Contextual Curriculum: Learning in the Matrix, Learning From the Matrix. *Academic Medicine* 2018; 93(11): 1645–1651.

Shermer M. *Patternicity*. Scientific American, December 2008: Online at https://michaelshermer.com/sciam-columns/patternicity/ accessed May 17 2022.

Smith G, Cordes J. *The Phantom Pattern Problem: The Mirage of Big Data*. Oxford: Oxford University Press: 2020.

Stanovich K. *How To Think Straight About Psychology*. Boston, MA: Pearson: 2007.

Tashakkori A, Teddlie C (eds.). *SAGE Handbook of Mixed Methods in Social & Behavioral Research*. (2nd ed). Thousand Oaks, CA: Sage: 2010.

Teo T. What is Epistemological Violence in the Empirical Social Sciences? *Social and Personality Psychology Compass* 4/5 (2010): 295–303. DOI: 10.1111/j.1751-9004.2010.00265.x

Tuckman BW. Developmental Sequence in Small Groups. *Psychological Bulletin* 1965; 63(6): 384–399.

Wenger E. *Communities of Practice: Learning, Meaning, and Identity*. Cambridge, UK: Cambridge University Press: 1998.

Wong G, Greenhalgh T, Westhorp G, Buckingham J, Pawson R. RAMESES Publication Standards: Meta-narrative Reviews. *BMC Medicine* 2013; 11: 20. DOI: 10.1186/1741-7015-11-20

10

A PATTERN MANIFESTO

In which I try to put everything in its right place.

I opened this volume with a critique of scientific and non-scientific uses of the concept of pattern, particularly its undefined use ("it just is") and its ambiguous use to refer to many different things (even within the same sentence). Noting a recurring (and apparently unnoticed) conflation of pattern with regularity, I distinguished between form (the specific morphology of a thing, action, idea, etc.), order (morphological principles perceived in the form of a particular thing), and regularity (the presence of repetition and periodicity, or the sense of something being controlled, directed, or constrained), and related each concept in turn to my provisional description of pattern. Noting that form, order, regularity, and pattern are all tied to perception and cognition led me to consider the neurological and cognitive basis of *pattern perception* and from this the nature of *pattern* and of *pattern thinking.* I then built on these ideas to outline a *pattern theoretical stance*, primarily articulated in terms of different aspects of pattern thinking. Acknowledging that, although our patterns are intrinsically cognitive, we regularly externalise and share them, I outlined concepts of *pattern instances* and *pattern expressions*, and then focused on *pattern representations* and their collection into *pattern systems* and *pattern languages*. I next explored some of the conceptual and philosophical implications of a pattern theoretical stance, both to test it and to provide more depth and substance. I then used pattern theoretical concepts to outline what I called *pattern inquiry*, and I explored practical applications thereof and their implications.

In this final chapter I will try to draw these threads together to present a coherent whole. I start by outlining the core tenets of pattern theory, grouped by the nature of patterns, the nature of pattern thinking, the nature of shared pattern thinking, and the nature of pattern inquiry – see Table 10.1. I then consider the place of a pattern theoretical approach in the context of contemporary trends in the social sciences, I reprise the matter of Wittgensteinian word games and the issue of what pattern as a word is and what it means, and I close with thoughts on the strengths and limitations of the ideas I have presented and where they may go next.

DOI: 10.4324/9781003543565-10

TABLE 10.1 Fifty core tenets of pattern theory

Patterns

1 Patterns are cognitive.
2 Patterns are pervasive.
3 Patterns are aggregative.
4 Patterns can include symbols.
5 Connections between a pattern's elements can vary.
6 Patterns are formed iteratively and selectively.
7 Patterns are dynamic.
8 Patterns are fuzzy.
9 Patterns are abstract.
10 Patterns are layered.
11 Patterns are lean.

Pattern Thinking

12 Pattern thinking is the use of patterns.
13 There are many neural correlates of pattern and pattern thinking.
14 Pattern thinking is inseparable from our cognitive makeup.
15 Perception is based on matching stimuli to pattern memory.
16 A pattern instance is our perception of something familiar
17 Pattern perception is primarily (but not exclusively) abductive and heuristic.
18 Pulling on a pattern's elements can invoke other elements and other associated patterns.
19 Pattern perception is shaped by anticipation and cueing.
20 Pattern perception works with partial information.
21 Pattern perception can convey meaning.
22 Pattern thinking is essential for reasoning and creativity.
23 Pattern thinking changes patterns.
24 Pattern and utility.
25 Humans are all pattern thinkers.
26 Pattern thinking and learning are intertwined.
27 Pattern thinking has strengths and limitations.
28 Other species also have pattern thinking minds.

Shared Pattern Thinking

29 Individuals can never fully share their patterns or pattern thinking.
30 Abstract pattern thinking capabilities allow individuals to share their patterns and pattern thinking.
31 Patterns and pattern thinking can be shared as pattern expressions, pattern representations, pattern systems, and pattern languages.
32 Sharing pattern thinking varies by individual, collective, and culture.
33 Shared pattern thinking underpins all human culture and society.
34 Shared pattern thinking is negotiated, fluid, and variant.
35 Participation and influence in shared pattern thinking tends to be asymmetrical.
36 Shared pattern thinking can persist beyond the mind of any participating individual.
37 The patterns of shared pattern thinking are implied, virtual.
38 Shared pattern thinking is reified in the actions and products of the pattern thinking collective.
39 Shared pattern thinking can be used to enable participants and to constrain them.
40 Different individuals in a shared pattern thinking collective will have different dispositions towards it.
41 There are both strengths and limitations of shared pattern thinking.

(*Continued*)

TABLE 10.1 (Continued)

Pattern Inquiry

42 Pattern inquiry is a pattern theoretical approach to structured scholarly inquiry.
43 Pattern inquiry is not a methodology, it is a meta-methodology.
44 Pattern inquiry makes use of pattern ontologies and epistemologies.
45 Pattern inquiry can serve many different purposes.
46 Pattern Inquiry may have many different foci.
47 Pattern inquiry requires certain investigator dispositions.
48 Pattern inquiry may involve many different methodologies, study designs, and methods.
49 There may be many different products of pattern inquiry.
50 There are both strengths and limitations to pattern inquiry.

Patterns

I started the development of pattern theory by outlining what patterns are, how they form, how they work, and what their core characteristics are:

1 *Patterns are cognitive.* Although they are often described as such, patterns do not exist in the world; patterns are entirely a product and function of mind. Individuals may perceive regularities around them (which they often call patterns), but patterns are and can only be nexuses of memories in their minds.

2 *Patterns are pervasive.* Pattern formation, recall, consolidation, and elaboration are fundamental characteristics of human (and likely many non-human) minds. Pattern is so pervasive and fundamental to the function of the mind that it largely goes unnoticed.

3 *Patterns are aggregative.* A pattern is not a singular cognitive construct; it is composed of interconnected memories of ideas, experiences, and feelings. Patterns can include other patterns, one pattern may serve as an element of many other patterns, and elements of one pattern may also be elements in other patterns. Connections may reflect shared or similar properties, referents (words, sounds, images) or contexts. Not every pattern element in a pattern is directly connected to every other element in a pattern.

4 *Patterns can include symbols.* Most words (nouns, verbs, adjectives, adverbs, etc.) refer to patterns such that learning a word establishes or elaborates a corresponding pattern. Using a word can invoke its associated pattern(s). Other symbols and symbol systems, such as numbers, shapes, graphical symbols, sounds, and gestures, can also be used in this way. Symbols and symbol systems are fundamentally articulated through patterns. Language is highly patterned and may be the primary patterning vehicle both in terms of what individuals use and in terms of what helps to develop their pattern thinking skills as children. A key part of this is the use of category as both an index of pattern and an aggregating capability of pattern thinkers.

5 *Connections between a pattern's elements can vary.* I have used the term "connected" memories for close or strong relationships within a pattern, and "associated" memories for more tenuous or weaker relationships between patterns, but this should be understood as a continuum rather than two distinct types. Connections and associations can reflect different kinds of relationships, such as "is a part of", "is a kind of", or "occurs with" (which are themselves patterns) that qualify the nature of pattern searching and recall.

6 *Patterns are formed iteratively and selectively.* Patterns are formed over time by connecting and associating memories. A pattern can be seeded around the connected memories

of a novel experience (what it looked like, how it behaved, what feelings it engendered, etc.). If subsequent experiences are matched to a pattern of past experiences, then, rather than forming new pattern nexuses, the existing pattern nexus is elaborated. This involves sorting and consolidation of memory such that only some experiences and thoughts are woven into patterns. Consolidation is facilitated by repeated pattern recall and elaboration such that a pattern grows each time it is recalled, particularly as new memories and associations are added to it or when multiple patterns are woven together into a single pattern. Although it can be influenced by deliberate thought, most pattern formation and elaboration occur without deliberate or conscious thought.

7 *Patterns are dynamic.* Patterns change over time. Patterns may be elaborated by thinking about them or using them. Patterns may atrophy or be lost, for instance, through injury or aging. Some patterns may be retained for a lifetime while others have much shorter duration. Pattern connections and associations may be added, strengthened, weakened, or lost over time, and pattern elements added, removed, or reconfigured. A change to one pattern nexus can impact the other pattern nexuses it is connected to such that patterns are always partial, situated, and incomplete. Memories of new experiences, thoughts, and feelings can always be added. Dynamic pattern building may reinforce a pattern if subsequent experiences align with it, or it may transform the pattern if experiences suggest alternative or divergent understanding and meaning.

8 *Patterns are fuzzy.* Patterns are not discrete units of memory; they are nexuses of thought that are deeply interconnected and constantly changing. Patterns can share common elements such that they seem to overlap. There are no distinct edges to a pattern, rather individual minds hold a continuity of pattern memory with denser nexuses of connections reflecting patterns. This pattern continuity is the substrate in which new patterns are woven, and in which patterns are accessed and organised.

9 *Patterns are abstract.* While a single memory relates to a particular experience of a phenomenon; patterns reflect abstractions and generalisations of similar phenomena such that individual minds can group them as instances of the same pattern both synchronously and asynchronously. Rather than assessing identity (whether a new experience is identical to a past experience), our minds assess similar differences and different similarities and flex the scope of patterns to accommodate new variations. Abstraction is greatly facilitated by symbolic thought of various kinds including language, logic, and category.

10 *Patterns are layered.* Patterns can involve multiple layers of aggregation and abstraction. This layering, aggregation, and abstraction reflects a developmental continuum from micropatterns (impressions of things around us and within us) through mesopatterns (perceptions and understandings of our micropatterns) to macropatterns (theories and abstractions of our mesopatterns). The development of macropatterns (macropatterning) can involve theorisation, generalisation, signification, formalisation, and systematisation. Macropattern development demonstrates ecological systems behaviours (adaptive cycles, panarchy, hysteresis)

11 *Patterns are lean.* Pattern formation and pattern elaboration reuse and aggregate patterns such that new experiences, feelings, or thoughts that match to similar memories are subsequently stored as elaborations of those memories. This is seen in the continuities of pattern nexus associations and in the aggregating nature of macropatterns and the symbols and symbol systems they employ. The result is that individual minds hold only one pattern nexus for any given aggregation of experiences, feelings, or thoughts.

Pattern Thinking

Having outlined what patterns are, I then explored the ways in which patterns can be used by our minds for various purposes and the consequences thereof:

12 *Pattern thinking is the use of pattern.* Pattern thinking in an individual mind both draws on and shapes the matrix of interwoven pattern nexuses in that mind. Not only could there be no pattern thinking without patterns to work with, there would be no point in developing patterns without the advantages of using them in pattern thinking.

13 *There are many neural correlates of pattern and pattern thinking.* The apparent structure and function of the cortex, its hemispheres, its lobes and centres, and the role of the hippocampus and amygdala all correlate with patterns and pattern thinking.

14 *Pattern thinking is inseparable from our cognitive makeup.* Pattern thinking is not something that happens in one part of the brain, nor is it one process among many. Rather, pattern thinking is a constantly active and connected part of an individual's mental being, both while they are awake and while they are asleep.

15 *Perception is based on matching stimuli to pattern memory.* Individual minds are constantly comparing stimuli to pattern memory and selectively activating or pulling putative pattern matches into active thought. Perceiving patterns in the world reflects a sense of recognition and meaning afforded by a matched pattern.

16 *A pattern instance is our perception of something familiar.* Pattern instances are ephemeral even if the phenomenon persists, and instances are personal and subjective such that different observers may perceive the same phenomenon differently. Individuals may collectively agree that something is an instance of a shared pattern based on consensus or social conventions.

17 *Pattern perception is primarily (but not exclusively) abductive and heuristic.* Pattern perception can work rapidly and deal with imperfect and ambiguous inputs to suggest provisional (satisficing) pattern solutions. These apparent pattern solutions may change as available information changes and better matches are found. Pattern solutions can satisfice without having to be exact or definitive.

18 *Pulling on a pattern's elements can invoke other elements and other associated patterns.* There is much evidence to suggest that we cannot recall individual independent memories, rather recall involves activating multiple interconnected memories. Pulling on a pattern part or element is how a pattern is evoked, matched, or otherwise triggered so that its understandings, meanings, and significances inform our perceptions. The stronger or longer the pull, the more connected and associated elements are activated and available to working memory and thought.

19 *Pattern perception is shaped by anticipation and cueing.* Patterns can include contextual associations. The activation of one pattern activates associated contextual patterns, which affords a degree of pattern anticipation in that these patterns are more readily available for matching. As a result, it can seem to a conscious mind that patterns are just "there" and their perception is fast and accurate. This can also bias perception by tending to match more readily available patterns.

20 *Pattern perception works with partial information.* Pattern perception need only match some elements or characteristics to activate a suitable pattern. Once active, the pattern can then help to complete or fill in the missing information to confirm what is there but

not directly available to the senses. Perceiving patterns thereby establishes and maintains an inferential understanding of the world based despite limited available information. Pattern thinking extends an individual's senses, it helps to make sense of senses, and it allows senses to be matched to memories such that the world becomes more comprehensible.

21 *Pattern perception can convey meaning.* Patterns can include memories of meanings, understandings, and feelings. Activating a pattern allows its meanings and understandings to be applied to current experiences. Perceiving a pattern can, therefore, include apparent causes, origins, explanations, and predictions. Affective pattern elements can convey a sense of significance such that an individual pays particular attention to the pattern match. This can lead to pattern anticipation where the mind and body can prepare for probable future pattern matches.

22 *Pattern thinking is essential for reasoning and creativity.* Reasoning is a patterned process; how an individual reasons, what they enter into their reasoning process, and the products of those processes are all drawn from patterns. Creativity is based on speculative reasoning around possible pattern variations and combinations. Reasoning and creativity can be significant contributors to macropattern development.

23 *Pattern thinking changes patterns.* An individual draws on their pattern memories to help them understand current experiences. Not only do patterns allow individuals to interpret, associate, and attach meaning to their perceptions and thoughts, they are how individuals build and retain understanding.

24 *Pattern and utility.* Patterns do not have to be useful in an instrumental way, but patterns that are more frequently matched, activated, and elaborated will tend to be more useful, more fully developed, and longer lasting than those that are not. Conversely, the longer a pattern persists the more likely it is to be elaborated and combined with other patterns. The utility of any given pattern can, therefore, be understood in terms of how often it is matched, activated, and elaborated.

25 *Humans are all pattern thinkers.* Fundamental human capabilities, such as the use of language the imagination, and mathematics, logic, and reason, are all based on pattern thinking. Active participation in human society requires pattern thinking; individuals who are disrupted in developing or using pattern thinking may experience disability or be perceived by others as disabled in the context of human society.

26 *Pattern thinking and learning are intertwined.* Early learning involves developing basic pattern thinking capabilities and a core repertoire of experiential and reasoning patterns to support later expansion of and sophistication in pattern thinking. Education involves the development of learners' pattern thinking and their acquisition and articulation of repertoires of patterns relevant to particular domains.

27 *Pattern thinking has strengths and limitations.* Pattern thinking makes up a considerable part of the capabilities of human minds. Each mind has unique pattern thinking strengths and weaknesses. There are many ways in which patterns and pattern thinking can be disrupted over the lifespan of individual humans; developmentally, through injury and oppression, and through aging and senescence. Pattern thinking is also subject to illusion, bias, manipulation, and irrationality.

28 *Other species also have pattern thinking minds.* Many other species demonstrate capabilities in pattern thinking, albeit (as far as we can tell) to a lesser degree than humans.

Shared Pattern Thinking

Having outlined pattern and pattern thinking in the context of individual minds, I next explored the many ways in which humans are able to share and exchange their patterns and pattern thinking:

29 *Individuals can never fully share their patterns or pattern thinking.* Individuals may have similar pattern experiences or ideas but the specific pattern nexus in one mind is not the same as any other nexus in any other mind even if they refer to the same phenomena. Humans nevertheless try to share their patterns and pattern thinking but can only do so in imperfect and reductive ways.

30 *Abstract pattern thinking capabilities allow individuals to share their patterns and pattern thinking.* The capacity for pattern sharing is greatly enabled and extended by language of various forms, as well as other symbolic systems such as those of mathematics and music. The capacity for sharing patterns is enshrined in human languages, cultures, and shared practices including those of sciences, arts, and philosophies.

31 *Patterns and pattern thinking can be shared in different ways*:

   a *Pattern expressions* are deliberately manufactured or performed instances of one or more patterns.

   b *Pattern representations* are deliberate attempts to record the elements and logics of a pattern such that they can be used to analyse or replicate the pattern or its use in pattern thinking.

   c *Pattern systems* are collections of pattern representations pertaining to a particular domain that cannot be encompassed with a single pattern representation.

   d *Pattern languages* are pattern systems with additional grammar and syntax that reflect how pattern representations relate to each other.

   Pattern representations, pattern systems, and pattern languages may be speculative, normative, naturalistic, or historical. Although a pattern's logic can be formalised and externalised such that it functions as an algorithm in lieu of a mind, the representation still refers back to its author(s) pattern thinking.

32 *Sharing pattern thinking varies by individual, collective, and culture.* Some individuals and groups are more adept at pattern sharing than others, examples include poets, writers, philosophers, and orators. Some individuals and groups are more adept at certain kinds of pattern sharing or sharing certain patterns than others. For instance, schoolteachers learn to develop the shared pattern thinking of their pupils in areas such as reading, writing, and arithmetic. Indeed, schooling (in whatever form it takes) plays a critical role in developing skills in the shared pattern thinking of particular subjects such as history, geography, physics, chemistry, and biology.

33 *Shared pattern thinking underpins all human culture and society.* Without shared pattern thinking human societies and cultures could not develop or survive. Not only do cultures and societies involve sharing pattern thinking, they also have many subcultures that have their own distinct pattern thinking, including those of its professions, religions, sciences, arts, and sports.

34 *Shared pattern thinking is negotiated, fluid, and variant.* Shared pattern thinking tends to change over time, although such change may be slow, and it may be sought or it may be

resisted. Generally, the values and tenets of shared pattern thinking vary between generations, between regions, and between the interpretations of individual participants.

35 *Participation and influence in shared pattern thinking tends to be asymmetrical.* Some minds may influence shared pattern thinking more than others and some minds may be influenced more than others. Some individuals may accept the shared pattern thinking of the community, society, or culture "as-is", while others may reject it or seek to change it. Dispositions change such that shared pattern thinking heretics and discontents may become more orthodox over time, while others may shift from orthodoxy to heterodoxy or even heresy.

36 *Shared pattern thinking can persist beyond the mind of any participating individual.* Shared pattern thinking does not falter when individual minds are lost, but if all participating minds are lost and the products of those minds are also lost then shared pattern thinking can indeed cease.

37 *The patterns of shared pattern thinking are implied.* There is no single substrate for shared patterns other than the community of minds that share them, which makes them somewhat emergent and fragile; there is no one definitive shared pattern, just a multitude of participant perspectives and articulations.

38 *Shared pattern thinking is reified in the actions and products of the pattern thinking collective.* Despite the implied nature of shared pattern thinking, it can be reified in the actions and products of a collective and these may persist long after its pattern thinkers have died out. For instance, the pattern thinking of ancient cultures can be seen in collections of ancient artefacts.

39 *Shared pattern thinking can be used to enable participants and to constrain them.* While shared pattern concepts and practices of freedom, justice, rights, and opportunities can be liberating and enabling, they may also be used to manipulate and control individuals, as can other shared pattern concepts such as those of obligation, duty, authority, and legitimacy.

40 *Different individuals in a shared pattern thinking collective will have different dispositions towards it.* Individuals may align with dominant or normative shared pattern thinking, they may contribute to it, or they may critique it, or even resist it.

41 *There are both strengths and limitations of shared pattern thinking.* Although shared pattern thinking provides the basis of science, art, and philosophy, it can also serve cruel, ignorant, and destructive ends. Patterns have positive and negative sides, and these may change according to context. Because pattern thinking can be manipulated, strengths in pattern thinking can sometimes become vulnerabilities and vice versa.

Pattern Inquiry

The last broad pattern theoretical area I considered was that of pattern inquiry:

42 *Pattern inquiry is a pattern theoretical approach to structured scholarly inquiry.* Pattern inquiry can both accommodate the patterned nature of investigator and subject/participant/population minds (individually and collectively) and focus on the phenomena of patterned minds or on those that are associated with patterned minds.

43 *Pattern inquiry is not a methodology, it is a meta-methodology.* There are many possible approaches and orientations to pattern inquiry. Rather than setting out procedural rules we can consider the application of pattern theoretical principles to different aspects of

pattern inquiry. These principles include: cognition, constitution, logics, uniqueness, awareness, dynamism, continuum, categorisation, formation, dependency, modalities, triggering, elaboration, weaving, micro-meso-macro, layering, similarity and difference, generation, limitations, sharing, latency, expressing and representing, and medium of expression.

44 *Pattern inquiry makes use of pattern ontologies and epistemologies.* By understanding phenomena as pattern constructs, pattern inquiry can accommodate the variance in nominally grouped phenomena. Pattern inquiry differs from traditional reductionism which breaks things apart into their functional pieces. Pattern inquiry deliberately considers similar differences and different similarities.

45 *Pattern inquiry can serve many different purposes.* These can include (but are not limited to): the pattern basis of human behaviours, skills, attitudes, beliefs, knowledge, competence, culture, etc.; exploring how minds and brains work, both human and non-human; exploring how shared pattern thinking can establish, sustain, or disrupt social structure and coherence; exploring similarities and differences in patterns and pattern thinking over a lifetime or between different individuals or groups of individuals; exploring the nature and limits of knowledge given its basis in pattern thinking; and testing the many claims of pattern theory.

46 *Pattern Inquiry may have many different foci.* Pattern inquiry can explore both patterns and pattern thinking in and of individual minds and the implied gestalt shared patterns and shared pattern thinking of collectives. Pattern inquiry may focus on people or on what they produce or create as a result of their individual and shared pattern thinking. Pattern inquiry may explore pattern formation, pattern recall, pattern elaboration, or pattern decay.

47 *Pattern inquiry requires certain investigator dispositions.* These include being attentive to investigators' pattern thinking (reflexivity and critical thinking of investigators, critical appraisal by others); and being attentive to subject/participant/population pattern thinking and its capacity to confound or bias acts of inquiry.

48 *Pattern inquiry may involve many different methodologies, study designs, and methods.* Pattern inquiry is not a methodological approach, it is a theoretical perspective on inquiry that can be used with other methodologies. These include the full gamut of social science approaches and paradigms, quantitative and qualitative methodologies, objectivities and subjectivities, and real and latent constructs, and varieties of mixed and multiple methods inquiry. A pattern theoretical perspective also affords a degree of reflexivity on methodological choices, adaptations, and their execution in acts of inquiry.

49 *There may be many different products of pattern inquiry.* These can include pattern representations, pattern systems, and pattern languages, as well as other products that reflect pattern characteristics, such as models, theories, and frameworks. A pattern theoretical perspective affords a degree of reflexivity on the macropatterning development of these products and on the ways in which these patterned products may be consumed by others.

50 *There are both strengths and limitations to pattern inquiry.* A pattern theoretical perspective affords new insights to the process of inquiry, to the rigour and integrity thereof, and to the patterned perspectives that are used in appraising the quality of acts of inquiry and of what they produce.

These fifty tenets of pattern theory are not exclusive or exhaustive. I fully accept the possibility that they may be revised or extended in future work or by other authors. Nevertheless, they provide a broad set of perspectives that outline pattern theory, at least as I have developed it throughout this volume.

Situating Pattern Theory

All knowledge is incomplete, all experience involves uncertainty, and all perceptions of fact can involve emotions. Our minds work with similarities and differences, parts and wholes, and connections and associations all the time. Pattern thinking is omnipresent such that all science, culture, politics, and faith are woven from patterns and pattern thinking.

On the one hand, I have argued for a definition of *pattern* that excludes its use as a synonym of regularity (or form or order) and that denies the existence of patterns outside minds and their interactions with other minds. On the other hand, I have suggested that pattern inquiry can be productively used in conjunction with different methodologies and paradigms. In this section I situate pattern theory within a range of broader scientific, philosophical, and sociological domains.

One domain is that of the neurosciences. Indeed, this book could be interpreted as being largely about brains and minds. In Chapter 4, I described a range of empirical findings and theoretical arguments that align with a pattern theoretical perspective. However, alignment is not the same as proof, nor is the level of understanding afforded by the neurosciences (at present) able to constitute a proof for pattern theory. From a neuroscientific perspective, pattern theory is just that, a theory that explains and connects ideas and understanding about brains and minds. The neurosciences continue to break boundaries and develop new insights, such that the evidence I have presented in support of my arguments may not age well and new findings may further support or confound the neurological bases of pattern theory. However, simply hoping that evidence will in some way advance pattern theory would be a rather passive position to take. Rather, an applied neurosciences might actively explore evidence for or against pattern theoretical perspectives. For instance, can pattern memory element (engram) formation be tracked? Can the weaving of these engrams into coherent pattern memories be tracked or observed? Can the topology of those pattern memories be mapped out? Can we intervene in these memory structures? Can we track pattern selection or searching, pattern activation, pattern perception, or pattern elaboration in a living human brain? Can we track or model the neurological basis of pattern thinking in its many modalities? I think it would be difficult to develop pattern theoretical perspectives without developing a foundational understanding of the basis of patterns and pattern thinking in the brain, and much of this would depend on the work of the neuropsychology community. There may also be much to be learned about patterns and pattern thinking from clinical neuropsychology in terms of neurodiversity, trauma, and neurodegeneration.

Another intersecting domain to note is that of artificial intelligence (AI). As I described in Chapter 5, AI was associated with to concepts of pattern by Hawkins and by Kurzweil (among others). Moreover, the work of Grenander and his colleagues on General Pattern Theory was primarily (although not exclusively) realised in work exploring machine perception and learning. Rather than replicating patterns and pattern thinking of human minds in terms of algorithms or semantic webs, the abductive cycles of reinforcement learning (in terms of what to pursue and what to avoid) and the layering and recursion of deep learning algorithms that are a key feature of contemporary artificial intelligence technologies seem to be a closer analogue of human pattern thinking than any previous generation or approach (Zai & Brown 2020). There may also be applications of artificial intelligence in analysing the artefacts of shared pattern thinking (such as discourses and structures) to identify convergences and divergences. Again, there is much potential for future research here.

A third domain is that of the psychological sciences. Given the central part that memory, recall, pattern recognition, reasoning, attention, and emotion play in pattern theory, psychology might be well-placed to establish, refute, or revise pattern theory. It is also the domain to which pattern theory might have the most relevance as part of structured inquiry of various kinds. There is much of pattern theory that aligns with psychological theory and empirical knowledge, not least because I developed it to follow psychological principles. But again, that does not establish proof. I would note that there are many other psychological theories that cannot be proved in any absolute way (such as dual processing, cognitive load, and schema theories). The value of these theories is that they provide useful working explanations that can render otherwise complex and elusive phenomena tractable. Pattern theory may be understood and used in this way in the context of psychology given that, although it has empirical and conceptual grounding, it is as much metaphorical as it is literal in what it offers.

To an extent this is a book about cognition, and as such it intersects with other cognitive models and theories. For instance, in describing the role of emotion in cognition, Feldman Barrett described three "inevitabilities" of human minds: affective realism, conceptual thinking, and social reality. The first, affective realism, is that emotions shape thinking including our perceived reality, even to the point of illusion:

> Affective realism, the phenomenon that you experience what you believe, is inevitable because of your wiring ... body-budget predictions laden with affect, not logic and reason, are the main drivers of your experience and behavior.
>
> *(Feldman Barrett 2017, p. 285)*

This parallels my descriptions of affective aspects of pattern memories being matched and activated and thereby shaping current perceptions. It also emphasises the importance of affect in pattern thinking more broadly. This brings me to the second of Feldman Barrett's inevitability, conceptual thinking, is that we readily develop concepts that then shape our perceived reality:

> ... you have concepts, because the human brain is wired to construct a conceptual system. You build concepts for the smallest physical details, like fleeting bits of light and sound, and for incredibly complex ideas.
>
> *(Feldman Barrett 2017, p. 287)*

This framing of conceptual thinking parallels my descriptions of the symbolising and categorising characteristics of patterns and pattern thinking. The third of Feldman Barrett's inevitabilities is that we think and act socially and our perception of and interaction with others shapes our thinking:

> Your personal experience ... is actively constructed by your actions. You tweak the world, and the world tweaks you back ... your movements, and other people's movements in turn, influence your own incoming sensory input. These incoming sensations, like any experience, can rewire your brain.
>
> *(Feldman Barrett 2017, p. 287)*

I would argue that this too aligns with pattern theory, particularly to concepts of shared pattern thinking and the products and consequences thereof. I would note though that, although the role of emotion would seem to play critical role in pattern thinking, I have only sketched out this space and much more work would be needed to truly do it justice.

This book has been much concerned with memory: pattern can be understood as the structure, organisation, and the nature, necessity, and consequence of how memory is developed, organised, and accessed. While many other theories of memory focus on the memories themselves and how they are translated from working memory to long-term memory and back again, pattern theory is more focused on the connections between memories and the connections that constitute memories then on the memories themselves, and it is more interested on the consequences and uses of these connections than on what they connect. That is why I have talked about pulling on networks of memory, that is why I have described pattern nexuses, that is why I have described pattern elements and the connections and associations rather than the memories themselves. Pattern theory is as much about the topology and function of memory, as it is about memories themselves.

Not only does pattern theory have a large overlap with psychology, it also has a large potential overlap with linguistics and neurolinguistics. I have argued that language is our ur-pattern thinking system in that it symbolises and abstracts, it affords the sharing of pattern thinking, and it provides a way to represent and abstract categories, groupings, and relationships. Pattern thinking can be understood as the basis of language and language can be understood as conditioning the mind to abstract pattern thinking. I have disagreed with (or perhaps reinterpreted) Chomsky's and Pinker's arguments that we have an innate mental disposition for language. Rather, I have suggested we have an innate disposition for pattern thinking and a cultural immersion in shared pattern thinking from birth that forms the substrate through which language develops. Indeed, pattern theory might be applied to many issues in linguistics such as the formation of language, the relationship between language and thought, and how the shared pattern thinking of linguistics changes according to language and dialect. There may even be applications in speech pathology. Going from linguistics to pattern theory, there might also be useful linguistic perspectives that might be applied to the grammars and syntaxes of pattern languages. I would also note the application of pattern theory to mathematics both as a form of language (acknowledging this argument is disputed) and to more directly explore the many claims made that mathematics is the science of pattern. Pattern theory can challenge such assertions based on its interpretation of what "pattern" means, and it may also provide the means to better understand and nuance the capabilities of mathematics in a pattern space (for instance, in modelling pattern topologies).

This is also a book about knowledge, a topic that I considered in Chapters 7 and 8. To an extent, pattern *is* knowledge since active knowledge is the interpretation of and assimilation of patterns with current experience, while stored knowledge is both woven through our and constituted in our pattern memories. This may be applied to different kinds of knowledge and knowledge use, but, as I explored in Chapter 8, it applies to both our knowledge about inquiry and the knowledge we create out of acts of inquiry. To that end, this is also a book about science and structured inquiry. Indeed, there are many apparent intersections between pattern theory and philosophy of science, and with philosophy more generally. I have used many philosophical constructs in advancing pattern theory, such as ontology, epistemology, and axiology, and there are many other philosophical implications that might be explored. For example, can we understand Kuhnian scientific paradigms or methodological or theoretical ideologies in terms of shared pattern thinking? As another example, how might we understand ethics or morality in the context of individual and shared pattern thinking? What are the relationships between material, personal, and social realities when understood as being mediated by pattern thinking?

That there are intersections between pattern theory and the social sciences should also not be unexpected, not least due to my own grounding in this domain of scientific inquiry. For one, shared pattern thinking (and all its manifestations and ramifications) would appear to provide a fertile ground for exploring social structures, processes, and thought, sequentially and in parallel. Indeed, pattern theory, in affording a phenomenological continuity from brain to mind to collaboration and interaction, may afford new ways to integrate social science theory and inquiry with other domains. Pattern theory can also serve in a meta-sociological context to examine the pattern thinking basis for different schools of and approaches to sociological inquiry. In Chapter 8 I also considered ways in which pattern inquiry might provide new insights and complementarities to existing social science approaches, including those of my own domain of professional education (such as its application to expertise theories and dual processing theories).

Pattern theory is not limited to applications in the psychological and social sciences. Indeed, I would argue that it has particular implications for the physical and biological sciences. For instance, in understanding the patterned and patterning nature of investigators' minds, the framing and design of the studies they conduct, and the findings they generate. This sits, I admit, on the cusp of philosophy of science but it does have significant implications for appraising and designing research that either embraces or seeks to step back from the inevitable pattern thinking of any domain that involves human actors. There are also fascinating opportunities afforded by pattern theory to explore the pattern thinking of species other than humans, something I have hinted at but have not pursued in this volume.

Pattern theory has implications for the arts both in terms of creative expression and in terms of the critical analysis of such expression. For instance, what can we say about the apparent pattern thinking of writers, painters, and composers? What can we say about the pattern thinking of architects, designers, and engineers? What can we say about the shared pattern thinking that constitutes artistic idiom, genre, and style? Indeed, potentially, pattern theory has implications for all human knowledge, human experience, and human undertakings.

Although pattern theory can provide ways to understand and analyse the nature of our patterned minds, our social pattern-based behaviours, and our pursuit of structured inquiry, we would not use a pattern theoretical approach to consider the physical world of, say, bosons, Hydrogen bonds, or dark matter. However, we would use a pattern approach to understand how scientists understand these things and how they communicate, explore, and act individually and collectively around this shared understanding. A pattern theoretical approach will not tell you how a nuclear power station works but it might tell you about the thinking that went into how and why the station was designed and how the station is operated.

Pattern theory is integrative in that it connects neuroscientific findings to cognition, and it connects the complex products of cognition in terms of individual and collective thinking, as well as sciences, cultures, etc. to both theory and evidence. As such, a pattern theoretical approach applies to phenomena of mind, things that are the product of mind (individual and collective), and to the ways we engage pattern in our thinking and behaviour in general. Pattern theory, like pattern itself, is generative in that it posits many new research questions and ways of approaching long-standing research problems. There is much more that could be said in that regard. However, pattern theory is also speculative in that while it explains and connects a great many phenomena, research findings, and theories, it has yet to be tested robustly across multiple settings and in different paradigms. Rather, I have woven together a theoretical construct whose implications I have perhaps only begun to outline. Indeed, I have suggested many jumping off points for empirical inquiry, and maybe these will be pursued and my theories disproved or elaborated. As Kant suggested, science can only progress in the dance between theory and experience.

Word Games

I opened the book by considering language games associated with the concept of pattern. Throughout this book I have continued to play word games regarding what is meant by perception, understanding, memory, or creativity? It only seems fitting, therefore, that I return to language games in this final chapter. As I described earlier, the word pattern in English conceals as well as affords so many different meanings and possibilities that that simple selection of the seven letters of PATTERN, hardly does justice to the breadth of phenomena, concepts, and theories that I have covered in this book. Indeed, is pattern a noun, an adjective, or a verb? Using pattern as a noun positions it as a thing or as an idea. Although we talk about patterns of weather or patterns in the landscape these are just regularities in which we may perceive patterns. If as a noun pattern refers to anything then it is to a nexus of associated memories and thoughts that constitute a pattern, although, as I discuss later in this volume, even then this is not a straightforward proposition. Using patterned as an adjective suggests it is a quality of something, although what it refers to may be ambiguous. For instance, saying that "this fabric is patterned" may refer either to its regularities or to the sense that those regularities were guided by patterns in the mind of the designer. Using "to pattern" as a transitive verb positions it as the result of some process, which is presumably driven by a mind ("she patterned her career on that of her mentor") although it might have some other causal basis ("the wind patterned the drifts of snow"). There is also the gerund of "patterning" ("those tiles show some beautiful patterning"), which again suggests but does not mandate the action or intent of a mind.

The issue becomes even more complicated in the context of translation and equivalency between languages. That said, I did not want to engage in neologistic games by inventing some other set of terminologies to replace that of pattern, but what are the choices when there has been so much confusion around this word and all its many possible meanings? Can we ever agree on what pattern means or should mean? Do we need to agree? Is pattern what we think, is it how we think, or is it a matter of our thinking being made tangible by externalising the patterns in our minds? Is it a matter of pattern thinking as a modality that underpins almost everything that we do and think? Or do we go back or stay with meanings to do with regularities? Do we stay with meanings of pattern as a synonym of regularity? Do we stay with the pattern language community's framing of pattern as being about problem-solving? Is a unified notion of pattern even necessary? In a poststructuralist age perhaps we should embrace multiple perspectives on what pattern means?

I really do not want to dispose of the word pattern, it is such a lovely word; simple, rich, and evocative. The solution I have used (mostly) in this book has been to extend and to adjectivise the term, and to work with compound terms involving pattern. I have talked about the difference between patterns and pattern instances, and I have identified components, relationships, variations, perimeters, connections, and associations within and between patterns. Using the word pattern in this way allows there to be a single unifying concept, something that ties all these different things together, and that is true to the essence of pattern, something that has associations and connections, something that ties parts into unified wholes. pattern is its concept of itself. Having pattern explaining itself is logically complex and potentially problematic given this elegant (albeit paradoxical) circularity.

Although I might have argued that pattern only needs to explain itself and as such does not need other reference points to exist, this is not the strongest defence. Rather, I would note that humans and other animals think in patterns, our cultures have been advanced and sustained through patterns. Patterns and pattern thinking can give us joy and sadness, patterns afford progress and regress, patterns are so much of what it is to be a human that maybe it is human beings ourselves (and other pattern thinking species) that defines what pattern is. If so then pattern is that which sentient life needs to exist as such, as well as to understand (albeit in a glass darkly) and thrive in a hostile and baffling world. A book on pattern that then disposes of the word pattern would, therefore, have failed its central thesis. I have, therefore, kept the word pattern and have woven my own patterns of meaning around the central concepts of pattern even as the concept of pattern itself has flexed and bent to accommodate these different meanings.

I do ask that the academic community use the term *pattern* in a more specific and informed way than has been the case. More specifically, I do ask scholars to stop using pattern to refer to regularities, at least in the context of scholarly discourse. If scholars say pattern but mean regularity, then call it regularity. If scholars mean pattern in a pattern theoretical sense, then by all means say pattern, or advance a different conceptual basis and an accompanying defence. Having said that, will the habits and conventions of the scholarly use of the concept of pattern change as a result of the arguments I have advanced, even if they are persuasive? My response to this flows from some of my earlier arguments regarding language in that: (1) pattern thinking precedes language and forms the basis for language development, (2) language acts as an accelerator and amplifier of the ongoing development of pattern thinking as a whole, and (3) language structures (in particular categories, logic, and abstract relationships) are essential in developing macropatterns and macropattern thinking.

Conclusions

Sociologist and philosopher Isaiah Berlin once argued that:

> To understand is to perceive patterns.
>
> *(Berlin 1997, p. 129)*

It is apparent from Berlin's subsequent text that he was thinking about regularities rather than cognition when he said this, but this statement works just as well when seen from a pattern theoretical stance. However, this is not where this book started from.

Writing this book was an abductive process. I did not know where it would end up when I started writing it and I had only a broad outline of the directions it might take. Indeed, my starting point was that I needed a book like this as a resource for my work as a social scientist, something that set out a substantial exploration of pattern as a cognitive and social phenomenon. Before starting this project, I had applied some pattern concepts in my research and publications, but I had found the literature thereof was both patchy (in regard of topics, methods, and products) and lacking in theoretical depth, and that meant that my earlier work in this area also struggled with depth and substance. Indeed, there have clearly been many other scholars who have been inspired by and who have built on the work of Christopher Alexander and colleagues, but I found very little critical reflection in this literature on the nature of pattern. Indeed, the work in this area has tended to be almost exclusively focused on developing and/or documenting normative representations to be used as tools in the hands of others. This begs the question of whether or not the knowledge from pattern inquiry is of value in its own right. Can pattern inquiry not explore things as they were or are rather than as they should be? Clearly, I have argued that pattern inquiry can have value and that there are many other approaches than normative utilitarian ones. Rather than pattern languages and their ilk remaining a somewhat fringe area of interest in the social sciences, I would like to see pattern inquiry taken more seriously and applied in many research contexts.

In reading this book you have followed the journey I took in developing what became pattern theory, and this reflects both the directions I took and what I considered to be essential parts of the puzzle that needed careful exposition and analysis. I wrote Chapters 1 and 2 very early on in the process and, other than tightening them up and adding more examples, they stand as the starting point; that pattern is a widely used concept, that it is used to mean many different things, that there is little coherence or agreement over what pattern is, and that there seem to be a multitude of missed opportunities to engage the concept of pattern in more conceptually grounded and procedurally rigorous ways. This led to a long trawl through philosophical and theoretical concepts of form, order, and regularity such that what is now Chapter 3 was originally two chapters. I still think that it is important to understand the meanings and connections between pattern, form, order, and regularity, but the latter did not provide the path I sought to grounding a theoretical basis of pattern; it helped to define what pattern is not but not necessarily what it is.

I took a very different approach in Chapter 4 by focusing on minds and brains as the basis of pattern. I would thank Kent Hecker and Fil Cortese for drawing my attention to developments and debates in the cognitive neurosciences, which I found readily aligned with my putative theories of patterns as cognitive phenomenon. Indeed, the alignment led me to be bolder than I had anticipated in asserting connections between neurological structures and processes and the theoretical nature of patterns. However, getting there took some time and if you felt an implied pause in my writing in around Chapter 4, you would not have been mistaken.

I approached Chapter 5 as a logical extension of Chapter 4; if pattern is cognitive then we also need to consider how patterns are used. This led to an exposition of what I called *pattern thinking* and its various modalities. A significant challenge I encountered in writing Chapter 5 was what I should include as modalities of pattern thinking. As a result, writing this chapter was a particularly abductive process going through multiple cycles of adding, merging, and consolidating subheadings and topics. I had wondered whether *pattern learning* deserved a chapter all its own (much of my day job is spent in health professions education), while I moved *pattern creativity* from Chapter 5 to Chapter 6 as it made little sense without first describing how we externalise our patterns and pattern thinking.

I must admit that pattern externalisation had not at first seemed to warrant a chapter of its own. However, I found that not only did it need substantive treatment, pattern externalisation needed unpacking into what became pattern theoretical concepts of instances, expressions, representations, systems, and languages. These concepts proved essential in advancing pattern inquiry in Chapter 8. However, having written Chapters 4–6, I found that I still had many loose threads, many ambiguities, and many questions that still needed resolving. This then was the purpose of Chapter 7 on pattern philosophy in which I grouped and explored a number of critical questions around the topics of pattern ontologies, pattern epistemologies, and pattern axiologies. Interestingly, I found I was better able to outline what patterns do and how they work than I was defining what patterns are. I ended with more questions than I started, which is not unusual in the social sciences, and suggests that this book is by no means the (or even a) final word in matters pattern theoretical.

This then brought me to Chapter 8 on pattern inquiry, a chapter that I had planned to write from the outset, although not at all in the way that I eventually did write it. Rather than outlining inquiry and the place of pattern theory within it, now that I had a body of pattern theory to work with, I could deductively outline the scope and philosophy of pattern theoretical approaches. To that end, I explored in turn purposes, foci, dispositions, processes, and products of pattern inquiry, and I provided a worked example of these principles in practice. This allowed me to consider issues of grammar and syntax, of scope, and of conceptual landscapes. This in turn allowed me to consider broader issues of pattern inquiry as a part of methodological and theoretical continua and of the potential strengths and weaknesses of a pattern theoretical approach to scientific inquiry. I close the book with this current manifesto chapter.

This feels like a long journey to me, one that took more than three years to complete, if it can be said to be complete. As much as I find great utility and broad applicability of pattern theory, I freely admit that much of this book has been predicated on a series of deductive interpretations backed up by empirical evidence and experience. However, most of my claims are still theories, ones that need to be put to the test. I must suggest, therefore, that this book is the start of pattern theory and not the definitive source. Indeed, there seems to be so much that can and should be explored further that this volume is no more than an overture to what I hope will follow.

I should also question whether I have done anything particularly new here. Clearly, I did not invent the term pattern; it is a concept/construct that a great many others have used in academic thought as I described in Chapter 2. I have, I think, been able to establish a sense of coherence around what pattern is and I have connected a large body of ideas in doing so. I did not come up with the first body of pattern theory. Grenander and colleagues' General Pattern Theory precedes my work by decades, and there have been other less well-developed theoretical positions taken. However, my approach differs in many respects to GPT and to these other models, and none that I have found have had such a wide scope or integrative focus. Others have explored whether pattern is no more than a product of perception or some more objective material reality. For example, Dennett argued that:

> Where utter patternlessness or randomness prevails, nothing is predictable. The success of folk-psychological prediction, like the success of any prediction, depends on there being some order or pattern in the world to exploit. Exactly where in the world does this pattern exist? What is the pattern a pattern of? Some have thought … that the pattern of

belief must in the end be a pattern of structures in the brain, formulae written in the language of thought. Where else could it be? ... When are the elements of a pattern real and not merely apparent?

(Dennett 1991)

What I have done, I think, is to answer some of Dennett's questions, albeit provisionally, and in doing so I have introduced new concepts and tools to continue this exploration. I think this applies too to Damasio's suggestions that:

Minds, based on the mapping of overt, multidimensional patterns, were a powerful advance that permitted, simultaneously, making images of the world outside the organisms and images of the world inside them.

(Damasio 2021, p. 191)

A pattern theoretical stance puts meat on these bones, it affords models and debates, and it connects so many otherwise loose threads. To this latter point, I would note parallels in schema theory, originally developed by Bartlett and later developed by others including Piaget. As Bartlett observed:

Remembering is not the re-excitation of innumerable fixed, lifeless and fragmentary traces. It is an imaginative reconstruction or construction, built out of the relation of our attitude towards a whole active mass of organised past reactions or experience.

(Bartlett 1932)

Not only does schema theory have some very pattern theoretical qualities, it has both a structural and a functional side. For instance, there are strong parallels between the pattern theoretical concept of pulling on a pattern and having its many associations being pulled into active thought and Chi and Ohlsson's observations that:

... it is essential to their hypothesized function that [schemas] are retrieved or activated as units ... if one part of a schema (relation or slot) is activated, there is a high probability that the rest of the schema complex declarative learning will also be retrieved.

(Chi & Ohlsson 2005, pp. 374–375)

As much as we can think about pattern theory as a generative approach to combining brain, mind, learning, creativity, reasoning, and research, is not a theory of everything simply a theory of nothing? Perhaps so. Nevertheless, I would argue that pattern allows us to break free of traditional boundaries and limitations of certain kinds of thought to consider thought as a unified phenomenon. Pattern theory allows us to think about previously unconsidered and unexplored kinds of thoughts, and it allows us to analyse those thoughts as a way of exploring the nature of inquiry as something woven from pattern thinking. Pattern is the way mind and matter interact. Pattern is the cave wall.

I am a scientist and yet I know I see things that are not there. I have a human tendency (that I try to keep in check) to sometimes over-interpret things in my eagerness to discover something new and unusual, or to find an answer to an important problem. These tendencies are not (I hope) a sign of a failing mind, but rather are those of a very typical human mind. That I can readily see things that are not there makes me think of the books I had as a child with illustrations by artists such as Oliver Rackham and Kay Nielsen who painted trees, rocks, water, and skies redolent with faces and limbs. They painted all manner of things that cannot possibly be there and yet seemed so familiar. I have from this (and from many other formative experiences) a sense of having been deeply socialised to pattern thinking, and I think it likely that you have been too. I interpret this as a priming of our minds to seek for meaning and appearance amongst noise and chaos, to find one thing in the presence of something else. I think it likely that you too *see* patterns all the time, some of which make sense, while others do not. Indeed, how could you have made sense of this book if you were not already an adept pattern thinker? I hope the journey has given you more insight into the workings of your own mind and into the possibilities afforded by a pattern theoretical approach. After all:

> When you are a Bear of Very Little Brain, and you Think of Things, you find sometimes that a Thing which seemed very Thingish inside you is quite different when it gets out into the open and has other people looking at it.
>
> *(Milne 1928)*

References

Bartlett FC. *Remembering: A Study in Experimental and Social Psychology*. Cambridge, UK: Cambridge University Press: 1932.

Berlin I. *The Proper Study of Mankind: An anthology of Essays*. London, UK: Chatto & Windus: 1997.

Chi MTH, Ohlsson S. Complex Declarative Learning. In: Holyoak KJ, Morrison RG (eds.). *The Cambridge Handbook of Thinking and Reasoning*. Cambridge UK: Cambridge University Press: 2005.

Damasio A. *Feeling and Knowing: Making Minds Conscious*. New York, NY: Pantheon: 2021.

Dennett DC. Real Patterns. *The Journal of Philosophy* 1991; 88: 27–51.

Feldman Barrett L. *How Emotions Are Made: The Secret Life of the Brain*. New York, NY. Houghton Mifflin Harcourt: 2017.

Milne AA. *The House at Pooh Corner*. London, UK: Methuen: 1928.

Zai A, Brown B. *Deep Reinforcement Learning in Action*. Shelter Island, NY: Manning: 2020.

INDEX

Printed in the United States
by Baker & Taylor Publisher Services

Printed in the United States
by Baker & Taylor Publisher Services